PLACES THROUGH THE BODY

It is relatively easy to imagine how bodies labour to make places and how the social conditions and physical form of a place might shape particular kinds of bodies. It is much more difficult to think about how bodies and places simultaneously and creatively make one another. Body parts, for example, are mapped onto spatial processes and places in culturally specific ways: in Western countries it is not unusual to say the bowels of the earth, the eye of the storm, the spine of a book, or the foot of a mountain. Similarly, spatial processes and landscape features are mapped onto human processes and bodies: an earthy person, a stormy marriage, a person who is a bookworm, or a person built like a rock. There is something remarkable about this transformation of bodies into places and places into bodies: something, that is, which is not easy to imagine.

The authors in this collection draw on novel, creative and often poetic ideas to show how bodies and places make one another. Together they reveal how places through the body and how bodies through places, come to be. The volume is divided into four sections. In part one, *FilteringPlacesBodies*, authors show how discourses, memories, and practices work to filter or screen experiences, shaping our symbolic and material understandings of the world. Part two, *ConfiningPlacesBodies*, demonstrates how various kinds of bodies have historically been denigrated and confined, and how confined bodies and places emerge simultaneously in cultural imaginaries. In contrast, part three, *ExcessingPlacesBodies*, shows how bodies and places always exceed their cultural confinements. The final section, *ProjectingPlacesBodies*, focuses on how bodies and places are materially and symbolically displaced onto one another, across different scales and through radically different forms of communication.

This exciting collection from and outstanding group of international authors opens up many new conversations on BodyPlace and reveals the surprising, as well as poignant, ways in which people move through places through the body.

Heidi J. Nast is an Assistant Professor in the International Studies Program at De Paul University, Chicago. She writes about sexuality and state formational processes and is the co-editor of *Thresholds in Feminist Geography* (1997). **Steve Pile** is a lecturer in the Faculty of Social Sciences, Open University. He is the author of *The Body and the City: Psychoanalysis, Space and Subjectivity* (1996) and co-editor, with Michael Keith, of *Geographies of Resistance* (1997).

PLACES THROUGH THE BODY

Edited by
Heidi J. Nast and Steve Pile

London and New York

First published 1998
by Routledge
11 New Fetter Lane, London EC4P 4EE

Transferred to Digital Printing 2004

Simultaneously published in the USA and Canada
by Routledge
29 West 36th Street, New York, NY 10001

Typeset in Garamond by Keystroke, Jacaranda Lodge, Wolverhampton

British Library Cataloguing in Publication Data
A catalogue record for this book is available from the British Library

Library of Congress Cataloguing in Publication Data
Places through the body / edited by Heidi J. Nast and Steve Pile.
p. cm.
Includes bibliographical references and index.
1. Human geography. 2. Body, Human–Social aspects. 3. Body,
Human–Symbolic aspects. 4. Spatial behavior. 5. Personal space.
I. Nast, Heidi J. II. Pile, Steve, 1961–
GF50.P57 1998
304.2′3–dc21 97–37994

ISBN 0 415 17904 1 (hbk)
ISBN 0 415 17905 x (pbk)

CONTENTS

CONTENTS

CONTENTS

FIGURES

FIGURES

CONTRIBUTORS

Stuart C. Aitken is Professor of Geography at San Diego State University, California. He has published articles and essays on sexuality and film in a variety of journals including *Gender, Place and Culture*, the *Transactions of the Institute of British Geographers*, and *Society and Space*. His recent books include *Family Fantasies and Community Space* (1998), *Putting Children in Their Place* (1994) and co-edited, with Leo Zonn, *Place, Power, Situation and Spectacle: A Geography of Film* (1994).

Emily Apter is Chair of the Department of Comparative Literature at Cornell University. She is the author of *Feminizing the Fetish: Psychoanalysis and Narrative Obsession in Turn-of-the-Century France* (1991) and co-editor with William Pietz of *Fetishism as Cultural Discourse* (1993). She has published articles in *October*, *Esprit Createur*, *Architecture New York*, *Modern Language Notes*, *Suitcase: A Journal of Transnational Traffic*, *Art Journal* and *Poétique*. Her book, *Continental Drift: From National Characters to Virtual Subjects*, is forthcoming.

Karen Bermann grew up in New York City, is active in writing, sculptural and architectural projects, and teaches at Iowa State University. Her writing, sculptural and architectural work has been published in journals including *Artforum*, *ANY Magazine*, *Assemblage*, and *Landscape Architecture*. She is interested in the invisible psychological and historical content of spaces and landscapes.

Virginia L. Blum is Associate Professor of English at the University of Kentucky. She is author of *Hide and Seek: The Child Between Psychoanalysis and Fiction* (1995). She is currently preparing a book on plastic surgery.

Susan Bordo is Professor of Philosophy and holds the Singletary Chair of Humanities at the University of Kentucky. She is the author of *The Flight to Objectivity: Essays on Cartesianism and Culture* (1987), *Unbearable Weight: Feminism, Western Culture, and the Body* (1993) and *Twilight Zones: the hidden life of cultural images from Plato to O.J.* (1997). *My Father's Body and Other Unexplored Regions of Sex, Masculinity, and the Male Body* is forthcoming.

Giuliana Bruno is Professor of Visual Studies at Harvard University. She is the author of *An Atlas of the Flesh: Journeys in Art, Architecture and Film* (1998) and *Streetwalking on a Ruined Map* (1993) and co-editor of *Off Screen: Women and Film in Italy* (1988) and *Immagini allo schermo* (1991). She has published widely on film, art and urban culture, feminist theory and cultural history.

Teresa de Lauretis is Professor of the History of Consciousness at the University of California, Santa Cruz. She is the author of numerous essays and books on feminist theory, film, semiotics, literature, and lesbian studies, including *Alice Doesn't* (1984), *Technologies of Gender* (1987) and *The Practice of Love* (1994). She edited *Feminist Studies/Critical Studies* (1986) and guest-edited the "Queer Theory: Lesbian and Gay Sexualities" issue of *Differences* (1991). She is currently working on *Fantasies of the Other*, a book of essays on film.

Michael L. Dorn is a doctoral candidate in Geography at the University of Kentucky. Applying the genealogical method of Michel Foucault to an in-depth examination of the historical geography of disability, Michael's dissertation examines the attempts of medical authorities in the early nineteenth century United States to discipline intemperate bodies and to improve public health.

Glen S. Elder is visiting Assistant Professor of Geography at the University of Vermont in Burlington, VT. His research interests focus on restructuring in contemporary South Africa where he was born. He has published on issues relating to sexuality, race, gender, and geography in particular.

J. K. Gibson-Graham is the pen name of Julie Graham and Katherine Gibson, feminist economic geographers who teach at the University of Massachusetts-Amherst and Monash University in Melbourne, respectively. Gibson-Graham has recently published *The End of Capitalism (As We Knew It): A Feminist Critique of Political Economy*.

Elizabeth Grosz is Director for the Centre for Critical and Cultural Studies at Monash University in Australia. She is the author of *Volatile Bodies* (1994) and *Space, Time and Perversion* (1995), and is currently working on a book on the virtual in architecture.

Ken Hillis is Assistant Professor of Communication Studies at the University of North Carolina at Chapel Hill. His work on virtual environments and information technologies looks to their ability to force reconceptualizations of the relationships between space and subjectivity. The author is a registered professional planner, and for many years worked with citizens' organizations dedicated to community development and planning. His book, *Digital Sensations: identity, embodiment, and space in virtual reality*, is forthcoming.

CONTRIBUTORS

Lynda Johnston is a contract lecturer and doctoral student in the Department of Geography at the University of Waikato. Her research interests include recent critical social theories on sexed and sexualized bodies, geographies of sexualities, poststructuralism, tourism and gay pride parades.

Binnie Klein is a licensed clinical social worker in private practice in New Haven, Connecticut. She is on the staff of Yale University Health Services. Binnie is a published poet and produces a music show for WPKN-FM radio.

Sandra Kryst completed her PhD in anthropology at the University of Kentucky in 1995 and has worked as a Visiting Assistant Professor in the Department of Sociology and Anthropology at Kalamazoo College. Her interests include medical anthropology, women's health, healing and chronic pain.

Christopher Lukinbeal is a PhD student in geography at San Diego State University. His recent articles and essays on film appear in *Progress in Human Geography* and the *Yearbook of the Association of Pacific Coast Geographers*.

Angela K. Martin is completing her dissertation in anthropology at the University of Kentucky and works as a part-time instructor in the Department of Sociology and Anthropology at DePaul University. Her interests include identity, embodiment and religious symbolism, and the study of transnational processes.

Heidi J. Nast is a feminist cultural geographer who teaches in DePaul University's International Studies Program (Chicago). She writes about sexuality and state formational processes and is a co-editor of *Thresholds in Feminist Geography* (1997).

Steve Pile is a lecturer in the Faculty of Social Sciences at the Open University. He has published work relating to identity, politics and geography. He is the author of *The Body and the City: Psychoanalysis, Space and Subjectivity* (1996) and co-editor, with Michael Keith, of *Geographies of Resistance* (1997).

Geraldine Pratt is Professor of Geography at the University of British Columbia and editor of *Environment and Planning D: Society and Space*. She writes about women and labour markets, and is co-author with Susan Hanson of *Women, Work and Space* (1995).

Marilyn K. Silverman, PhD, is a clinical psychologist with a private practice in psychotherapy in Wilton, Connecticut. She is a consultant in Pediatrics and Psychiatry at Norwalk Hospital and holds academic appointments at Yale University and Vanderbilt University.

Matthew Sparke is an assistant professor in Geography and International Studies at the University of Washington. He has published on topics relating to gender, capitalism and post-coloniality in *Antipode*, *Society and Space*, *Gender Place and Culture*, and *Cartographica*. He is author of *Negotiating Nation-States: North American geographies of culture and capitalism* (1988). He is currently researching the reconfiguration of borders in transnational regions.

Gregory A. Waller teaches film and cultural studies at the University of Kentucky. He was a Fulbright lecturer in New Zealand in 1993 and has published four books, including *Main Street Amusements: Film and Commercial Entertainment in a Southern City, 1896–1930*, which won the 1995 Theatre Library Association Award for best book in film and television studies.

Mabel O. Wilson is assistant professor of architectural design at the University of Kentucky. She is currently a doctoral student in the American Studies program at New York University and a partner in the architectural collective Architecture et al. Her architectural designs were included in the exhibition House Rules and her writings on racism and architectural discourse have appeared in *Assemblage*, *ANY Magazine*, and *Practices*.

ACKNOWLEDGMENTS

One kernel of this book was a special session organized by ourselves at the annual meeting of the Association of American Geographers in Chicago. During this time, Virginia Blum, Angela Martin and Mabel Wilson traveled from the University of Kentucky, Lexington, despite the fact that none of them are geographers, none had funding, and none had ever attended a geography meeting previously. We thank them for investing so heartily in this endeavor. Subsequently, conversations began with colleagues at the University of Kentucky, specifically Susan Bordo with Binnie Klein and Marilyn Silverman, Michael Dorn, Sandra Kryst and Greg Waller, all of whom we thank for writing original works for this collection. All those who took the time and effort to produce drafts and yet more drafts of chapters deserve appreciation for their patience and endurance – including those not yet mentioned: Stuart Aitken, Glen Elder, Katherine Gibson, Julie Graham, Ken Hillis, Lynda Johnston, Chris Lukinbeal, Gerry Pratt and Matthew Sparke. Similarly, we thank Emily Apter, Karen Bermann, Giuliana Bruno, Teresa de Lauretis and Elizabeth Grosz for their enthusiasm and cooperation in the production of this book and, subsequently, for allowing us to reprint their work. Because of the people involved, the entire endeavour of "collecting" and "editing" has been not only intellectually broadening, but thoroughly enjoyable.

We would like to acknowledge the following for granting permission to reprint work published elsewhere. J. K. Gibson-Graham's chapter is a revised version of a paper which originally appeared under the title "Querying Globalization" in *Rethinking Marxism*, Volume 9, Part 1 (Spring 1996/7). "Bodies-Cities" by Elizabeth Grosz is reprinted, with kind permission, from B Colomina, ed., 1992, *Sexuality and Space* (Princeton, Princeton Architectural Press), pages 241–53. Emily Apter's chapter originally appeared, in both Italian and English, in *Ottagono*, 1991, Number 97, pages 97–107 and is reprinted with kind permission. Karen Bermann's chapter is reprinted, with kind permission, from *Architecture New York*, 1994, Volume 4, pages 16–21. Kind permission has been given to reprint Teresa de Lauretis's article by *Australian Feminist Studies*, 1991, Volume 13, pages 18–26; a revised and extended version of this article appears in her book, *The Practice of Love: Lesbian Sexuality and Perverse Desire* (Routledge, 1994). Matthew Sparke's chapter is a revised version of a paper which originally appeared under the title "Between demythologizing and deconstructing the map" in *Cartographia*, Volume 32, number 1 (Spring 1995). It is reprinted with kind permission from the University of Toronto Press. Kind permission has been given by Princeton University Press to reprint Chapter 17 of Giuliana Bruno's book, *Streetwalking on a Ruined Map* (1993).

We could not have completed this collection in such a timely (or fun) fashion without funding from the University Research Council of DePaul University, for which we are very grateful. We also thank both of our respective departments, and heads of department, for allowing us to tap into numerous institutional resources.

THE COVER ILLUSTRATIONS

A word from the artist

When first approached about doing the artwork for the book jacket of *Places through the Body*, Heidi asked me if I remembered ever seeing an illustration of Alice in *Alice in Wonderland*, bursting through the roof of a house after consuming some mushrooms that said, "Eat me." It seems we both had seen a similar illustration as children, but were unable to find it in any current copies of the story. After making an initial sketch, I began thinking about how our bodies are defined, controlled and more often, transformed by our surrounding spaces. Using Alice as a starting point, I began to conceptualize a fantastical drawing that would illustrate these ideas. It occurred to me that there was another well known character – Dorothy from the Wizard of Oz – who suffered a similar fate of being brought into a world unfamiliar to her and having to make her own way.

Both Alice and Dorothy were at odds with their spaces. Poor, dreaming, misfit Alice, while pursuing the time conscious white rabbit, was either too big or too small for her space; too sane for the Mad Hatter's Tea Party and dangerously outspoken in the Queen's Court. Plucky Dorothy was limited by her dreary rural existence, dreaming of places faraway and happy over the rainbow, only to be unceremoniously transplanted there via a tornado and a house which unexpectedly lands on the body of the Wicked Witch of the West's sister: ("Who killed my sister . . . "). As the evil witch's sister slowly expires under the house, the ruby red shoes she wears magically appear on Dorothy's feet, making her the witch's sworn enemy. Thereafter, Dorothy is directed to the Wizard of Oz who, everyone tells her, has the power to bring her back home to Kansas. Only later does she realize that she did not need the Wizard, that her powers to leave lay in her own two feet. Clicking her rubied heels together, Dorothy returns home.

Thinking about the two characters and their places, I found they shared many allegorical similarities. We, too, in our search for sense of place in society and among even family relationships, often find ourselves misfits in spaces in which we are "supposed" to feel comfortable. Consequently, we are driven to make our own place, in the process, transforming ourselves. These experiences throughout one's years, from birth to childhood through adulthood are the rooms through which we pass, each one leaving its mark on both body and place. (And aren't we always mystified or surprised at how we turn out as adults, when we so carefully craft and are crafted by every experience along the way?) So, too, Alice and Dorothy, shaped by trials and tribulations, adjust and redefine their way at each bend in the road. Dorothy manages to pick up a few friends along the way. Alice pretty much went on through Wonderland confused and alone.

The front and back covers are meant to be viewed in order. They represent a mixed up new tale of "Places through the Body". The cover drawing depicts Alice just as she lands at the end of the rainbow after tasting that mushroom. Bang! She lands on the path to Oz, and the white rabbit, trapped with Alice, flees in terror, still worried about time. She is out of place in the surroundings, fast becoming out of space, as she experiences the first surge of her body bursting through the roof. Bending Dali-like watches symbolize three measure of time – one counts melting years; another is hourly; the last a stopwatch, hands bent.

The second drawing depicts Alice as she completely obliterates the house in a huge burst of growth. She has lost control. Emerald City and Oz go up in flames. The rabbit is flung against us in the explosion. It is a redefining, meltdown moment for Alice in Dorothy's space.

It is in the essays of *Places through the Body* that we catch glimpses of ourselves and others, some folks looking backwards to attempt an understanding; others looking forward, being refined and retrofitted for new BodyPlaces that lie ahead.

Cindy Davies

1

INTRODUCTION

MakingPlacesBodies

Heidi J. Nast and Steve Pile

Very few things are universal. Bumper stickers across the United States of America proclaim only taxes and death. Of course, many escape taxes, but no-one sadly has eluded death (to our knowledge!). On the brighter side: death presumes life. And, since we are alive, we must have bodies. And, since we have bodies, we must be some place. Here we are then. Before this book has even started, it seems that it is over – we all have bodies and we are all some place. But let us not be so hasty, so quick to assume that this just about covers the story of bodies and places. We might also take a little time to wonder about the stories that we have failed to tell. Sure, we all have bodies, but the idea that we *have* bodies – that bodies are a possession that the individual has – is culturally, historically and geographically specific. Further, the impression that the individual is located *in* a body and that being in a body is also about being in a place warrants further scrutiny. It turns out that our universals – the body, the body in place, being in place – are actually unique, specific, singular. Paradoxically, then, at the same time that we all have bodies, none of us has the same body as anyone else; conversely, at the same time as we live in a particular place, no place is completely isolated from everywhere else (even Robinson Crusoe's island was connected to other parts of the world – just not very often!).

Our bodies are unique, yet everyone else has a body too. If our bodies and places are unique, then this implies that only we can experience the world in the way we do – but, since other people have bodies and can live in the same places, our experiences cannot after all be unique. The argument is moving in circles. Both bodies and places need to be freed from the logic that says that they are either universal or unique. Instead, it would be better to think of the ways in which bodies and places are understood, how they are made and how they are interrelated, one to the other – because this is how we live our lives – through places, through the body.

There is an urgent need to look at the relationship between bodies and places, not because of an academic requirement to sort out paradoxes, but

1

because the ways in which we live out body/place relationships are political. In her thought-provoking article, "Notes towards a Politics of Location" (1984), Adrienne Rich begins to doubt Virginia Woolf's claim that women have no country and that women's country is the whole world. Instead of "globalizing" the experience of "woman," Rich argues that she does have a country and that, much as she might wish it otherwise, she cannot simply divest herself of her geographical location. If an appropriate feminist politics is to be devised, Rich says, "I need to understand how a place on the map is also a place in history within which as a woman, a Jew, a lesbian, a feminist I am created and trying to create" (1984: 212).

She urges that an analysis of these intersections of gender, race, sexuality and politics *in place* begins with an analysis of "the geography closest in – the body. Here at least," she continues, "I exist" (1984: 212). Of course, there have been many studies of the body in a wide range of disciplines,[1] and geographers themselves have sought to incorporate the body into their studies of lived place and spatial relationships.[2] Nevertheless, cultural geographers have recently been intrigued by the idea of the geography closest in. For Rose, Rich's arguments suggest that the female body is a site of struggle. Rich's cartography of the female body maps out a particular political terrain:

> The politics of pregnability and motherhood. The politics of orgasm. The politics of rape and incest, of abortion, birth control, forcible sterilization. Of prostitution and marital sex. Of what had been named sexual liberation. Of prescriptive heterosexuality. Of lesbian existence.
>
> (Rich 1984: 212–13; see also Rose 1993: 29)

For Rose, the body is placed "geopolitically": its location is marked by its position with specific historical and geographical circumstances. It matters, to Rich, that she is a citizen of the United States of America. At the time of writing, she was arguing that the Cold War was reaching new heights, as American foreign (and domestic) politics was dominated by the fear of a communist take-over, prompting insidious interventions in central America and beyond. Meanwhile, black politics at home made Rich acutely aware of her whiteness, while in Nazi Germany she would not have been white enough. For Rich, it is not enough to assert some kind of universal feminist struggle, but to recognize the specificities of women's struggles in their situatedness, their location in history, on the map.

For Soja, Rich's recognition of the "geopolitics of the body" suggests a spatial hierarchy of scales of oppression, from the body outwards, to the global (1996: 36). However, Kirby has argued that Rich moves "fluidly through a number of spatial registers" (1996: 19). For Kirby, Rich's project is to identify the connecting points between the individual and place. These connecting points are variable, for the individual occupies, not one, but many positions.

2

"This body. White female; or female, white" (Rich 1984: 215). The distinction may seem trivial, but Rich talks about the ways in which white and black babies were separated into separate wards in the hospital where she was born. In the first instance, she was marked by skin colour, by blood. She was white and female: children marked by race and sex.

> To locate myself in my body means more than understanding what it has meant to me to have a vulva and clitoris and uterus and breasts. It mean recognizing this white skin, the places it has taken me, the places it has not let me go.
>
> (Rich 1984: 215–16)

The body is both mobile and channeled, both fluid and fixed, into places. It is not only the "geopolitics of the body" but also the politics of connection and disconnection, of rights over the body, of the body as a site of struggle. Rich maps out her connections to the world, maps out its territories, and shows where she is within these connections and territories. This cartography of places through her body reveals that ways in which she is positioned through her body, but also how her body becomes capable of imagining these connections and territories differently. Thus, Rich moves fluidly through spatial registers because her mappings of connections, of territories, subvert commonplace understandings of places as bounded, sealed areas. She shows that these spatial registers – these geopolitics of the body – are produced through unequal power relations, between men and women, between blacks and whites, between heterosexuals and homosexuals, between one nation and another. Beginning her analysis with the body allows her to map her place, to map out a history of spatial registers: points, connections, dislocations, boundaries, territories, countries, regions, power blocs. Her aim: to bring politics down to earth, to create a ground for struggle (1984: 218–19).

The meaty body is where Rich wishes to ground politics. However, it remains an uncomfortable place. Rich's geography closest in is meant to stretch out to incorporate others in struggle and, because the geography of the body is closest in, it is also the ground on which to fight for women's rights. In this, Rich herself marks the body through exactly the same power relations she is hoping to overthrow: she is "woman," "Jew," "lesbian" – there are not many other places to be. Rich's use of fluid spatial registers suggest other ways of thinking places and bodies. The body is not simply the bearer of some pre-given cultural categories. For Grosz, "the body cannot be understood as a neutral screen, a biological *tabula rasa* onto which masculine and feminine could be indifferently projected" (1994: 18).

Rich argues that her white skin has taken her places and stopped her going to other places, but Grosz would also insist that the privileging of skin or of whiteness or of white skin is a particular relationship, interlocked with other

3

hierarchical forms of power, that make the body in other ways. For Grosz, the body also exists beyond social relations and the categories that social relations impose on the body. While medical science has been dominated, since the end of the eighteenth century, by the idea that the sexes are opposite (see Laqueur 1990), Grosz wishes to argue that bodies are unstable and indeterminate. Such medical knowledge might have permitted the better treatment of diseases specific to women, but it has also perpetuated a particular kind of understanding of the relationship between men and women: they are opposite, opposed. Yet men and women have more in common with each other than with any other "thing". For Grosz, this suggests an alternative understanding of embodied difference: rather than being opposites, people might instead be "neighbors". Not one sex, nor opposite sex, but neighborhood sex: a thousand sexes, a conurbation of sexes.

From this perspective, sex, gender, race, skin, blood are indeterminate and unstable signifiers of the differences and similarities between bodies. This understanding provokes questions, not about the real make-up of bodies, but about how bodies are really made-up. More and more, it seems as if the relationship between bodies and places is like not only Alice's journey through Wonderland[3] but also Dorothy's trip down the yellow brick road to the Wizard of Oz (see also Cindy Davies's discussion of the cover illustrations, pp. xvii–xviii of this volume). Alice's body was never stable enough to qualify as being that of a little girl: she shrunk, grew, was in place and out of place. Alice, in Wonderland, never quite fitted in. Though she tried to understand, her bodies and their places never stabilized long enough to make any real sense, though sense there was of a kind. Dorothy was transported in a whirl-wind of dreams to a strange land where she never follows its strange logic, partly maybe because all the male characters appeared to be missing something vital. As she crossed from one place to another, Dorothy refigured and unmasked the charade of bodies. Eventually she gets to her goal, but even the Wizard turns out to be a sham; seemingly all powerful, but only while he was hidden. Alice and Dorothy, hand in hand, point to the fragile illusions through which the endurances and solidities of bodies and places are built.

Bit by bit, bodies become relational, territorialized in specific ways. Indeed, places themselves might be said to be exactly the same: they, too, are made-up out of relationships between, within and beyond them; terri-torialized through scales, borders, geography, geopolitics. Bodies and places, then, are made-up through the production of their spatial registers, through relations of power. Bodies and places are woven together through intricate webs of social and spatial relations that are made by, and make, embodied subjects. It is these intricate webs that this book is concerned to trace out.

While scholarly work frequently notes body–space associations, nowhere have these associations been systematically teased out or explicitly thought through across a number of social, spatial and cultural fields. Nor have the motivation and potentialities for thinking of bodily-spatial displacements been theorized or explored in any cross-disciplinary detail. What is being suggested here is that it is not enough simply to treat the body–place relationship as if it was either universal (non-specific) or unique (too specific). Instead, the particular ways in which spatial relationships come together to make bodies and places, through the body and through places, needs to be exemplified, demonstrated and clarified, in places, through the body. Thus, this collection privileges the embedded social practices that are constitutive of particular bodies and places; places–bodies as seemingly wide apart as those of capitalism, cities, museums, gymnasiums and harems. Body–place relationships are thus not only delivered from the fixed coordinates of social relationships and their constitutive spatial registers to *places through the body* (as they are produced, inspired, dreamed, born, borne); but they are also released from static, reified notions of bodies to bodies that make, and are made through, the practices and geography of places.

Though this collection privileges the relationship between bodies and places, this does not result in either becoming some arbitrary fixed point or point of departure for analysis. Instead, the contributors to the book come from a wide range of disciplinary backgrounds – revealing their particular takes on bodies and places. Accordingly, the collection speaks of "places through the body" in differently disciplined ways, allowing their styles and contents to be voiced one with another. These differences of style and content allow for a creative interplay across the book as a whole. However, it presents readers – and editors! – with some challenges: one of which is how to read the diverse contributions. To help "place" the chapters, we outline the thinking behind the "running order" of the book below. But we do this, not as a "final script" nor to prevent other readings, but just to situate our take on the chapters and on how we envisage the themes running through them.

In the end, we had to settle on a sequence of chapters in the book – but we hope this does not seem too final, too closed, for this would be to make-up the body, once again, into something too set, too fixed, too material. In this respect, we have tried to leave room for thoughts which might be a little surprising, ideas that are not immediately obvious, because we hope that one "message" of the book will be that bodies and places remain as indeterminate and unstable as they are distinct and enduring; open to analysis, but always beyond the limits of categories or orders to finally seal them up and close them down, like bodies in body bags, or places conquered by the Empire of Reason.

* * * * *

Part 1 deals with *FilteringPlacesBodies*. It shows how discursive practices, materiality (what is considered to be "material") and spatiality (the spatial organization/relationality of the social world) collectively and mutually, figuratively and practically, filter through one another. Filtering, for us, evokes thoughts about keeping out, keeping back, defensiveness, interchanges, flows between, across, over, through places, through the body. However, the theme also inspires images of cleansing, contaminating, fertilizing, washing bodies and places. So, in this part of the book, chapters give clues as to how flows filter through places through bodies (and back again).

Drawing upon some of the metaphors of filtering and fluidity, then, Gibson-Graham (Chapter 2) reveal how the images and practices of heterosexuality filter through normative scriptings of capitalism. "Capital", for example, is proto-typically cast as a perfect phallic hardness able to penetrate foreign and/or virgin markets at will. They go on to present an alternative scripting of capitalism as an indiscriminate and fluid process spurting out of control. Indiscriminate processes of credit card issuance, for example, leaks capital's fluids into production arenas that negatively interfere with, or infect, capitalism's exploitative processes. In doing so, credit cards paradoxically support modes of production and social relations antithetical and corrosive to capitalist interests. Lastly, Gibson-Graham suggest ways for changing the heterosexual economy and discourse of capitalist desire, production, and exchange, using some of the language and tactics of feminism and queer theory to wrest agency out of the body of capitalism.

At a different scale, Grosz's work (Chapter 3) deals with how bodies and cities co-filter and constitute (rather than causally produce or representationally prefigure) one another. Her piece can also be taken to question what happens when filtering processes are ruptured, scarred and, possibly, healed, asking what happens, for example, when "a body inscribed by one cultural milieu finds itself in another involuntarily"? How is filtering (whereby a body sensually and knowledgeably ingests the city as the city, in another sense, ingests it) to be stabilized when the city (for whatever reasons) changes (a move to a new city? changes in zoning laws or investment patterns?) or becomes unintelligible, becomes a "wall" – or when a body-subject refuses (for whatever reasons – illness or disability? trauma? unemployment?) to consume the city? How, she asks, will computers and cyberspace – key material and labor anchors in the global capitalist economy – affect, rupture, or transform the spatial economy of bodies-cities as we know them? Are the majority of us to be refused, extruded, filtered out by the city's architectural economy (the non-unionized worker looking in through the high rise windows of corporate computer headquarters as he washes them for little pay?). Are we going to be passively absorbed (the telecommuting woman who "does" data entry, locked in her home, her key strokes – and those of others – automatically tabulated by a master-computer elsewhere)? Or will the woman funnel capital into her home, eventually to form an independent site of auto-surplus extraction? Is she then the same worker? Or is she now a cyborg actor within a city-global circuitry of accumulation – wresting agency, productive power and organizational capital out of passivity?

Recovering the bodily in cyberspace is the concern of Hillis (Chapter 4). Cyberspace is highly embodied, he argues; it's just that the bodies are forced into invisibility through the ways in which we create and enforce a coding of cyberspace: it is a spatial void; it is abstract, neutral, solitary, fake. Here, we might tie in some of the urban arguments made by Sassen about cities (1994, 1996a, 1996b). She argues that post-industrial landscapes are not as clean and privileged as we would like to think: havens of white collar workers working in tidy, white, computerized environments. Instead, the landscape is littered with the refuse of a lost middle class: a growing army of non-unionized maintenance workers whose bodies, while filling the landscape (janitors, window-washers, secretaries, and those with deadening data entry jobs), are abjected. No one wants to imagine their presence, just as previously, nineteenth century industrial capitalists refused to think about the body/spaces of their workers (see Engels 1844). Hillis asks us to think through the fact that cyberspace depends upon a hard technology run by hard bodies located in hard places. In this sense, "software" is a disembodying euphemism that removes our imaginaries away from the material (hard) realities of labor processes. In this sense, computers are screening devices that displace, filter out, and disallow "real" places and bodies, ideologically replacing them with virtual ones.

The filtering of bodies through memories and space is, poignantly, also the subject matter of Susan Bordo, Binnie Klein and Marilyn Silverman's emotional exploration of "places through the body" (Chapter 5). These three sisters together and separately explore their often agonizing, always partial, memories of childhood spaces and of their experiences of their parents. They talk, one by one, from their own perspectives, about how they remember their father and mother, about how they remember the different parts of the homes they inhabited. What becomes apparent is that their pain and anger bound up in the emotional truth of these spaces of family and home do not "map" directly onto the physical spaces of parent and home. As they make the journey through memory and place, they come to realize that something is missing: they can't quite remember the kitchens, and in this they recognize that they have forgotten something important about their mother. Their mother and father, their homes, becomes a series of absences and presences, simultaneously there and not there. Their mapping of their own truths of father, mother, home, are produced through a dissolving and resolving of their senses of self within senses of place. However, these senses of place are marked, not by a security of identity and place that tells them who they are and where they are, but by a series of painful dislocations that made it almost impossible to find a location for themselves.

The difficulties of finding a proper place are considered in Nast's work (Chapter 6). By reworking the notion of "reflexivity" through her experiences of cross-cultural fieldwork in Nigeria, she shows how everyday practices act as socio-spatial filters that regulate and constrain us into understanding, obeying, and/or resisting cultural norms: those who do not fit in are disciplined and/or excreted, both processes only ever partially accomplishing what is intended. In particular, she anecdotally discusses how she learned to recognize her place amongst several very different social groups by allowing her body to be a kind of "place" where others registered their subjectivities, showing her what socio-spatial and bodily norms were acceptable. Through strategically placing herself in ways that facilitated particular socio-spatial exchanges, or by being placed against her will into positionings with which she was culturally unfamiliar or uncomfortable, she realized how bodily "fieldings" of difference made her negotiate difference in ways that less embodied, "academic" exercises in reflexivity (especially literary exercises) did not allow for.

Bodily and spatial constrictions, proscriptions, and confinements are a recurring theme in Part 2: *ConfiningPlacesBodies*. Here, the contributions demonstrate not only the ways in which bodies are restricted and enclosed by the places which they inhabit, but also how bodies themselves incarcerate and bind. These confinements shape a social architecture of the psyche: someone somewhere confines, while someone elsewhere is confined. These restraints on bodies and places are not simply material or discursive, but fantasized and lived – sometimes with barbaric and deadly consequences.

Apter (Chapter 7) begins by showing the key place which harem/bodies have occupied in Western psyches, being deployed culturally to structure not only Freudian psychoanalytic imaginings and discourse, but nineteenth and twentieth century European art forms, such as painting, literature and architecture. The harem is a site where desire for the maternal is recognized at the same time that it is distanced (being romanticized as oriental), sequestered, viewed, and called to modern order (inner home/vaginal spaces define and confine the modern identity/place of Woman). Metaphors of movement prevail as Apter traces the harem/body through a number of different material and scalar contexts – from the hymen, to the veil, to the domestic interior of a home, to masks and the feminine masquerade, to feminine cornucopia, to the feminine as secret. Apter charts the harem's dual function: it unabashedly "prompts forays" into the unconscious, as it creates a subjugated, domesticated female sexuality. The harem/body is "what the West most desires and fears in the Other as well as what it most wants and detests in itself': incestuous desire called to order through colonization. Colonization's instrumentality is understood not only in terms of figures of speech (e.g. Apter notes how harem tropes are "obligatory" in colonial French fiction), but geographically: imperialist conquests at once prefigure and displace desire onto faraway lands, bodies, and places. The harem-body allows, for example, for Algeria to be encoded as feminine, while the colonizers become the masculine master (paralleling the work of Fanon; see Gordon, Sharpley-Whiting and White 1996, and Read 1996). Apter calls for a critical "spatial ethnology" of harem/bodies to help the West remove itself from its "psychosis of coloniality" – that is, from a thoroughly racialized space.

Uncovering the racialization of space informs Wilson's analysis of modern architecture, particularly that produced by Le Corbusier (Chapter 8). Tacking back and forth between his sketches, writings, building plans, and even his own (disavowed) body, Wilson shows how Taylorism was the confining conceptual grid through which Le Corbusier imagined an ideal city. His thoughts channeled through the heterosexist, racist spirit of industrial capitalism to imagine a utopian, Taylorized urban form in perfect synchronization with the Metropolis' mechanical workings: women sited in a perfected, domestic home bliss; spatially efficient transportation nodes carrying energetic men to work sites; enclaves of green nature dispersed in controlled ways to efficiently bring healthful benefits to urbanites; gleaming white towers ("white cathedrals") erected for, and full of, city managers. Blackness, on the other hand, is what the ideal city rises above; it is the savage frenzy inhering in the Machine Age, a frenzy that must (like women) be called to order. "Grinding street cars", "the unchained madness of subways", and "pounding machines" are metonymically linked (via experiences in New York) with black tap dancers, and jazz: all are scripted as matter and spirit needing mastery. New York is too-full of blackness, its lack of trees a sign of an unnatural remove from a natural "white", clean state of being. Like Apter's linking of the harem to Algeria, Wilson shows how America's savage blackness and evil femininity plays Other to France's refinement. Like Heidegger, who called for an Aryan embrace of the German language as the means for recovering a pure state of being (for a recent discussion, see Dear 1997), so too Le Corbusier called for an architectural language that would bring the world to a pure and racist Oedipal order.

Elder (Chapter 9) works from a different spatial direction, the apartheid state of South Africa, to expose similar associations between racism and heterosexism which work to confine. In particular, he shows how apartheid was not only regulated through spatial directives and proscriptions, but through miscegenation laws and laws defining what constitutes legitimate sexual acts and sexuality. Compulsory heterosexuality (a phrase coined by Adrienne Rich, 1980), while technically to be embraced by all men, in practice, was enforced only with respect to "white" male terrains/bodies – black men's homosexuality was of little relevance to a state concerned with managing and imagining purity through epidermic "whiteness." (Perhaps this reflects an additional eugenic glee that male homosexuality allowed for black seed to be squandered.) State concerns over lesbianism, moreover, encoded female–female sexual relations according to heterosexualized morphological norms: an apparent fascination with who plays "husband" and who plays "wife," as well as anxiety over the size of dildos (is it as big – or bigger – than the real thing?). Elder thus shows how the confinement of apartheid bodies and places are co-constituting, sexuality traversing and informing both domains.

Poignant stories of Jews seeking to escape confinement in Nazi-occupied Europe are told by Bermann (Chapter 10): for years a woman hides her husband in a specially fitted blanket chest daily before leaving for work; a woman in a concentration camp hides her infant behind her shawl, only to have it learn how to walk, one day slipping away to run to its death in front of a soldier. Bermann mostly works with different scales of confinement, showing how bodies and places make each other, through evocative engagements with Anne Frank's diary (whose original title, *het achterhuis*, can be translated as "Behind the House") and her life-in-hiding. Bermann shows how bodies continually exceed the rigid confines and borders of the house where Anne and her family live: food peelings need to be burned or thrown out; toilets need to be flushed; bones and flesh need exercise, warmth, and fresh air. These excesses become potentially traceable pathways to betrayal. But Bermann's text is just as much about mapping out how Anne's diary is a real, alternative space through which she can live, even as it serves as a "paradoxical record of self-obliteration." Bermann compares reading the diary to the reading of the axonometric drawing of the building in which Anne lived, its exposures allowing for a painful looking. In the end, bodily excesses, confined, spell death.

If this excess of bodyspaces has already prefigured in the bodies of those in hiding places described by Bermann, then in Part 3 these excesses are less dangerous, more troubling, more open. *ExcessingPlacesBodies* evokes senses of excess more forcefully, that the body always slips from the categories and materialities that are used to describe and delineate it, to control and oppress it. Moreover, this part also shows how excess is sometimes rewarded with accommodations embedded within political and social struggle.

Perhaps the most dramatic example of cultural excess and accommodation is found in Dorn's chapter (11) on the life and space/times of Patty Hayes, confined to a wheelchair since 1980. In particular, he charts how Patty's relationship to her chair/body changed over a fifteen-year period, in the process marking out what he sees as four paradigmatic shifts in body/place relations: first, feeling immobilized and oppressed by space and ableism; second, accepting her deteriorating physical condition, allowing "experts" to help insert her into the normalizing bodily/spatial order of things; third, coming to the "Copernican" realization that she could make space and wheelchairs work for her, consequently redesigning her home and chair; and, finally, realizing the importance of political action to change the normative body/place. Her journey is a fascinating one, leading Dorn to invent a vocabulary of "spatial dissidents" (persons who refuse normative body/space) and "geographical maturity" (the ability to constantly change one's libidinal economy and cultural practice as bodily and spatial contours shift and change). He also uses Patty's life to critique spatially facile, postmodernist notions of nomads – persons apparently able to traverse a number of landscapes without bodily struggle and without leaving any trace.

Patty's fourth stage speaks of a kind of "place contagion" whereby her efforts to change the ableist order of her homeplace leak out to infect the ableist ordering of the streets – e.g. civic law, architectural codes, and cultural support networks and lobby groups. The term "place contagion," however, is coined by Martin and Kryst (Chapter 12) in their analysis of the remarkable, recent proliferation of Virgin Mary ("Marian") apparition sites (200 in the last 60 years alone!). Focusing on several sites in Kentucky, Ohio, and Georgia, the authors weave a captivating tale of bodily and spatial mimesis involving one of the world's premier Marian sites, Medjugorje. Drawing creatively upon, and working beyond, the work of Bourdieu on body habitus, we are shown how Marian contagion is carried from a site in the former Yugoslavia to several sites in the United States via pilgrims who re-create Mary's sacred presence. In one instance, a new site is established on a former farm with a Mary grotto built to face a distant ridge which is, in turn, made to mimic Calvary by the emplacement of large crosses. Objects brought from Medjugorje, ranging from stones to rosaries, become conduits of sacred authenticity, the blessed visionaries themselves also conduits, in each case having visited Medjugorje. But we are not merely left at the level of place/body mimesis. Rather, the authors show how the apparitions fit into larger body/places. Mary, for example, is cast as the passive, suffering recipient of the Father's will. Ethnographic material show that the Marian pilgrims are fascinated with bodily suffering, suffering registered not only at the level of their own bodies, but at the level of social bodies, especially in terms of a felt dissolution of the heterosexual nuclear family. Mary, then, becomes iconic of what is required (Motherly submission and suffering) to reproduce the Holy Family – a social body which, as the authors point out, is increasingly difficult to reproduce in a post-industrial Western world.

The creative possibilities of such excessive bodily sufferings are brought into the realm of adult desire by de Lauretis (Chapter 13). De Lauretis inverts Freud's theory of perversion ("inversion") by using Radclyffe Hall's novel, *The Well of Loneliness*, to show how and why lesbian desire exists. Counter to Freud, she posits a form of lesbian fetishism whereby the fetish is not solely a defensive male gesture of disavowing the mother's (read all women's) symbolic castration. Rather, a lesbian can also fetishize – but in terms of the missing maternal body; what she disavows is her individuation, placing difference itself (and thus desire) onto whatever feminine object, body part, or body recalls (and refuses) that loss. Her re-reading of Freud is important not only in terms of re-figuring lesbianism (and therefore a different kind of body) but in terms of re-figuring how excesses in economies of desire are transferred into the world itself. Thus, de Lauretis's work prompts new *spatial* figurations and imaginings of bodily displacement and desire.

Creating new space for desiring and desirable bodies, deemed too transgressive to accommodate in normative body/place, is the subject of Johnston's work (Chapter 14) on female body builders and gyms. Female body builders disrupt biologistic binaries of masculinity and femininity. Breasts reduced in size (or eliminated), menstruation stopped, fat content nil, muscles hard and powerful and strong, protruding muscular forms – all of these produce dis-ease and disgust. Steroids for women are vilified: they make women not-women; steroids for men, make men into "more" men. Accordingly, women are encouraged into fitness centres and discouraged from the spaces of weight training, Johnston carefully exposing the carefully engineered gym divisions of gender-differentiated body/spaces. Nonetheless, these transgressive body/spaces exceed the confines of the gym, spilling out into public arenas (e.g., competitions), forcing an awareness of the political and cultural investments which inhere in normative body/places.

Blum (Chapter 15) shows how the gendered shaping and spatial channeling of bodies – along with the refusals to accommodate body/space transgressions – begin at birth and are never-ending. Gendered differentiations occur through forceful dichotomies of movements, signaled by such unspoken proscriptions as those spelled out on the doors of toilets – Ladies or Gentlemen. Blum carries forward this analysis of toilet-bound sex-encoding using two stories of stories: Lacan's account of two children on a train who mistakenly assume that the names affixed on the entrance doors to the public toilets of the train station spell out the name of the "country" in which they have stopped (the boy only sees the word Gentleman; the girl, only the word Ladies); and a contemporary cyber-game called *Leather Goddesses of Phobos*, which begins by asking players to choose and enter a bathroom door. In either case, these doors (channeling subjects into separate lands of experience) at-once suggest thresholds, choice, confinement, and movement. The door we choose irrevocably establishes foundations for further gendered movement (the ladies fitness centre versus the men's weight room). The doors are not supposed to function as two-way valves, allowing both backward and forward movement; one must choose and (without looking back) go on. Nonetheless, as Blum points out, using other psychoanalytic accounts, children continually exceed the limits of sexual (and other modes of) identification through, for example, blurring boundaries between themselves and others, between interior and exterior.

Part 4, *ProjectingPlacesBodies* demonstrates that the body is woven into many layers of signification, through projection of images onto seemingly blank surfaces, onto the unwritten surfaces of the skin; through the writing of meanings onto bodies and places by intersecting fields of discourse. The body is told, and acted out, through the stories that are folded into it. These stories are never singular, never complete. However, the body itself speaks its own language – and never remains silent for long.

Writing cultural codes onto the bodies of migrant female Filipina "domestic workers" (nannies) are discussed by Pratt (Chapter 16). For her, these inscriptions are produced not just by the meanings that are placed on bodies, but also through form-filling governmental regulations, and also through various kinds of Canadian cultural refusals. The Canadian government requires, for example, that domestic workers live inside their employers' homes *and* that they do not increase their formal educational level! Thus, former skills and professional identities are denied and debased at the same time that the women are forcibly and spatially familiarized in necessarily heterosexualized ways: the woman is hired to serve, but she is also assigned a family bedroom. Like a sister or daughter, the room can be opened and used by family members and guests; like the "wife," she cooks and cares for the family; and she is to eat and drink with, and be loyal to, the family. This dependent insertion into family space brings with it heterosexualized tensions, invasions of privacy, and a painful erosion of cultural identity. In one of the most poignant discussions, about food, Pratt shows how "ethnic" and "bodily" cultural registers extend simultaneously into bodies and into place. Employers complain that a nanny consumes too much food, while the nanny is afraid to cook foods that smell, taste, or look "ethnic." The nannies find ways of not sticking out: some sit by windows as they sneak-eat their own food; upon spotting an employer, rushing to throw out or hide their food; others go to relative's homes to eat. Afraid that they might be accused of stealing, some workers wear gloves when handling jewellery; afraid of accusations of husband-theft, some women make themselves physically unattractive (not combing hair, wearing oversized t-shirts); afraid to become the images projected onto them, they find ways of resisting and surviving the highly circumscribed places they inhabit by projecting their own bodies and meanings into their world. The women, thereby, resist their experience of being all too visible by actively inscribing themselves into safe(r) spaces, into new forms of associations, such as domestic worker organizations and through their own socio-spatial "ethnic" networks.

It is with the idea of inscription, and a sense that struggle will involve ways of negotiating surveillance and erasure, that Sparke begins his interpretation of the maps of a Beothuk woman, Shawnadithit, living in what was to become Canada (Chapter 17). Sparke argues that colonial practices required and produced specific ways of mapping the world. These cartographic techniques assumed either that native maps were effectively useless or that the land was somehow unmapped and unpeopled. On the contrary, Sparke shows that the first nations in Canada used maps, though these were not constructed on the rational spatial grids that the colonizers were using. Instead, maps – such as those of Shawnadithit – were evocative of stories, the dreams, that underlay them. In this way, native maps came to represent not only the story of colonization, but also the heart-rending genocide of native peoples. Like Pratt's domestic workers, Shawnadithit lies in-between impossible positions: between a dying colonized people and a now dominant colonizer. Her maps trace out a fast-fading way of life, but they also embody that world. Her maps survive. Yet, Sparke worries that historical recoveries of these maps are too systematic, produced by the same impulse to control and know the native. These projections disembody Shawnadithit's dream-maps, desiccated by

13

historical scrutiny. Instead, the spaces of colonization are painful memories of disappearance and re-emergence.

Waller's work (Chapter 18) also draws upon a contemporary setting of colonization, the Maori in New Zealand, but this time as read through the highly acclaimed 1993 New Zealand film, *Once were Warriors*. Here, Maori identity is depicted as a collection of urban survival sites occupied by members of a single family. An abusive father whose second home is an isolated bar, an abused mother who stays at home, one son striving to become a member of a Maori gang (their turf defined by barbed wire and its own security system; their bodies defined by face tattoos), the second son who comes of age in prison, and a daughter who is raped and hangs herself, taking up only marginal life-spaces in the film. Waller analyzes the bodies and places in these images to question what "traditional" Maori identity might be in a postmodern urban world of juxtaposed cultural styles (traditional and modern tattooing, hip hop, reggae, doo-wop, electric blues, techno-pop) and places ("traditional village," barbed-wire gang enclave, the pub, "home").

In the films of Gus van Sant, the instability of male identities are represented as hysterical, as somehow "out of place." It is this male hysteria that Lukinbeal and Aitken take as the pivotal theme of their interpretation of *To Die For*, *Drugstore Cowboy* and *My Own Private Idaho* (in Chapter 19). In these films, they suggest, Van Sant dramatizes the ways in which patriarchal social relationships simultaneously rely on, and are resisted by, masculinities that are marginal, whether they be "emasculated" or "effeminate." Lukinbeal and Aitken suggest that it is precisely because hysterical masculinities can, or are authorized to, dominate women that patriarchy can be sustained – at every scale of analysis, from the body to geopolitics. Nevertheless, they see the possibility in these hysterias that new trangressive paces for sexual and political identities can be – and are being – opened up. However, they recognize that the aberrant sexual practices and identities are "rewarded" in Van Sant's films with death.

In Chapter 20, Bruno deals with the ways in which sexuality and death are written through the female body.[4] She situates the striking images that the early twentieth century Italian film-maker Elvira Notari uses in her films within the social context of Italian Catholic iconography. Just as Martin and Kryst (Chapter 13) show how the images and sensibilities of Marian appearances are contagious across different sites, so Bruno shows how images from one context can be taken up and projected onto another screen, with contagious effect. In particular, Bruno argues that the production of images of the female body were designed to install a sense of awe and worship in the audience – the films deliberately painted the bodies of ordinary working class women into situations symbolically associated with the saints and with angels. In these depictions, the wounds of everyday life were to be suffered through reference to the suffering of saints (thus, adding to Martin and Kryst's story). A moral economy of desire and devotion was mapped through the body – and this body is significantly female (see also de Lauretis). Notari's films do not simply animate Catholic imagery, nor merely represent the female body, they are active in the production of topographies of female desire and the regulative fictions of femininity, in the straining of desire through place and through the body. In this sense, these images (Notari's, Catholicism's) are also places through the body.

* * * * *

These are the images, themes, that spoke to us personally, as geographers, as people: the lines of argument we chose to draw out from each work. But, like pick-up-sticks, many lines could be chosen, juxtaposed, or heaped upon one another, the heaping being necessarily both purposeful and contingent. Maybe this image does not work for you and you might have fun imagining other schemes, other ways to imagine – as you read – places through the body.

Notes

1 A representative list of studies of the body would stretch to many pages and would, therefore, be long and useless. Instead, we would like to filter this list by dealing with what might be called the body politic. Studies of the body have been provoked, arguably, by politics associated with social movements such as feminism, civil rights and sexual liberation. These have, in one way and another, seen oppression and liberation as starting with the body. From this perspective, the following references might provide a flavor of what is out there. We are all too aware of the brevity of this list, but we hope that it gives a starting point for further explorations of the literature.

On various aspects of contemporary sociology and the body, see Giddens (1992), Shapiro and Alker (1996), Shilling's classic study (1993), Synnott (1993) and Turner (1996). The body has been thoroughly examined in art practices, for example, see Foster (1995), Mirzoeff (1995), Pollock (1996), Salaman (1994) and Schneider (1997). One of the leading historians of the body is Michel Foucault, see for example Foucault (1975), and Jones and Porter (1994). Histories of the body have been provided, also, both by Gallagher and Laqueur (1987) and Laqueur (1990), and by historians of science, see Turner (1992), and Jacobus, Keller and Shuttleworth (1990). Anthropologists have noted differences in bodily practices, see Douglas (1966) and Lingis (1994), including the emotions, see Harré and Parrott (1996) and Lutz (1988). The cultural production and regulation of bodies has been demonstrated in Epstein and Straub (1991), Featherstone, Hepworth and Turner (1991) and Stallybrass and White (1986). In many texts on the body, the gendering and sexing of bodies is seen to be of great significance. On this, see Butler (1993), Grosz and Probyn (1995) and Russo (1994). The relations of power associated with the gendered and sexed body are a central concern of corporeal feminisms, see Bordo (1993), Butler (1990), Davis (1997) and Grosz (1994). This has prompted discussions of body ethics by feminists, see Gatens (1996) and Irigaray (1977, 1993). Racialized and racist corporeal schemas produced in colonial practices have been explored in Gordon, Sharpley-Whiting and White (1996), Low (1996), McClintock (1995), Read (1996) and Young (1995). In these texts, intersections of race, sexuality and gender become increasingly important. These intersections have been explored, through the situatedness of bodies in particular spaces and places, by Bruce Pratt (1984), Diprose and Ferrell (1991) and, in a series of essays, by Grosz (1995). One particular site, where bodies and spaces/places are seen as co-constitutive, is the city, see Sennett (1994) and Vidler (1992, 1993).

2 Like many other disciplines, the body has been an "absent presence" in much of the literature, but geographers have also sought to explore lived experiences of space and place. For different styles of analysis, see for example Lefebvre (1974), but also see Stewart's discussion of the body in Lefebvre's work (1995) and Nast and Wilson's use of Lefebvre's work on spaces and bodies (1994); the humanistic approaches of Rodaway (1994) and Seamon (1979), and the shift from time geography to a geography of practices in the work of Thrift (1996). A critical review of the geographical literature is provided by Longhurst (1995). More recently, cultural geographers have increasingly addressed themselves to discussions of the body and gender and sexual identity, see Bell and Valentine (1995), Duncan (1996), Jones, Nast and Roberts (1997) and Rose (1993). Further discussions of the body, space and subjectivity are provided by Pile (1996) and in Pile and Thrift (1995). While the imposition of boundaries as a way of delimiting bodies and their spaces have been explored by Kirby (1996) and Sibley (1995).

3 One might follow the looking glass to other, more academic, wonderlands: see Grosz (1994: 175–6), but also Irigaray's Alice in *The Sex Which is Not One* (1977: chapter 1) and Deleuze (1990).

4 This chapter is taken from Bruno's book, *Streetwalking on a Ruined Map* (1993), where she is concerned to map out the imaginative geographies seen in the films of Elvira Notari. Notari (1875–1946) was the motivating force behind Dora Films, which made over 60 feature films and 100 documentaries in Naples, Italy, between 1906 and 1930. Bruno shows that the body, place and social context were the organizing elements, allowing the plot of the film to literally "take place." She argues that the narrative tensions, often dramatized through desire and fear, are also spatial tensions: desire is presented as a distance between bodies, and especially the (un)attainability of the female body.

References

Bell, D. and Valentine, G., eds, 1995, *Mapping Desire: geographies of sexuality* (London, Routledge).

Bordo. S., 1993, *Unbearable Weight: Feminism, Western Culture and the Body* (Berkeley, University of California Press).

Bruce Pratt, M., 1984, "Identity: skin, blood, heart" in E. Bulkin, M. Bruce Pratt and B. Smith, eds, *Yours in Struggle: three feminist perspectives on anti-semitism and racism* (New York, Long Haul Press) pp. 11–63.

Butler, J., 1990, *Gender Trouble: feminism and the subversion of identity* (London, Routledge).

Butler, J., 1993, *Bodies That Matter: on the discursive limits of "sex"* (New York, Routledge).

Davis, K., ed., 1997, *Embodied Practices: feminist perspectives on the body* (London, Sage).

Dear, M., 1997, "Identity, Authenticity and Memory in Place-Time" in S. Pile and M. Keith, eds, *Geographies of Resistance* (London, Routledge) pp. 219–35.

Deleuze, G., 1990, *The Logic of Sense* (New York, Columbia University Press).

Diprose, R. and Ferrell, R., 1991, *Cartographies: poststructuralism and the mapping of bodies and spaces* (Sydney, George Allen & Unwin).

Douglas, M., 1966, *Purity and Danger: an analysis of the concepts of pollution and taboo* (London, Routledge).

Duncan, N., ed., 1996, *BodySpace: destabilizing geographies of gender and sexuality* (London, Routledge).

Engels, F., 1844, *The Condition of the Working Class* (1969, Harmondsworth, Penguin).

Epstein, J. and Straub, K., eds, 1991, *Body Guards: the cultural politics of gender ambiguity* (London, Routledge).

Featherstone, M., Hepworth, M. and Turner. B. S., eds, 1991, *The Body: social process and cultural theory* (London, Sage Publications).

Foster, S. L., ed., 1995, *Corporealities: dancing knowledge, culture and power* (London, Routledge).

Foucault, M., (1975) *Discipline and Punish: the birth of the prison* (1977, Harmondsworth, Penguin).

Gallagher, C. and Laqueur, T., eds, 1987, *The Making of the Modern Body: sexuality and society in the nineteenth century* (Berkeley, University of California Press).

Gatens, M., 1996, *Imaginary Bodies: ethics, power and corporeality* (London, Routledge).

Giddens, A., 1992, *The Transformation of Intimacy: sexuality, love and eroticism in modern societies* (Cambridge, Polity Press).

Gordon L. R., Sharpley-Whiting T. D. and White R. T., eds, 1996, *Fanon: a critical reader* (Oxford, Basil Blackwell).

Grosz, E., 1994, *Volatile Bodies: toward a corporeal feminism* (Bloomington, Indiana University Press).

Grosz, E., 1995, *Space, Time and Perversion* (London, Routledge).

Grosz, E. and Probyn, E., eds, 1995, *Sexy Bodies: the strange carnalities of feminism* (London, Routledge).

Harré, R. and Parrott, W. G., eds, 1996, *The Emotions: social, cultural and physical dimensions* (London, Sage).

Irigaray, L., 1977, *This Sex Which is Not One* (1985, Ithaca, New York, Cornell University Press).

Irigaray, L., 1993, *An Ethics of Sexual Difference* (London, Athlone Press).

Jacobus, M., Keller E. F. and Shuttleworth, S., eds, 1990, *Body/Politics: women and the discourses of science* (London, Routledge).

Jones, C. and Porter, R., eds, 1994, *Reassessing Foucault: power, medicine and the body* (London, Routledge).

Jones, J. P., Nast, H. J. and Roberts, S., eds, 1997, *Thresholds in Feminist Geography: difference, methodology, representation* (Lanham, Rowman & Littlefield).

Kirby, K., 1996, *Indifferent Boundaries: exploring the space of the subject* (New York, Guilford Press).

Laqueur, T., 1990, *Making Sex: body and gender from the Greeks to Freud* (Cambridge, Mass., Harvard University Press).

Lefebvre, H., 1974, *The Production of Space* (1991, Oxford: Basil Blackwell).

Lingis, A., 1994, *Foreign Bodies* (London, Routledge).

Longhurst, R., 1995, "The body and geography" *Gender, Place and Culture* 2(1) pp. 97–105.

Low, G. C.-L., 1996, *White Skins, Black Masks: representation and colonialism* (London, Routledge).

Lutz, C., 1988, *Unnatural Emotions: everyday sentiments on a Micronesian Atoll and their challenge to Western theory* (Chicago, University of Chicago Press).

McClintock, A., 1995, *Imperial Leather: race, gender and sexuality in the colonial context* (London, Routledge).

Mirzoeff, N., 1995, *Bodyscape: art, modernity and the ideal figure* (London, Routledge).

Nast, H. J. and Wilson, M. O., 1994, "Lawful transgressions: this is the house that Jackie built . . . " *Assemblage: a critical journal of architecture and design culture* 24 pp. 48–56.

Pile, S., 1996, *The Body and the City: psychoanalysis, space and subjectivity* (London, Routledge).

Pile, S. and Thrift, N., eds, 1995, *Mapping the Subject: geographies of cultural transformation* (London, Routledge).

Pollock, G., ed., 1996, *Generations and Geographies in the Visual Arts: feminist readings* (London, Routledge).

Read, A., ed., 1996, *The Fact of Blackness: Frantz Fanon and visual representation* (London, Institute of Contemporary Arts and Institute of International Visual Arts).

Rich, A., 1980, "Compulsory Heterosexuality and Lesbian Existence" in A. Rich, *Blood, Bread, and Poetry: selected prose 1979–1985* (1986, London, Virago Press) pp. 23–75.

Rich, A., 1984, "Notes toward a Politics of Location" in A. Rich, *Blood, Bread, and Poetry: selected prose 1979–1985* (1986, London, Virago Press) pp. 210–231.

Rodaway, P., 1994, *Sensuous Geographies: Body, Sense and Place* (London, Routledge).

Rose, G., 1993, *Feminism and Geography: the limits of geographical knowledge* (Cambridge, Polity Press).

Russo, M., 1994, *The Female Grotesque: risk, excess and modernity* (New York, Routledge).

Salaman, N., ed., 1994, *What She Wants: women artists look at men* (London, Verso).

Sassen, S., 1994, *Cities in World Economy* (Thousand Oaks, CA., Pine Forge Press).

Sassen, S., 1996a, "Rebuilding the Global City: economy, ethnicity and space" in A. D. King, ed., *Re-Presenting the City: ethnicity, capital and culture in the 21st-Century Metropolis* (London, Macmillan) pp. 23–42.

Sassen, S., 1996b, "Analytic Borderlands: race, gender and representation in the new city" in A. D. King, ed., *Re-Presenting the City: ethnicity, capital and culture in the 21st-Century Metropolis* (London, Macmillan) pp. 183–202.

Schneider, R., 1997, *The Explicit Body in Performance* (London, Routledge).

Seamon, D., 1979, *A Geography of the Lifeworld: movement, rest, and encounter* (London, Croom Helm).

Sennett, R., 1994, *The Flesh and the Stone: the body and the city in western civilisation* (London, Faber & Faber).

Shapiro, M. J. and Alker, H. R., eds, 1996, *Challenging Boundaries: global flows, territorial boundaries* (Minneapolis, University of Minneapolis Press).

Shilling, C., 1993, *The Body and Social Theory* (London, Sage).

Sibley, D., 1995, *Geographies of Exclusion: society and difference in the west* (London, Routledge).

Soja, E., 1996, *Thirdspace: journeys to Los Angeles and other real-and-imagined places* (Oxford, Basil Blackwell).

Stallybrass, P. and White, A., 1986, *The Politics and Poetics of Transgression* (London, Methuen).

Stewart, L., 1995, "Bodies, visions, and spatial politics: a review essay on Henri Lefebvre's *The Production of Space Environment and Planning D: Society and Space* 13 pp. 609–18.

Synnott, A., 1993, *The Body Social* (London, Routledge).

Thrift, N., 1996, *Spatial Formations* (London, Sage).

Turner, B. S., 1992, *Regulating Bodies: essays in medical sociology* (London, Routledge).

Turner, B. S., 1996, *The Body and Society: explorations in social theory* (second edition, London, Sage).

Vidler, A., 1992, *The Architectural Uncanny: essays in the modern unhomely* (Cambridge, Mass., MIT Press).

Vidler, A., 1993, "Bodies in space/subjects in the city: psychopathologies of modern urbanism" *Differences* 5(3) pp. 31–51.

Young, R., 1995, *Colonial Desire: hybridity in theory, culture and race* (London, Routledge).

1

FilteringPlacesBodies

2

QUEER(Y)ING GLOBALIZATION

J. K. Gibson-Graham

It was an article on rape by Sharon Marcus that first "drove home" to me the force of globalization. The force of it as a discourse, that is, as a language of domination, a tightly scripted narrative of differential power. What I mean by "globalization" is that set of processes by which the world is rapidly being integrated into one economic space via increased international trade, the internationalization of production and financial markets, the internationalization of a commodity culture promoted by an increasingly networked global telecommunications system. Heralded as a "reality" by both the right and the left, globalization is greeted by the right with celebration and admiration. On the left, it tends to be viewed in a more sinister light as the penetration (or imminent penetration) of capitalism into all processes of production, circulation and consumption, not only of commodities but also of meaning:[1]

> [T]he prodigious new expansion of multinational capital ends up *penetrating* and colonizing those very precapitalist enclaves (Nature and the Unconscious) which offered extraterritorial and Archimedean footholds for critical effectivity.
>
> (Jameson 1991: 49, emphasis mine)

> Multinational capital formation . . . no longer makes its claims through direct colonial subjugation of the subject, but rather by the hyperextension of interpellative discourses and representations generated with and from a specifically new form of capital *domination*. Thus, it is important to recognize that *domination* occurs intensively at the levels of discourse, representation, and subjectivity.
>
> (Smith 1988: 138, emphasis mine)

The dynamic image of penetration and domination is linked to a vision of the world as already or about to be wholly capitalist – that is, a world "rightfully owned" by capitalism.

It was in the light of this pervasive discourse of capitalist penetration and of assertions that the Kingdom of Capitalism is *here* and *now* that Marcus's argument about rape spoke to me so directly. Marcus challenges the inevitability of the claim that rape is one of the "real, clear facts of women's lives" (1992: 385). In her suspicion of claims about "reality" and the un-mediated nature of women's "experience," indeed, in her refusal to recognize "rape as the real fact of our lives" (1992: 388), Marcus is led to explore the construction of rape as a cultural artifact, and to suggest on the basis of her exploration that we might begin to imagine an alternative "script of rape" (1992: 387). This project inspired me to interrogate globalization as a narrative of capitalist domination, hoping to become more active in the face of it, less prey to its ability to map a social terrain in which I and others are relatively powerless.

Becoming subject, unbecoming victim

Marcus understands a script as a narrative – "a series of steps and signals" (1992: 390) – whose course and ending is not set. In her conception, a script involves the continual making and remaking of social roles by soliciting responses and responding to cues, and in this sense she highlights the ways in which a script can be challenged from within (1992: 391, 402). When she narrates the common rape script she draws our attention to the choices that are made, through which unset responses become set and woven into a standardized story.

The standardized rape script is embedded in a language of social representation that permits the would-be rapist to constitute feelings of power and causes the woman to experience corresponding feelings of terror and paralysis. In the gendered grammar of violence, men are the subjects of violence and aggression. Their bodies are hard, full, projectile, and bio-logically endowed with the strength to commit rape. Women are naturally weaker than men. They can employ empathy, acquiescence or persuasiveness to avert (or minimize the violence of) rape, but they cannot physically stop it. In the gendered grammar of violence, women are the subjects of fear. Their bodies are soft, empty, vulnerable, open.

There are many obvious points of connection between the language of rape and the language of capitalist globalization. In the globalization script, especially as it has been strengthened and consolidated since 1989, only capitalism has the ability to spread and invade. Capitalism is represented as inherently spatial and as naturally stronger than the forms of noncapitalist economy (traditional economies, "third world" economies, socialist economies, communal experiments) because of its presumed capacity to universalize the market for capitalist commodities. In its most recent guise, "the market" is joined by the operations of multinational corporations (MNCs) and finance capital in the irreversible process of spreading and spatializing capitalism.

Capitalist social and economic relations are scripted as penetrating "other" social and economic relations but not vice versa. (The penis can penetrate or invade a woman's body, but a woman cannot imprint, invade or penetrate a Man.) After the experience of penetration – by commodification, market incorporation, proletarianization, MNC invasion – something is lost, never to be regained. All forms of noncapitalism become damaged, violated, fallen, subordinated to capitalism:

> [C]apital, through its command over the logic of social cooperation, envelops society and hence becomes social: capital has to extend its logic of command to cover the whole of society. . . . (Footnote: It is not being claimed here that "real subsumption" precludes the existence of noncapitalist formations. Rather, in "real subsumption" a form of social cooperation exists that enables even noncapitalist formations to be inserted into a system of accumulation that renders them productive for capital.)
>
> (Surin 1994: 13, interpreting Negri)

To paraphrase Marcus (1992: 399), the standardized and dominant globalization script constitutes noncapitalist economic relations as inevitably and only ever sites of potential invasion/envelopment/accumulation, sites that may be recalcitrant but are incapable of retaliation, sites in which cooperation in the act of rape is called for and ultimately obtained.

For the left, the question is, how might we challenge the dominant script of globalization and the victim role it ascribes to workers and communities in both "first" and "third" worlds? To break from the victim role prescribed for her in the rape script, Marcus suggests, a woman might change the script, use speech in unexpected ways, "resist self-defeating notions of polite feminine speech as well as develop physical self-defence tactics" (1992: 389). She might remember that an erection is fragile and quite temporal, that men's testicles are certainly no stronger than women's knees, that

> a rapist confronted with a wisecracking, scolding, and bossy woman may lose his grip on his power to rape; a rapist responded to with fear may feel his power consolidated.
>
> (Marcus 1992: 396)

Reading this I found myself asking, how might we get globalization to lose its erection – its ability to instill fear and thereby garner cooperation? I was drawn to an alternative reading of the globalization script that highlights contradictory representations of multinational corporations.

Like the man of the rape script, the MNC is positioned in the standard globalization script as inherently strong and powerful (by virtue of its size and presumed lack of allegiance to any particular nation or labor force):

TNCs are unencumbered with nationalist baggage. Their profit motives are unconcealed. They travel, communicate, and transfer people and plants, information and technology, money and resources globally. TNCs rationalize and execute the objectives of colonialism with greater efficiency and rationalism.

(Miyoshi 1993: 748)

But is this representation of a powerful agent able to assert and enact its will the only possible picture? Could we see the MNC in a different light – perhaps as a sometimes fragile entity, spread out and potentially vulnerable? The current literature on MNCs offers some promising counter-images.[2] On the basis of these, might it be possible to offer different (non-standard) responses to the set cues of the globalization script?

And I grabbed him by the penis, I was trying to break it, and he was beating me all over the head with his fists, I mean, just as hard as he could. I couldn't let go. I was determined I was going to yank it out of the socket. And then he lost his erection . . . pushed me away and grabbed his coat and ran.

(Bart and O'Brien, quoted by Marcus 1992: 400)

A non-standard response to the tactics of globalization was offered by the United Steel Workers of America (USWA) during the early 1990s when Ravenswood Aluminum Company (RAC) locked members of Local 5668 out over the negotiation of a new contract (Herod 1995). Research into the new ownership of RAC (which had been brought about by a leveraged buyout) established that the controlling interest was owned by a global commodities trader who had been indicted by the US Department of Justice on 65 counts of tax fraud and racketeering and was now residing in Switzerland to avoid his jail sentence (Herod 1995: 350). The union surmised that the "key to settling the (contract) dispute . . . lay in identifying (the trader's) vulner-abilities and seeking to exploit them internationally" (p. 351). They decided (amongst other strategies) to allege that the company was a corporate outlaw " . . . thereby potentially damaging its public image and encouraging government examination of its affairs" (pp. 351–2) and, through the channels of several truly international labor organizations, persuaded unions in a number of countries to lobby their own governments against dealing with such a corporate operator. Their interventions ended up ranging from Switzerland, Czechoslovakia, Rumania, Bulgaria and Russia to Latin America and the Caribbean, all in all 28 countries on five continents. This inter-national strategy – combined with a domestic end-user campaign to boycott RAC products – resulted (20 months after initiating the action) in the approval of a new three year contract and in placing restrictions on the trader's expansion into Latin America and Eastern Europe. Terrier-like, the USWA pursued the company relentlessly around the globe yanking and pulling at it until it capitulated. Using their own globalized networks, workers met internationalism with internationalism and eventually won.

Reading globalization as Marcus reads rape, as a scripted series of steps and signals, allows me to see the MNC attempting to place regions, work forces and governments in positions of passivity and victimization and being met by a range of responses – some of which play into the standard script and others that don't. It helps me to challenge representations of the superior power of the MNC, to recognize how that power is constituted through language as well as in action, or more importantly, in a complex interaction between the two. As Ernesto Laclau has argued, we do ourselves an injury, and promote the possibility of greater injury, by accepting a vision in which "absolute power has been transferred . . . to the multinational corporations":

> A break must be made with the simplistic vision of an ultimate, conclusive instance of power. . . . One instance . . . is presented as if it did not have conditions of existence, as if it did not have a constitutive outside. The power of this instance does not therefore need to be hegemonically and pragmatically constituted since it has the character of a ground.
>
> (Laclau 1990: 58–9)

Inscribing sexual/economic identity

Stories of resistance, of cases where women averted rape or fought back, provide empowering images that might help women to "rewrite" the standard rape script. But Marcus is concerned to go beyond the tactics of reversal and individual empowerment. This leads her to another kind of interest in the metaphor of a script.

One of the powerful things about rape in our culture is that it represents an important *inscription* of female sexual identity. Marcus argues that we should

> view rape not as the invasion of female inner space, but as the forced *creation* of female sexuality as a violated inner space. The horror of rape is not that it steals something from us but that it *makes* us into things to be taken. . . . The most deep-rooted upheaval of rape culture would revise the idea of female sexuality as an object, as property, and as inner space.
>
> (Marcus 1992: 399, emphasis mine)

The rape act draws its legitimacy (and therefore the illegitimacy of resistance) from a very powerful discourse about the female body and female and male sexuality. Thus the subject position of victim for the woman is not created merely by the strength and violence of the rapist but also by the

27

discourse of female sexual identity that the rape script draws upon in its enactment:

> Rape *engenders* a sexualized female body defined as a wound, a body excluded from subject–subject violence, from the ability to engage in a fair fight. Rapists do not beat women at the game of violence, but aim to exclude us from playing it altogether.
>
> (Marcus 1992: 397 emphasis mine)

In Marcus's view, the creation of alternatives to the standard rape script is predicated upon a significant revision of the powerful discourses of sexual identity that constitute the enabling background of all rape events.

The feminist project of rewriting and reinscribing the female body and sexuality has borne much theoretical fruit in recent years, prompted especially by the work of French philosophers Irigaray, Kristeva and Cixous. Marcus's challenge and encouragement to marry this rethinking with strategizing a new (poststructuralist) feminist politics of rape prevention has prompted me to further thoughts about globalization and the politics of economic transformation. In particular, they have led me to consider how a rewriting of the male body and sexuality might affect views of capitalism and its globalizing capacities.

> Men still have everything to say about their sexuality, and everything to write. For what they have said so far, for the most part, stems from the opposition activity/passivity, from the power relation between a fantasized obligatory virility meant to invade, to colonize, and the consequential phantasm of woman as a "dark continent" to penetrate and "pacify."
>
> (Cixous 1980: 247)

Feminist theorists have generated many new representations to replace that of the vacant, dark continent of female sexuality. However, as Grosz remarks, the particularities of the male body that might prompt us to challenge its naturalized hard and impermeable qualities have largely remained unanalyzed and unrepresented (1994: 198). Discussion of bodily fluids, for example, is rarely allowed to break down the solidity and boundedness of the male body:

> Seminal fluid is understood primarily as what it makes, what it achieves, a causal agent and thus a thing, a solid: its fluidity, its potential seepage, the element in it that is uncontrollable, its spread, its formlessness, is perpetually displaced in discourse onto its properties, its capacity to fertilize, to father, to produce an object.
>
> (Grosz 1994: 199)

Grosz's suggestive words offer a brief glimpse of how we might differently conceive of the body of capitalism, viewing it as open, as penetrable, as weeping or draining away instead of as hard and contained, penetrating, and inevitably overpowering. Consider the seminal fluid of capitalism – finance capital (or money) – which has more traditionally been represented as the lifeblood of the economic system whose free circulation ensures health and growth of the capitalist body. As seminal fluid, however, it periodically breaks its bounds, unleashing uncontrollable gushes of capital that flow every which way, including into self-destruction. One such spectacle of bodily excess, a wet dream that stained markets around the globe, occurred in October 1987, when stock markets across the world crashed, vaporizing millions of dollars in immaterial wealth (Wark 1994: 169). The 1987 crash was for many the bursting of a bubble of irrationality, "a suitable ending, fit for a moral fable, where the speculators finally got their just deserts" (Wark 1994: 171–2). But the growth of activity on international financial markets represents an interestingly contradictory aspect of the globalization script.

One of the key features of globalization has been the complete reorganization of the global financial system since the mid 1970s:

> The formation of a global stock market, of global commodity (even debt) futures markets, of currency and interest rate swaps, together with an accelerated geographical mobility of funds, meant, for the first time, the formation of a single world market for money and credit supply.
>
> (Harvey 1989: 161)

On the one hand this growth has been seen to facilitate the rapid internationalization of capitalist production, and the consolidation of the power of finance capital is seen to represent the "supreme and most abstract expression" of capital (Hilferding, quoted in Daly 1991: 84). On the other, this growth has unleashed money from its role as a means of circulation and allowed the rampant proliferation of global credit.[3] Money has become

> a kind of free-floating signifier detached from the real processes to which it once referred. Through options, swaps and futures, money is traded for money. Indeed since much of what is exchanged as commodities are future monetary transactions, so what is traded in no sense exists.
>
> (Lash and Urry 1994: 292)

The whole relation between signifier, signified and referent has been ruptured, unleashing capitalism's "third nature," "the spectacle" (Debord 1983), "the enchanted world" (Lipietz 1985) in which the "economic real" is buried under the trade of risk, information, image, futures, to be revealed only in momentary displays such as occurred when stock markets crashed in October 1987 (Wark 1994). Globalization, it seems, has set money free of the "real economy" and allowed capital to seep if not spurt from the productive system. But the implications of this unboundedness, this fluidity, for the identity of capitalism remain unexplored. Having set the signifier free from the referent, theorists of the global economy are loath to think about the effects of seepage, porosity, uncontrollability, that is, to feminize economic identity (Grosz 1994: 203). The global economy may have been opened up by international financial markets, but nothing "other" comes into or out of this opening. It would seem that the homophobia that pervades economic theorizing places a taboo on such thinking.[4]

> Part of the process of phallicizing the male body, of subordinating the rest of the body to the valorized functioning of the penis, with the culmination of sexual activities occurring ideally at least, in sexual penetration and male orgasm, involves the constitution of the sealed-up, impermeable body. Perhaps it is not after all flow in itself that a certain phallicized masculinity abhors but the idea that flow moves or can move in two-way or indeterminable directions that elicits horror, the possibility of being not only an active agent in the transmission of flow but also a passive receptacle.
>
> (Grosz 1994: 200–1)

How might we confront the economic "unthinkable," engendering a vision of the global economy as penetrable by noncapitalist economic forms? Perhaps the very same financial system that is represented as both the origin of seepage and the agent of capital's assertion of identity might yield a further surplus of effects. It might, for example, be possible to see the proliferation of credit and deregulation of financial markets as creating opportunities for the growth of noncapitalist class relations as well as capitalist ones. The huge expansion of consumer credit over the past two decades (including credit card financing with large maximum limits, home equity loans, and a variety of other instruments almost forced upon "consumers") is often assumed to promote personal indebtedness associated with a culture of consumption. Yet, given the growth in self-employment and of home-based industries – some of which is associated with the downsizing and streamlining of capitalist firms – it is clear that much of what is seen as consumer credit is actually (or also) producer credit, in other words it is used to buy means of production (including computers and other equipment) and other inputs into the production processes of self-employed workers. Historically such loans have been notoriously difficult to obtain from traditional financial institutions like local and regional banks, but with the growth of new international credit markets they have become quite instantaneous and straightforward. This has contributed to an increase in small businesses that are sites of noncapitalist class processes of individual and collective surplus appropriation (as well as providing a source of credit to small capitalist firms). From this perspective, then, the financial sector can be seen as an *opening* in the body of capitalism, one that not only allows capital to seep out but that enables noncapitalism to invade.

The script of globalization need not draw solely upon an image of the body of capitalism as hard, thrusting and powerful. Other images are available, and while we cannot expect the champions of globalization to express pleasure in leakage, unboundedness and invasion, it is important to draw upon such representations in creating an anti-capitalist imaginary and fashioning a politics of economic transformation. If the identity of capitalism is fluid and malleable, able to penetrate and be penetrated, then the process of globalization need not constitute or inscribe "economic development" as inevitably *capitalist* development. Globalization might be seen as liberating a variety of different economic development paths. In fact, the script of globalization may already (without explicit instances of opposition) be engendering economic differences.[5]

Marcus encourages a rejection of the fixed sexual identity *inscribed* upon the body of women by the rape script. Given that this identity is rooted in a dominant discourse of heterosexuality (in which the bodies of men and women are distinguished by rigid and fixed gender differences and in which male and female behavior can only be understood in terms of opposition, complementarity or supplementarity), one of the implications of Marcus's argument is that rape (along with marriage) is a recognized and accepted (if not acceptable) practice of heterosexuality. Challenging the legitimacy our culture implicitly grants to rape becomes, in this formulation, a challenge to heteronormativity itself.

My desire to reject globalization as the inevitable inscription of capitalism prompts me to take Marcus's implication one step further and to explore the ways in which discourses of homosexuality might liberate alternative scripts or inscriptions of sexual/economic identity:

> [M]any gay men . . . are prepared not only to send out but also to receive flow and in this process to assert other bodily regions than those singled out by the phallic function. A body that is permeable, that transmits in a circuit, that opens itself up rather than seals itself off, that is prepared to respond as well as to initiate, that does not revile its masculinity . . . or virilize it . . . would involve a quite radical rethinking of male sexual morphology.
>
> (Grosz 1994: 201)

Queering globalization/"the universality of intercourse"

Rethinking capitalist morphology in order to liberate economic development from the hegemonic grasp of capitalist identity is indeed a radical project. Yet resources for such a project are already available in the domain of social theory, especially within queer theory, where a rethinking of sexual morphology is taking place. For queer theorists, sexual identity is not automatically derived from certain organs or practices or genders but is instead a space of transitivity (Sedgwick 1993: xii).

> [O]ne of the things that "queer" can refer to [is] the open mesh of possibilities, gaps, overlaps, dissonances and resonances, lapses and excesses of meaning when the constituent elements of anyone's gender, of anyone's sexuality aren't made (or *can't be* made) to signify monolithically.
>
> (Sedgwick 1993: 8, emphasis in original)

Sedgwick's evocation of the way in which things are supposed to "come together" at certain social sites – she speaks of the family, for example, where a surname, a building, a legal entity, blood relationships, a unit in a community of worship, a system of companionship and caring, a sexual dyad, the prime site of economic and cultural consumption, and a mechanism to produce, care for, and acculturate children[6] are "meant to line up perfectly with each other" (1993: 6) – captures the oppressiveness of familiar identity-constituting/policing discourses and provides a glimpse of the productiveness of fracturing and highlighting dissonances in seemingly univocal formations. Through its challenges to such correspondences and alignments, queer theory has encouraged me to attempt to rupture monolithic representations of capitalism and capitalist social formations.

One key conflation or "coming together" that participates in constituting a capitalist monolith is the familiar association of capitalism with "commodification" and "the market." When it is problematized, which is not very often, the presumed overlap between markets, commodities and capitalism will often be understood as the product of a capitalist tendency to foster what Marx calls "the universality of intercourse" (1973: 540). This tendency is central to the representation of the capitalist body as inherently capable of invading, appropriating and destroying:

> [W]hile capital must on one side strive to tear down every spatial barrier to intercourse, i.e. to exchange, and conquer the whole earth for its market, it strives on the other side to annihilate this space with time, i.e. to reduce to a minimum the time spent in motion from one place to another. . . . There appears here the universalizing

tendency of capital, which distinguishes it from all previous stages of production.

(Marx 1973: 539–40)

The world market is the site of economic reproduction of the global capital relation, as well as of the political organization of hegemony. An opening to the world market is thus synonymous with integration into the global process of economic reproduction and a historically determined system of hegemony.

(Altvater 1993: 80–1)[7]

In discussions of the bodily interactions (that is, market transactions) between capitalism and its "others," the metaphor of infection sometimes joins those of invasion and penetration:

[C]apital is the virus of abstraction. It enters into any and every social relation, corrupts it, and makes it manufacture more relations of abstraction. It is a form of viral relations which has a double aspect. It turns every qualitative and particular relation into a quantitative and universal one.

(Wark 1994: xii)

It is perhaps not surprising that capitalism is represented as a body that invades and infects, but is not itself susceptible to invasion or infection. In the shadow of the AIDs crisis queer theorists have shed light upon the homophobia that pervades social theory, if not the specific heteronormativity of economic theory (Sedgwick 1993). The infection metaphor suggests an "immensely productive incoherence" (to use Eve Sedgwick's phrase (1993: xii)) that may help to disrupt the seamlessness of capitalist identity, reconfiguring capitalism's morphology in ways that our earlier rethinking of heterosexual rape and penetration could not.

The locus and agent of universalizing intercourse is "the market," which continually seeks new arenas/bodies in which to establish a medium, or cir-cuitry, through which contamination by capitalism may flow. Globalization discourse highlights the one-way nature of this contamination and the virtual impossibility of immunity to infection. But markets/circuits cannot control what or who flows through them. The market can, in fact, communicate many diseases, only one of which is capitalist development. Consider as an example the increasingly international labor market, another character in the standard script of globalization. The marked increase in international migration and the establishment of large immigrant "underclasses" in the metropolitan capitals (world cities) of the global economy since the 1960s is attributed both to economic and political destabilization in "donor" countries and to the voracious appetite of "first world" capitalism for cheap labor, particularly in the growing service industries and low-wage manufacturing (Sassen 1988). The loss of labor from source countries is usually portrayed as yet another outcome of the penetration of capitalism, via a process in which "the market" or "commodification" has destroyed the traditional, often agricultural, economy and forced people into proletarianization.

When incorporated into a script of globalization, the extraordinary economic diversity of immigrant economies in "host" countries is subsumed to, or depicted as an "enclave" within, the capitalist totality, located on a trajectory of homogenization or synthesis with the host society. The noncapitalist nature of immigrant entrepreneurial activity is documented for its cultural interest, but is rarely allowed the autonomy afforded to "capitalist enterprise." Yet immigrant economies made up of self-employed, feudal and communal family-based enterprises (as well as small capitalist enterprises) operate their own labor and capital markets, often on a global scale (Collins *et al.* 1995; Waldinger 1986). They maintain complex economic as well as political and cultural connections with other diasporic communities as well as with their "home" countries.

Even immigrant workers who may be wage laborers in metropolitan capitalist economies cannot be seen as "subsumed" in any complete sense. Rouse (1991) documents how Mexican immigrants working as wage laborers in service industries of US cities manage two distinct ways of life, maintaining small family farms or commercial operations in Mexico. Money received as wages in the US is siphoned into productive investment in noncapitalist activity in the Mexican economy.[8]

> Aguilillans have come to link proletarian labor with a sustained attachment to the creation of small-scale, family based operations. . . . Obliged to live within a transnational space and to make a living by combining quite different forms of class experience, Aguilillans have become skilled exponents of a cultural bifocality that defies reduction to a singular order.
>
> (Rouse 1991: 14–15)

Globalization, it would seem, has not merely created the circuitry for an increased density of international *capital* flows: "Just as capitalists have responded to the new forms of economic internationalism by establishing transnational corporations, so workers have responded by creating transnational circuits" (Rouse 1991: 14). Labor flows have also grown, and with them a variety of noncapitalist relations.

This case is a small but suggestive example of the productive incoherence that can be generated through a metaphor of infection – an infection whose agent is the market (in this instance the international labor market). It challenges the imperial nature of capitalism, depriving capitalism of its role as sole initiator of a spatially and socially expansive economic circuitry of infection. If capitalist globalization is an infection, it can be said to coexist with many other types of infection. Were we not bedazzled by images of the superior morphology of global capitalism, it might be possible to theorize the global integration of noncapitalist economic relations and non-economic relations and to see capitalist globalization as coexisting with, and even facilitating, the renewed viability of noncapitalist globalization.

Another productive incoherence introduced by the metaphor of infection arises from the possibility of immunity. Lash and Urry note that "(g)oods, labour, money and information will not flow to where there are no markets" such as into eastern Europe after the devastation of state economic governance or into the "black ghetto in the USA" (1994: 18). The globalization of the capitalist market is a spotty affair – characterized by cobweb-like networks of density and sparseness (p. 24), areas of infection and areas of immunity. Within the areas of sparseness/immunity the economic transactions that take place are seen to be of no consequence, and yet it is easy to imagine that in the interstices of the (capitalist) market, other markets do exist. And in these interstices noncapitalist commodities are exchanged.[9]

But the market is notoriously catholic in its tastes and desires for the exotic, the noncapitalist, the unclean. And, so goes the standard script, sooner or later, these areas of immunity, of capitalist vacuity, of noncapitalist commodity production, will come under the power of capitalism. It is only a matter of time before capitalism reaches out via "the market" into the unknown, infecting subjects with the desire for capitalist commodities and gathering to itself objects for exchange.

> Capitalist culture industries typically seek out the "new" in the cultures of various "others." . . . African American, Third World, gay and lesbian, working-class, and women's (sub)cultures frequently serve the function of research and development for cultural capitalists.
>
> (MacNeill and Burczak 1991: 122)

> [I]t is not the people who have nothing to lose but their chains, but the flow of qualitative information about people, their places, and the things they produce, in short their "culture." Culture . . . ceases in the process to be culture, and becomes instead a postculture or a transculture. . . . Culture is something that will be overcome – whether we like it or not.
>
> (Wark 1994: xiii)

Contact, via the market, with noncapitalist commodities or "other cultures" never permits the transmission of infection back into the capitalist body. Instead, in the wake of the capitalist market, comes the commodification of culture and the death of "authentic culture."

The homogenizing claims made for the global market and capitalist commodification are, in the eyes of Grewal and Kaplan, insufficiently interrogated:

> What is not clear in such debates is what elements of which culture (including goods and services) are deployed where, by whom, and for what reason. Why, for instance, is the Barbie doll sold in India but not the Cabbage Patch doll? Why did the Indian Mattel affiliate choose to market Barbie dressed in a sari while Ken remains dressed in "American" clothes? . . . Which class buys these dolls and how do children play with them in different locations? . . . Getting away from . . . "transnational centrism" acknowledges that local subjects are not "passive receptacles" who mechanically reproduce the "norms, values, and signs of transnational power."
>
> (1994: 13)

The market is not all or only capitalist, commodities are not all or only products of capitalism,[10] and the sale of Barbie dolls to Indian girls or boys does not at all or only presage the coming of the global heterosexist capitalist kingdom. A queer perspective can help to unsettle the consonances and coherences of the narrative of global commodification. That there need not be a universal script for this process is suggested by Mary Gossy's discussion of the use to which a Barbie-like doll is put in a series of photographs entitled "Gals and Dolls":[11]

> Barbie is not a dildo, not a penis substitute, but rather a figure of lesbian eroticism, a woman in the palm of a woman's hand. In fact, playing with Barbies and Kens teaches girls lesbian butch/femme roles – and thus in puberty the doll becomes a representative of lesbian desire.
>
> (1994: 23)

In the same spirit, might we not wonder exactly what is being globalized and how infection is moving when Barbie dolls are marketed in India or, for that matter, Indiana?[12] What power and potentialities do we relinquish when we accept the univocality of the market/commodity/global capitalist totality?

The severing of globalization from a fixed capitalist totality is enabled by a queering of economic identity, a breaking apart of the monolithic significations of capitalism (market/commodity/capital) and a liberation of different economic beings and practices. A space can be made for thinking of globalization as many, as other to itself, as inscribing different development paths and economic identities.[13] Globalization need not be resisted only through recourse to the local (its other within) but may be redefined discursively, in a process that makes room for a host of alternative scriptings, capable of inscribing a proliferation of economic differences.

Conclusion: rewriting the body

Making connections between the body and the social or economic totality, and using one as a metaphor for the other, is not a new way of thinking. Haraway (1991) reminds us, for example, that the concept of the body politic can be traced to the ancient Greeks. Interestingly, though, the now prevalent conception of the body as an "organism" seems to have its roots not only in antiquity but in modern economic discourse, specifically in the writings of Adam Smith. When Smith theorized the social division of labor as the most productive form of economic organization, he was envisioning a set of separate and specialized parts working together in efficient harmony to reproduce the social whole. This vision of a systemic totality, divided yet integrated, self-regulating and bent on reproduction, became a fruitful metaphor for eighteenth century students of animal nature, who even sometimes referred to the animal body as the "animal economy." In the physiological economy, various organs operated in a functional division of labor for the "common good" of the animal body.

Through a complex process of interaction and translation, what began as an economic conception developed into the fundamental physiological concept of the "organism," now a familiar feature of body discourse and representation.[14] This history suggests an interesting possibility. Perhaps our own reworking of globalization could coalesce, in generative and mutually reinforcing ways, with other bodily rethinkings.

In the shadow of the global (capitalist) economy we have tried to unthink the economic organism: to undermine the monism associated with organicist visions of an economic system; to erode the solidity conferred upon bounded entities, which in turn confers their ability to penetrate other less solid spaces; to distance ourselves from the heterosexual masculinism of nonreciprocal penetration; to return to the body economic what has been erased, excluded, negated, scorned – the noncapitalist activities that have become invisible or inconsequential in the presence of capitalism. What might it mean to theorize "the body economic" as *not* an organism? Not to theorize it, that is, as bounded and unified, hierarchically ordered, animated by a life force, and governed by a telos of reproduction? How might a different type of economic theory valorize and empower diverse figurings of the body?

Our "economy" is a space of difference where noncapitalist economic activities are not subordinated to a capitalist unity (as parts of a capitalist totality, or differences that are ultimately the same.) This economic body is neither the locus of a dominant drive (like expansion or reproduction) nor governed by a mentality (like the financial sector or the calculus of profitability). Nor is it the domain of a singular subject (like global capitalism), or solid or self-continuous (like the global market for capitalist commodities), or subsumed under a principal logic (like capital accumulation), or organized hierarchically (starting, say, with capital or capitalist production). In fact, in the organismic sense of subsumption to a unity, it is not organized at all. In many ways it recalls and reinforces the Body without Organs (BwO) of Deleuze and Guattari. Unlike the organism, which is understood in terms of what it contains and how it is organized, the BwO is specified in terms of surfaces, intensities and flows, connections, movements, lines of flight, bodily products, fragments of subjectivity, capabilities (unrealized by the organism because unimagined). It is a body that does not reflect other bodies or relate to them via negation or analogy. It is becoming but not becoming unity. No telos governs and completes it. Self-realization is not its trajectory. Destratified and demassified, it is not an emptiness or an amorphousness but a multiplicity.[15]

Such a body is undoubtedly queer. And just like the queer economy, where things don't line up and come together, where the market and the commodity do not necessarily signify capitalism and capitalism does not necessarily signify dominance, the queer body is a place of uncertainty and possibility. Gender, for example, does not necessarily accord with male or female embodiment, clothing, object choice, preferences for certain organs and orifices (or neither), identification, forms of dominance, self-presentation and manner (feminine or effeminate), SM top or bottom, or even with itself. Is a leatherdyke daddy a man? Is a dyke faggot (a butch who desires other butches) a woman? Are leatherdyke play parties – where partygoers have to certify that they are women at the door in order to enter a space where they can explore and develop their masculinities – male or female spaces? Or are these questions themselves part of the regulatory apparatus of heterosexual normativity (Hale 1996)?

At one and the same time, or in one and the same space, bodies may be gendered male and female, for different purposes and in different cultural contexts. (A male-embodied crossdresser may use the women's bathroom but may not be allowed to take a man in marriage, a legal process that requires a birth certificate.) It is difficult to presume a unitary and stable gender status across cultural (or subcultural) contexts and purposes (Hale 1996). What this suggests is not only that the gendered body is socially constructed but that it is fluid and multiple as well. This recognition can reinforce our doubts about calling an economy capitalist (or noncapitalist) in some definitive and permanent way. When a women's producer collective in South East Asia sells a commodity to a multinational corporation, is that a capitalist or noncapitalist transaction? Is it local or global? All one can say is that penetration (of metaphors and economies) seems to go both ways.

Acknowledgments

We gratefully acknowledge the helpful comments of Heidi Nast, Steve Pile and David Ruccio on earlier drafts of this paper.

Notes

1 Despite the criticisms of globalization that emanate from the left, there is an undercurrent of leftist desire for "penetration." Consistent with the often expressed view that the one thing worse than being exploited by capital is not being exploited at all, there is a sense that not to be penetrated by capitalism is worse than coming into its colonizing embrace. The ambivalent desire for capitalist penetration is bound up with the lack of an economic imaginary that can conceive of economic development which is not *capitalist* development (with its inherent globalization tendency), just as conceptions of sexuality that are not dominated by a phallocentric heterosexism (in which the act of penetration, whether called rape or intercourse, defines sexual difference) are difficult to muster.

2 See Gibson-Graham (1996, Chapter 6).

3 Mitchell (1995: 369) refers to the credit system characteristic of "casino capitalism" as a "feral credit system" running wild and uncontrolled, which "seems completely disconnected from the productive economy" in which the movement of goods has become "more and more negotiated and regulated" throughout the 1980s (Thrift 1990: 1136).

4 Indeed, as Whitford suggests, "(t)he house of the male subject is closed . . . whereas one characteristic of woman's sexual bodies is that they are precisely not closed; they can be entered in the act of love, and when one is born one leaves them, passes across the threshold. (One might argue that men's bodies can be entered too, but no doubt Irigaray would argue that the massive cultural taboo on homosexuality is linked to men's fear of the open or penetrable body/houses)" (1991: 159).

5 Kayatekin and Ruccio (1996: 14) point to the many forms of noncapitalist production and provisioning that are enabled by revenues that could be said to "flow out of global capitalism." These include wages paid by multinational corporations that are used to pay for inputs to "family" businesses which may be organized within individual, communal, feudal, or even slave relations of surplus appropriation (not to mention wages that support the household itself, which is a major locus of noncapitalist production).

6 All these features and more are listed in Sedgwick (1993: 6).

7 By contrast, Daly notes: "There is nothing . . . which is essentially capitalist about the market. Market mechanisms pre-exist capitalism and are clearly in operation in the socialist formations of today. It is clear, moreover, that a whole range of radical enterprises exist within the sphere of the market." (1991: 88).

8 Here we are making the not fully warranted assumption that family-based enterprises are noncapitalist (i.e., that family members are not hired as wage laborers by a family capitalist who appropriates surplus value from his/her family members). Presumably when Rouse speaks of "different forms of class experience" he is referring to the communal, feudal and even slave relations that may characterize family businesses; but since he does not actually specify the forms of class relations, we cannot exclude the possibility of capitalist exploitation nor can we assume that noncapitalist family relations are necessarily less exploitative than capitalist ones. The point is simply that capitalism does not necessarily breed capitalism.

9 Commodities (goods or services for sale) can of course be produced within a variety of class relations – slaves may produce commodities as may independent and collective commodity producers who appropriate their own surplus labor.
10 See note 9.
11 The photographs show a Barbie-like doll emerging from or entering a vulva.
12 Rand concludes her fascinating queer exploration of Barbie memories and practices within the US with this reflection: "Surprisingly often, the stories I heard were about how Barbie turned people into cultural critics and political activists – about how seeing activists queer Barbie, or remembering their own Barbie queerings, or hearing about my Barbie work, induced them to move from Barbie anecdotes to thinking about cultural politics, ideology, oppression, and resistance, and sometimes to political plotting and practice" (1995: 195).
13 Theorists such as Haraway have been concerned to make visible the many and heterogeneous globalized organizations that build alliances across situatednesses and establish "webs of systematicity" between locales (Harvey and Haraway 1995).
14 For an extended discussion of organismic metaphors of capitalism and economy, see Gibson-Graham (1996, Chapter 5).
15 This discussion relies on Grosz (1994, Chapter 7).

References

Altvater, E. 1993. *The Future of the Market: An Essay on the Regulation of Money and Nature After the Collapse of "Actually Existing Socialism."* Trans. P. Camiller. London: Verso.

Cixous, H. 1980. "The Laugh of the Medusa." In *New French Feminisms*, eds E. Marks and I. de Courtivron, 245–64. Amherst: University of Massachusetts Press.

Collins, J., Alcorso, C., Castles, S., Gibson, K., and Tait, D. 1995. *A Shopful of Dreams: Ethnic Small Business in Australia.* Sydney: Pluto.

Daly, G. 1991. "The Discursive Construction of Economic Space: Logics of Organization and Disorganization." *Economy and Society* 20(1): 79–102.

Debord, G. 1983. *Society of the Spectacle.* Detroit: Black and Red.

Gibson-Graham, J.K. 1996. *The End of Capitalism (As We Knew It): A Feminist Critique of Political Economy.* Oxford: Blackwell.

Gossy, M. 1994. "Gals and Dolls: Playing with Some Lesbian Pornography." *Art Papers*, November and December: 21–4.

Grewal, I. and Kaplan, C. (eds) 1994. *Scattered Hegemonies: Postmodernity and Transnational Feminist Practices.* Minneapolis: University of Minnesota Press.

Grosz, E. 1994. *Volatile Bodies: Toward a Corporeal Feminism.* Bloomington: Indiana University Press.

Hale, C.J. 1996. "Dyke Leatherboys and Their Daddies: How to Have Sex Without Men or Women." Paper presented at the Berkshire Conference on the History of Women, Chapel Hill NC, June.

Haraway, D. 1991. *Simians, Cyborgs, and Women: The Reinvention of Nature.* New York: Routledge.

Harvey, D. 1989. *The Condition of Postmodernity.* Oxford: Blackwell.

Harvey, D. and Haraway, D. 1995. "Nature, Politics and Possibilities: A Debate and

Discussion with David Harvey and Donna Haraway." *Environment and Planning D: Society and Space* 13: 507–27.

Herod, A. 1995. "The Practice of International Solidarity and the Geography of the Global Economy." *Economic Geography* 71(4): 341–63.

Jameson, F. 1991. *Postmodernism, or the Cultural Logic of Late Capitalism.* Durham NC: Duke University Press.

Kayatekin, S. and Ruccio, D. 1996. "Global Fragments: Subjectivity and Politics in Discourses of Globalization." Paper presented at the Allied Social Science Association Meetings, San Francisco, January.

Koechlin, T. 1989. "The Globalization of Investment: Three Critical Essays." PhD dissertation, University of Massachusetts, Amherst.

Laclau, E. 1990. *New Reflections on the Revolution of Our Time.* London: Verso.

Lash, S. and Urry, J. 1994. *Economies of Signs and Space.* London: Sage.

Lipietz, A. 1985. *The Enchanted World: Inflation, Credit and the World Crisis.* London: Verso.

MacNeill, A. and Burczak, T. 1991. "The Critique of Consumerism in 'The Cook, the Thief, His Wife and Her Lover.'" *Rethinking Marxism* 4(3): 117–24.

Marcus, S. 1992. "Fighting Bodies, Fighting Words: A Theory and Politics of Rape Prevention." In *Feminists Theorize the Political*, eds J. Butler and J. Scott, 385–403. London: Routledge.

Marx, K. 1973. *Grundrisse: Foundations of the Critique of Political Economy.* Trans. M. Nicolaus. New York: Random House.

Mitchell, K. 1995. "Flexible Circulation in the Pacific Rim: Capitalisms in Cultural Context." *Economic Geography* 71(4): 364–82.

Miyoshi, M. 1993. "A Borderless World? From Colonialism to Transnationalism and the Decline of the Nation State." *Critical Inquiry* 19(Summer): 726–51.

Rand, E. 1995. *Barbie's Queer Accessories.* Durham NC: Duke University Press.

Rouse, R. 1991. "Mexican Migration and the Social Space of Postmodernism." *Diaspora* 1(1): 8–23.

Sassen, S. 1988. *The Mobility of Labor and Capital: A Study in International Investment and Labor Flow.* Cambridge: Cambridge University Press.

Sedgwick, E.K. 1993. *Tendencies.* Durham NC: Duke University Press.

Smith, P. 1988. "Visiting the Banana Republic." In *Universal Abandon: The Politics of Postmodernism*, ed. A. Ross, 128–48. Minneapolis: University of Minnesota Press.

Surin, K. 1994. "Reinventing a Physiology of Collective Liberation: Going 'Beyond Marx' in the Marxism(s) of Negri, Guattari, and Deleuze." *Rethinking Marxism* 7(2): 9–27.

Thrift, N. 1990. "The Perils of the International Financial System." *Environment and Planning A* 22: 1135–40.

Waldinger, R. 1986. *Through the Eye of the Needle: Immigrants and Enterprise in New York's Garment Trades.* New York: New York University Press.

Wark, M. 1994. *Virtual Geography: Living with Global Media Events.* Bloomington IN: Indiana University Press.

Whitford, M. 1991. *Luce Irigaray: Philosophy in the Feminine.* London: Routledge.

3

BODIES-CITIES

Elizabeth Grosz

Congruent counterparts

For a number of years I have been involved in research on the body as sociocultural artifact. I have been interested in challenging traditional notions of the body so that we can abandon the oppositions by which the body has usually been understood – mind and body, inside and outside, experience and social context, subject and object, self and other, and underlying these, the opposition between male and female. Thus "stripped," corporeality in its sexual specificity may be seen as the material condition of subjectivity, that is, the body itself may be regarded as the locus and site of inscription for specific modes of subjectivity. In a "deconstructive turn," the subordinated terms of these oppositions take their rightful place at the very heart of the dominant ones.

Among other things, my recent work has involved a kind of turning *inside out* and *outside in* of the sexed body, questioning how the subject's exteriority is psychically constructed, and conversely, how the processes of social inscription of the body's surface construct for it a psychical interior. In other words, I have attempted to problematize the opposition between the inside and the outside by looking at the outside of the body from the point of view of the inside, and looking at the inside of the body from the point of view of the outside, thus re-examining and questioning the distinction between biology and culture, exploring the way in which culture constructs the biological order in its own image, the way in which the psychosocial simulates and produces the body as such. Thus I am interested in exploring the ways in which the body is psychically, socially, sexually, and discursively or representationally produced, and the ways, in turn, bodies reinscribe and project themselves onto their sociocultural environment so that this environment both produces and reflects the form and interests of the body. This relation of introjections and projections involves a complex feedback relation in which neither the body nor its environment can be assumed to form an organically unified ecosystem. (The very notion of an ecosystem implies a kind of higher-order unity or encompassing totality that I will try to problematize in this chapter.) The body and its environment, rather, produce each other as forms of the hyperreal, as modes of simulation which have overtaken and transformed whatever reality each may have had into the image of the other: the city is made and made over into the simulacrum of the body, and the body, in its turn, is transformed, "citified," urbanized as a distinctively metropolitan body.

One area that I have neglected for too long – and I am delighted to have the opportunity to begin to rectify this – is the constitutive and mutually defining relation between bodies and cities. The city is one of the crucial factors in the social production of (sexed) corporeality: the built environment provides the context and coordinates for most contemporary Western and, today, Eastern forms of the body, even for rural bodies insofar as the twentieth century defines the countryside, "the rural," as the underside or raw material of urban development. The city has become the defining term in constructing the image of the land and the landscape, as well as the point of reference, the centerpiece of a notion of economic/social/political/cultural exchange and a concept of a "natural ecosystem." The ecosystem notion of exchange and "natural balance" is itself a counterpart to the notion of a global economic and informational exchange system (which emerged with the computerization of the stock exchange in the 1970s).

The city provides the order and organization that automatically links otherwise unrelated bodies. For example, it links the affluent lifestyle of the banker or professional to the squalor of the vagrant, the homeless, or the impoverished without necessarily positing a conscious or intentional will-to-exploit. It is the condition and milieu in which corporeality is socially, sexually, and discursively produced. But if the city is a significant context and frame for the body, the relations between bodies and cities are more complex than may have been realized. My aim here will be to explore the constitutive and mutually defining relations between corporeality and the metropolis, if only in a rather sketchy but I hope suggestive fashion. I would also like to project into the not-too-distant future some of the effects of the tech-nologization and the technocratization of the city on the forms of the body, speculating about the enormous and so far undecidable prosthetic and organic changes this may effect for or in the lived body. A deeper exploration would of course be required to elaborate the historico-geographic specificity of bodies, their production as determinate types of subject with distinctive modes of corporeality.

Before going into any detail, it may be useful to define the two key terms I will examine *body* and *city*.

By *body* I understand a concrete, material, animate organization of flesh, organs, nerves, muscles, and skeletal structure which are given a unity, cohesiveness, and organization only through their psychical and social inscription as the surface and raw materials of an integrated and cohesive totality. The body is, so to speak, organically/biologically/naturally "incomplete"; it is indeterminate, amorphous, a series of uncoordinated potentialities which require social triggering, ordering, and long-term "administration," regulated in each culture and epoch by what Foucault has called "the micro-technologies of power."[1] The body becomes a *human* body, a body which coincides with the "shape" and space of a psyche, a body whose epidermal surface bounds a psychical unity, a body which thereby defines the limits of experience and subjectivity, in psychoanalytic terms, through the intervention of the (m)other, and, ultimately, the Other or Symbolic order (language and rule-governed social order). Among the key structuring principles of this produced body is its inscription and coding by (familially ordered) sexual desires (the desire of the other), which produce (and ultimately repress) the infant's bodily zones, orifices, and organs as libidinal sources; its inscription by a set of socially coded meanings and significances (both for the subject and for others), making the body a meaningful, "readable," depth-entity; and its production and development through various regimes of discipline and training, including the coordination and integration of its bodily functions so that not only can it undertake the general social tasks required of it, but so that it becomes an integral part of or position within a social network, linked to other bodies and objects.

By *city*, I understand a complex and interactive network which links together, often in an unintegrated and de facto way, a number of disparate social activities, processes, and relations, with a number of imaginary and real, projected or actual architectural, geographic, civic, and public relations. The city brings together economic and informational flows, power networks, forms of displacement, management, and political organization, interpersonal, familial, and extra-familial social relations, and an aesthetic/economic organization of space and place to create a semipermanent but ever-changing built environment or milieu. In this sense, the city can be seen, as it were, as midway between the village and the state, sharing the interpersonal inter-relations of the village (on a neighborhood scale) and the administrative concerns of the state (hence the need for local government, the preeminence of questions of transportation, and the relativity of location).

Body politic and political bodies

I will look at two pervasive models of the interrelation of bodies and cities, and, in outlining their problems, I hope to suggest alternatives that may account for future urban developments and their corporeal consequences.

In the first model, the body and the city have merely a de facto or external, contingent rather than constitutive relation. The city is a reflection, projection, or product of bodies. Bodies are conceived in naturalistic terms, predating the city, the cause and motivation for their design and construction. This model often assumes an ethnological and historical character: the city develops according to human needs and design, developing from nomadism to sedentary agrarianism to the structure of the localized village, the form of the polis through industrialization to the technological modern city and beyond. More recently, we have heard an inverted form of this presumed relation: cities have become (or may have always been) alienating environments, environments which do not allow the body a "natural," "healthy," or "conducive" context.

Underlying this view of the city as a product or projection of the body (in all its variations) is a form of humanism: the human subject is conceived as a sovereign and self-given agent which, individually or collectively, is responsible for all social and historical production. Humans *make* cities. Moreover, in such formulations the body is usually subordinated to and seen merely as a "tool" of subjectivity, of self-given consciousness. The city is a product not simply of the muscles and energy of the body, but the conceptual and reflective possibilities of consciousness itself: the capacity to design, to plan ahead, to function as an intentionality and thereby be transformed in the process. This view is reflected in the separation or binarism of design, on the one hand, and construction, on the other, of the division of mind from hand (or art from craft). Both Enlightenment humanism and marxism share this view, the distinction being whether the relation is conceived as a one-way relation (from subjectivity to the environment), or a dialectic (from subjectivity to environment and back again). Nonetheless, both positions consider the active agent in social production (whether the production of commodities or in the production of cities) to be the subject, a rational or potentially rational consciousness clothed in a body, the "captain of the ship," the "ghost in the machine."

In my opinion, this view has at least two serious problems. First, it subordinates the body to the mind while retaining a structure of binary opposites. The body is merely a tool or bridge linking a nonspatial (i.e., Cartesian) consciousness to the materiality and coordinates of the built environment, a kind of mediating term between mind on the one hand and inorganic matter on the other, a term that has no agency or productivity of its own. It is presumed to be a machine, animated by a consciousness. Second, at best, such a view only posits a one-way relation between the body or the subject and the city, linking them through a causal relation in which body or subjectivity is conceived as the cause, and the city its effect. In more sophisticated versions of this view, the city can have a negative feedback relation with the bodies that produce it, thereby alienating them. Implicit in this position is the active causal power of the subject in the design and construction of cities.

Another equally popular formulation proposes a kind of parallelism or isomorphism between the body and the city. The two are understood as analogues, congruent counterparts, in which the features, organization, and characteristics of one are reflected in the other. This notion of the parallelism between the body and social order (usually identified with the state) finds its clearest formulations in the seventeenth century, when liberal political philosophers justified their various allegiances (the divine right of kings, for Hobbes; parliamentary representation, for Locke; direct representation, for Rousseau, etc.) through the metaphor of the body-politic. The state parallels the body; artifice mirrors nature. The correspondence between the body and the body-politic is more or less exact and codified: the King usually represented as the head of the body-politic,[2] the populace as the body. The law has been compared to the body's nerves, the military to its arms, commerce to its legs or stomach, and so on. The exact correspondences vary from text to text, and from one political regime to another. However, if there is a morphological correspondence or parallelism between the artificial commonwealth (the "Leviathan") and the human body in this pervasive metaphor of the body-politic, the body is rarely attributed a sex. If one presses this metaphor just a little, we must ask: if the state or the structure of the polis/city mirrors the body, what takes on the metaphoric function of the genitals in the body-politic? What kind of genitals are they? In other words, does the body-politic have a sex?

Here once again, I have serious reservations. The first regards the implicitly phallocentric coding of the body-politic, which, while claiming it models itself on the *human* body, uses the male to represent the human. Phallocentrism is, in my understanding, not so much the dominance of the phallus as the pervasive unacknowledged use of the male or masculine to represent the human. The problem, then, is not so much to eliminate as to reveal the masculinity inherent in the notion of the universal, the generic human, or the unspecified subject. The second reservation concerns the political function of this analogy: it serves to provide a justification for various forms of "ideal" government and social organization through a process of "naturalization": the human body is a natural form of organization which functions not only for the good of each organ but primarily for the good of the whole. Similarly, the body politic, whatever form it may take,[3] justifies and naturalizes itself with reference to some form of hierarchical organization modeled on the (presumed and projected) structure of the body. A third problem: this conception of the body-politic relies on a fundamental opposition between nature and culture, in which nature dictates the ideal forms of culture. Culture is a supercession and perfection of nature. The body-politic is an artificial construct which replaces the primacy of the natural body. Culture is molded according to the dictates of nature, but transforms nature's limits. In this sense, nature is a passivity on which culture works as male (cultural) productivity supercedes and overtakes female (natural) reproduction.

But if the relation between bodies and cities is neither causal (the first view) nor representational (the second view), then what kind of relation exists between them? These two models are inadequate insofar as they give precedence to one term or the other in the body/city pair. A more appropriate model combines elements from each. Like the causal view, the body (and not simply a disembodied consciousness) must be considered active in the production and transformation of the city. But bodies and cities are not causally linked. Every cause must be logically distinct from its effect. The body, however, is not distinct, does not have an existence separate from the city, for they are mutually defining. Like the representational model, there may be an isomorphism between the body and the city. But it is not a mirroring of nature in artifice. Rather, there is a two-way linkage which could be defined as an *interface*, perhaps even a cobuilding. What I am suggesting is a model of the relations between bodies and cities which sees them, not as megalithic total entities, distinct identities, but as assemblages or collections of parts, capable of crossing the thresholds between substances to form linkages, machines, provisional and often temporary sub- or microgroupings. This model is a practical one, based on the practical productivity bodies and cities have in defining and establishing each other. It is not a holistic view, one that stresses the unity and integration of city and body, their "ecological balance." Instead, I am suggesting a fundamentally disunified series of systems and interconnections, a series of disparate flows, energies, events or entities, and spaces, brought together or drawn apart in more or less temporary alignments.

The city in its particular geographical, architectural, spatializing, municipal arrangements is one particular ingredient in the social constitution of the body. It is by no means the most significant. The structure and particularity of, say, the family is more directly and visibly influential, although this in itself is to some extent a function of the social geography of cities. But nonetheless, the form, structure, and norms of the city seep into and effect all the other elements that go into the constitution of corporeality and/as subjectivity. It effects the way the subject sees others (domestic architecture and the division of the home into the conjugal bedroom, separated off from other living and sleeping spaces, and the specialization of rooms are as significant in this regard as smaller family size[4]), as well as the subject's understanding of, alignment with, and positioning in space. Different forms of lived spatiality (the verticality of the city, as opposed to the horizontality of the landscape – at least our own) effect the ways we live space, and thus our comportment and corporeal orientations and the subject's forms of corporeal exertion – the kind of terrain it must negotiate day by day, the effect this has on its muscular structure, its nutritional context, providing the most elementary forms of material support and sustenance for the body. Moreover, the city is, of course, also the site for the body's cultural saturation, its takeover and transformation by images, representational systems, the mass media, and the arts – the place where the body is representationally reexplored, transformed, contested, reinscribed. In turn, the body (as cultural product) transforms, reinscribes the urban landscape according to its changing (demographic, economic, and psychological) needs,

47

extending the limits of the city, of the sub-urban, ever towards the country-side which borders it. As a hinge between the population and the individual, the body, its distribution, habits, alignments, pleasures, norms, and ideals are the ostensible object of governmental regulation, and the city is a key tool.[5]

Body spaces

Some general implications

First, there is no natural or ideal environment for the body, no "perfect" city, judged in terms of the body's health and well-being. If bodies are not culturally pregiven, built environments cannot alienate the very bodies they produce. However, what may prove unpleasant is the rapid transformation of an environment, such that a body inscribed by one cultural milieu finds itself in another involuntarily. This is not to deny that some city environments are forbidding, but there is nothing intrinsically alienating or unnatural about the city. The question is not simply how to distinguish life-enhancing from life-denying environments, but to examine how different cities, different sociocultural environments actively produce the bodies of their inhabitants as particular and distinctive types of bodies, as bodies with particular physiologies, affective lives, and concrete behaviors. For example, the slum is not inherently alienating, although for those used to a rural or even a suburban environment, it produces extreme feelings of alienation. However, the same is true for the slum dweller who moves to the country or the suburbs. It is a question of negotiation of urban spaces by individuals/groups more or less densely packed, who inhabit or traverse them: each environment or context contains its own powers, perils, dangers, and advantages.

Second, there are a number of general effects induced by cityscapes, which can only be concretely specified in particular cases. The city helps to orient sensory and perceptual information, insofar as it helps to produce specific conceptions of spatiality, the vectorization and setting for our earliest and most ongoing perceptions. The city orients and organizes family, sexual, and social relations insofar as the city divides cultural life into public and private domains, geographically dividing and defining the particular social positions and locations occupied by individuals and groups. Cities establish lateral, contingent, short- or long-term connections between individuals and social groups, and more or less stable divisions, such as those constituting domestic and generational distinctions. These spaces, divisions, and interconnections are the roles and means by which bodies are individuated to become subjects. The structure and layout of the city also provide and organize the circulation of information, and structure social and regional access to goods and services. Finally, the city's form and structure provide the context in which social rules and expectations are internalized or habituated in order to ensure social conformity, or position social marginality at a safe or insulated and bounded distance (ghettoization). This means that the city must be seen as the most immediately concrete locus for the production and circulation of power.

I have suggested that the city is an active force in constituting bodies, and always leaves its traces on the subject's corporeality. It follows that, corresponding to the dramatic transformation of the city as a result of the information revolution will be a transformation in the inscription of bodies. In his paper, "The Overexposed City," Paul Virilio makes clear the tendency toward hyperreality in cities today: the replacement of geographical space with the screen interface, the transformation of distance and depth into pure surface, the reduction of space to time, of the face-to-face encounter to the terminal screen:

> On the terminal's screen, a span of time becomes both the surface and the support of inscription; time literally . . . surfaces. Due to the cathode-ray tube's imperceptible substance, the dimensions of space become inseparable from their speed of transmission. Unity of place without unity of time makes the city disappear into the heterogeneity of advanced technology's temporal regime.[6]

The implosion of space into time, the transmutation of distance into speed, the instantaneousness of communication, the collapsing of the work-space into the home computer system, will clearly have major effects on specifically sexual and racial bodies of the city's inhabitants as well as on the form and structure of the city. The increased coordination and integration of microfunctions in the urban space creates the city not as a body-politic but as a political machine – no longer a machine modeled on the engine but now represented by the computer, facsimile machine, and modem, a machine that reduces distance and speed to immediate, instantaneous gratification. The abolition of the distance between home and work, the diminution of interaction between face-to-face subjects, the continuing mediation of interpersonal relations by terminals, screens, and keyboards, will increasingly affect/infect the minutiae of everyday life and corporeal existence.

> With the advent of instantaneous communications (satellite, TV, fiber optics, telematics) arrival supplants departure: everything arrives without necessarily having to depart. . . . Contributing to the creation of a permanent present whose intense pace knows no tomorrow, the latter type of time span is destroying the rhythms of a society which has become more and more debased. And "monument," no longer the elaborately constructed portico, the monumental passageway punctuated by sumptuous edifices, but idleness, the monumental wait for service in front of machinery:

everyone bustling about while waiting for communication and telecommunication machines, the lines at highway tollbooths, the pilot's checklist, night tables as computer consoles. Ultimately, the door is what monitors vehicles and various vectors whose breaks of continuity compose less a space than a kind of countdown in which the urgency of work time plays the part of a *time center*, while unemployment and vacation time play the part of the periphery – *the suburb of time*: a clearing away of activity whereby everyone is exiled to a life of both privacy and deprivation.[7]

The subject's body will no longer be disjointedly connected to random others and objects according to the city's spatio-temporal layout. The city network – now vertical more than horizontal in layout – will be modeled on and ordered by telecommunications. The city and body will interface with the computer, forming part of an information machine in which the body's limbs and organs will become interchangeable parts with the computer and with the technologization of production. The computerization of labor is intimately implicated in material transformations, including those which pose as merely conceptual. Whether this results in the "cross-breeding" of the body and machine – that is, whether the machine will take on the characteristics attributed to the human body ("artificial intelligence," automatons) or whether the body will take on the characteristics of the machine (the cyborg, bionics, computer prosthesis) remains unclear. Yet it is certain that this will fundamentally transform the ways in which we conceive both cities and bodies, and their interrelations.

Notes

1 See, in particular, *Discipline and Punish* (New York: Vintage, 1979) and *The History of Sexuality*, Vol. 1: *An Introduction* (New York: Pantheon, 1978).
2 The king may also represent the heart. See Michel Feher *et al.* eds, *Fragments of a History of the Human Body*, Vol. 1 (New York: Zone, 1989).
3 There is a slippage from conceptions of the state (which necessarily raise questions of legal sovereignty) and conceptions of the city as a commercial and cultural entity:

> The town is the correlate of the road. The town exists only as a function of a circulation and of circuits; it is a singular point on the circuits which create it and which it creates. It is defined by entries and exits; something must enter it and exit from it. It imposes a frequency. It effects a polarization of matter, inert, living or human. . . . It is a phenomenon of transconsistency, a network, because it is fundamentally in contact with other towns. . . .
>
> The State proceeds otherwise: it is a phenomenon of ultraconsistency. It makes points resonate together, points . . . very diverse points of order – geographic, ethnic, linguistic, moral, economic, technological particulars. The State makes the town resonate with the countryside . . . the central

power of the State is hierarchical and constitutes a civil-service sector; the center is not in the middle but on top because [it is] the only way it can recombine what it isolates . . . through subordination.

(Gilles Deleuze and Félix Guattari, "City/State," *Zone* 1/2 (1986): 195–7)

4 See Jacques Donzelot, *The Policing of Families* (New York: Pantheon, 1979).
5 See Foucault's discussion of the notion of biopower in the final sections of *The History of Sexuality*.
6 Paul Virilio, "The Overexposed City," *Zone* 1/2 (1986): 19.
7 Ibid., pp. 19–20.

References

Clifford, S. 1988. "Common Ground." *Meanjin* 47 (4), Summer: 625–36.

Cook, P. 1988. "Modernity, Postmodernity and the City." *Theory, Culture and Society* 5 (1988): 475–493.

De Landa, M. 1986. "Policing the Spectrum." *Zone* 1/2: 176–93.

Deleuze, G. and Guattari, F. 1986. "City/State." *Zone* 1/2: 194–9.

Donzelot, J. 1979. *The Policing of Families*. New York: Random House.

Feher, M., Nadoff, R. and Tazi, N. eds. 1989. *Fragments for a History of the Human Body*, Vols 1–3. New York: Zone Books.

Foucault, M. 1978. *The History of Sexuality*, Vol. 1: *An Introduction*. Trans. Robert Hurley. New York: Pantheon.

Foucault, M. 1979. *Discipline and Punish: The Birth of the Prison*. Trans. Alan Sheridan. New York: Vintage.

Jerde, J. 1988. "A Philosophy for City Development." *Meanjin* 47 (4), Summer: 609–14.

Kwinter, S. 1986. "La Città Nuova: Modernity and Continuity." *Zone* 1/2: 80–127.

Sky, A. 1988. "On Site." *Meanjin* 47 (4) Summer: 614–25.

Virilio, P. 1988. "The Overexposed City." *Zone* 1/2, Summer: 14–39.

Yencken, D. 1988. "The Creative City." *Meanjin* 47 (4), Summer: 597–609.

4

HUMAN.LANGUAGE.MACHINE

Ken Hillis

A middle-aged, Toronto man of average appearance subscribed to a regional on-line chat service through which he had made contact with five women all living in a nearby city. He planned to visit them during his upcoming vacation. Two of the women had agreed to put him up at their homes on separate evenings, sight unseen. Each had described herself as very beautiful and great in bed. In the pursuit of his fantasies, the man believed the idealizations and seemed oblivious to any unmediated potential that, in the flesh, these women might be only as human as he.

The Toronto gentleman, in concert with a text-based and non-immersive Information Technology (IT), had imaginatively substituted abstract alpha-numeric characters for physical realities – the voices, gestures, gazes, stances, inflections and other indicative aspects of corporeal intelligence of the women in question. In this chapter my interest extends to immersive virtual environments (VEs), both as off-line and soon to be on-line practices, that are permitted by a particular form of IT popularly called Virtual Reality (VR). I distinguish between virtual technologies and virtual environments. On-line virtual technologies are instances of telematics – the synthesis of telephony and digital computation.[1] VEs are made possible by virtual technologies; they are the computer-generated environments or "virtual worlds" one seems to enter, and they rely upon a naturalized "picture language" that is more conducive to collapsing experiential differences between symbols and referents, or the virtual and the real, than any yet available in text-based applications.

My interest is twofold: namely *why* a cultural desiring now exists for these kinds of psycho-technologies (see de Kerckhove 1991), which have the potential to remap the modern experiential "distances" – metaphoric and concrete – between subject and object, mind and body, meaning and desire; and *how* immersive technologies might influence contemporary self-perception and bodily experience. I examine the intersection of language conceived as a technology, with the spatial relationships set up between VEs and users' bodies, in hopes that the result will be of interest to cultural and political theorists.

The place of these technologies' digital spaces – often referred to as a cyberspace that runs parallel to embodied reality – is really one of language and code. No user's body can enter this metaphoric space. Visually oriented contemporary culture, however, generally equates seeing with knowing. This conflation helps privilege suggestions made by industry players and academics that VEs offer a space into which postmodern sub-identities of a previously more unitary subjectivity might profitably relocate. Paul Smith's (1988: xxxv) argument that the contemporary subject occupies a series of *positions* – scholar, employee, partner, consumer and so forth – into which he or she momentarily is called suggests precisely the kinds of fractured identities and disincorporating subjectivities for which VEs seem tailor-made, and which are privileged by certain schools of academic postmodernism. The increased computational power that lies behind virtual technologies promises an opportunity to *simultaneously* act out or perform these plural sub-identities as an array of iconic representations positioned within VEs. Referent and symbol, subjectivities and icons, seem to merge as spatialized positions within the coordinated visual gridworld of cyberspace.

Virtual space?

Physicist Roger Jones (1982: 17) observes that science's assumption of a "separate, quantifiable, objective world" has led to a conception of "absolute space" based on operational definitions and physical measure. Jones also argues that this assumption has produced a metaphysical belief that accords space an independent agency. For Jones, verifiable measure and representation matter little to "the essence and experience of space itself the basic mystery in which we all participate, which permeates our every act and thought, and whose ubiquitous presence we accept unconsciously as synonymous with our very existence in the world" (1982: 17). It is worth considering that an exteriorized and idealized conception of absolute space – based on thinking that there is a knowable concrete reality (even if this reality often has been theorized as an Ideal and formally separate from our own materiality) – has mutated to inform current metaphoric digital spaces such as VEs. VEs hold out the eventual picture-perfect promise of rejoining an image of our own embodied materiality with the image of a larger collective information space or electronic on-line environment.

Western understandings of space implicitly equate space with distance. In the case of cultures privileging mobility, space, therefore, becomes a distance to be crossed. If, on some level, the implicit equation space = distance raises "difficulties" to be overcome by transportation and communication technologies, the equation has been solved by replacing it with a new one – space = movement – where movement means *continual* transit across the old space of distance. What is understood as movement across or through space is increasingly conceptually displaced on to electronic networks which merge conceptions of "absolute" and "relative" space. This merger – made to perform on an implicit plane of continuous circulation – collapses distances and distinctions between objects, and between figure and ground. However, as Jones notes, the experience of space is synonymous with our very existence, and I accept his premise that we are unable to conceive of existence except in space. To exist derives from the Latin to stand out. What we stand out from is space (1982: 69). *To stand* immediately invokes our bodies. Bodies stand out from space. Without them there could be no meaningful geography, which "emerges" from the complex dialectic instantiated within the relationship we establish with space. Standing out from the space around our bodies, we prove we exist.

In on-line environments, "proof" of our existence depends on communicated images. "Proof" of our physicality depends on communicating that fact across space itself. In a kind of double recursivity, motion or information flow understood as communication stands in for the broader existential reality of which communication is only ever a part, along the way managing to assert that concepts and ideas are more real than material reality, and certainly more noble – a replaying of the old intellectual parlour trick in which the head swallows its own body, a trick at least as old as Plato's Cave in which reality is the true illusion, a dim reflection of the Ideal (see Heim 1993: 91–2, 117).

Philosopher Lorenzo Simpson (1995) makes a critical point about the use of electronic communications technology that directly speaks to the relationship between human bodies and space. He writes of the relationship between social recognition and self-confirmation. I am acknowledged in the street by the shopkeeper, who recognizes me because I purchase things at her store on a daily basis. This interrelationality "reminds" me who I am. However, in the emerging electronic "communities of representation," that which gets recognized by other people is my *"constructed* identity, which lacks the vulnerability of my primary identity" (Simpson 1995: 158). What is recognized by others, then, doesn't confirm me to myself, but possibly confirms only an ideal construction of myself, or even less than this, for the other "knows it is only a persona" (Simpson 1995: 158). The validating dimension of social interaction is less available as a result of a confusion between the brute fact of what embodied presence allows others to confirm to me about myself, and an idealized self-image projected outward into a communications Net that is made to operate as both a metaphor for space and a "space-collapsing" technology.

Bringing the metaphors together creates a situation that has the power to suggest that movement and circulation constitute an acceptable ontological ground "within" which to perform one's constructed identity. There is no cause to pause in cyberspace, and its speedy dynamics constituted in information flow have the ability to suggest to users such as the Toronto gentleman noted at the beginning of this chapter that bodily confirmation of representations might no longer matter.

The academy of cyberspace

I want to introduce certain theoretical points made by Norbert Elias and Mikhail Bakhtin that are useful in relating VEs and telematics generally to language and space, and to the way we understand our bodies. Elias offers a useful spatial metaphor for theorizing why the modern subject would come to see imaginative extension beyond its bodily boundaries, for example, into a VE, as a desirable state. Bakhtin's stance on language and bodies is ironic. I introduce it as an example of how an unproblematized use of necessarily unstable metaphors can unwittingly promote, in this instance, a trend toward living virtually that I suspect Bakhtin might have deplored.

In *The Civilizing Process*, Elias (1968) posits the "construction" of a fictional and invisible wall – a spatial metaphor underpinning modern self-identity and which demarcates the interiorized individual from public life. In Elias's account, over the past seven hundred years this "wall" has helped insert an actual pause between the brain's command and the hand's carrying out of the order. This separation (something of an eye-hand *dis*articulation) permits modern interdependent social relationships and capitalist economies. It is not feasible, for example, in an economy based on the division of labour for individuals dependent on one another for goods and services to kill one another spontaneously during heightened emotional states. Such a wall helps create a critical distance between the individual and the "outside" world. It permits a cooling off period that allow angers to dissipate, thus minimizing bodily harm and social disruptions between progressively isolated individuals who, nonetheless, depend on one another's productive contributions within a modern organization of economic activity and labor. This wall, Elias argues, appears to modern individuals

> as an eternal condition of spatial separation between a mental apparatus apparently locked "inside" man . . . and the objects "outside" and divided from it by an invisible wall. . . . The act of conceptual distancing from the objects of thought that any more emotionally controlled reflection involves . . . appears to self-perception . . . *as a distance actually existing* between the thinking subject and the objects of his thought.
>
> (1968: 256–7, emphasis added)

The popularity of virtual technologies today in part reflects a desire that these machines might represent, in commodity form, an acceptable commons in which fragmented but nonetheless highly individuated modern subjectivities might achieve virtual reunification with other similar entities. Such an "electronic commons" or digital "public sphere" heralds a postmodern imaginative vaulting of Elias's wall, yet one which, however ironically, continues to maintain and even intensify the critical spatial separation of users' bodies in absolute space via the dialectic of technology established by interface devices.

I am sympathetic to the aim of Bakhtin's (1984) project, as set forth in *Rabelais and His World*, to introduce contemporary readers to the liberatory potential of medieval carnivals. Bakhtin's theory of the "carnivalesque" medieval marketplace brings together poor and rich, peasant and lord. The collective belly laugh of the crowd – the public performance of gross bodily functions, oral utterances, and speech practices – levels social stratification, thereby offering the oppressed a temporary taste of political agency. In passing, I would note that carnivals were also subtly sanctioned spatio-temporal means of venting steam. Whatever hegemonic state of feudal affairs existed, and however gruesome it might be, could be resumed after a brief abeyance. However, I introduce Bakhtin's theory of carnival for a different reason. It exemplifies what may happen when the referent – here Bakhtin's text describing the embodied carnivalesque material reality – comes to stand in for that to which it refers. Following this, language "speaks" only to itself, and is positioned merely as a "social technology." Along the way, links between language and embodiment are forgotten. Yet Bakhtin links the carnival's belly laugh to the interlocutory possibilities contained in the form of the modern novel. Minus embodied speech, Bakhtin's exhumed medieval carnival is readily transubstantiated into the novel's abstract textual representations.

By idealizing medieval carnival, Bakhtin aims to recapture a Rabelaisian moment of hope, one that he wants us to remember, even as he understands this moment is no longer an embodied part of humanity. When, perhaps because he grasps the modern impossibility of Pantagruel's Utopia, Bakhtin then fuses the embodied carnivalesque – the belly-laugh and the fart in the face of *noblesse oblige* – with the dialogic and formal possibilities of print, he employs paralogism. He relies upon an excessive use of metaphor that skirts issues of form and scale – and suggests the interchangeability not only of historical epochs, but of physical embodied realities and their representations.

I want to suggest that, *voided of embodied speech as spoken oral language*, Bakhtin's medieval carnival is readily made available for conceptual relocation not only to the print-dependent novel whose abstract, modern textuality and physical inertness in the reader's hands may allow for a temporal and critical self-reflexivity. Carnival is also even more easily relocatable to metaphoric and virtual public spheres made possible by telematics and ITs – precisely because of their seemingly greater iconographic and linguistic naturalness.

Neil Smith and Cindi Katz (1993) offer a useful cautionary in this regard. They argue against the ill-considered taking up of spatial metaphors by cultural theorists. In the rush to the city-as-text and other virtual phenomena (such as carnival-as-novel), the authors note that real cities and urban exploitation continue, often under cover of metaphor which allows concepts to be isolated from the active world to which they refer. Smith's and Katz's point is well taken, in that paralogism influences the cultural understandings under which equally "metaphoric" virtual technologies get imagined, built and used.

For purposes of the present discussion, Frances Barker's (1984) study of the effects of the Restoration on the social and political imaginary of 1660s England offers a useful corrective in considering the implications of Bakhtin's theory. During the Restoration, the subject largely abandoned the performance of its subjectivity in public. Instead, it came to terms with performing "itself" in text made possible by print technology. The silent watching of theatre as spectacle exemplified this withdrawal of the subject from an earlier more "carnivalesque" performance in which the line between actors and audience had been socially and spatially demarcated less clearly. At the same time, the bourgeois subject found itself opened to new forms of manipulation by virtue of its spatial and mental isolation. Traces of this isolation are found in the extreme forms of spatial privacy that develop during this period, and within which the private individual gestates behind locked doors. The cartesianized self retreated from public display, finding its expression increasingly through texts – a spatial (re)move that temporally parallels the establishment of Elias's wall.

If, for Elias, the modern division of labour depends on this wall, Barker identifies the silent sublimation and channeling of physical energies into textual productions as necessary for modern capital formations and industrialization to have taken place. In a way Bakhtin never considers, Barker recognizes the synergy between changing forms of communications and social relations, from the performative song and dance of the communicating public body to its representational remove to a fully commodifiable media form consumed "behind closed doors." Novels and human bodies, however, are at least similar in that both take specifically material forms. Even this similarity need not be the case with virtual bodies in cyberspace. The body and the book are formed on a "durable material substrate. Once encoding [on either] has taken place, it cannot easily be changed. . . . [E]lectronic media . . . receive and transmit signals but do not permanently store messages, books carry their information in their bodies" (Hayles 1993: 73).[2]

Bakhtin privileges voice as a political act. He wants to demonstrate the commonality between embodied speakers and listeners. Yet his theory also suggests a threefold and slippery interchangeability or formal equality among (1) utterance and vocal sounds; (2) the words we speak aloud to one another as part of our presentations of self; and (3) these words subsequently codified within communication technologies such as print and ITs. However, there is a difference between physically performing in carnival's boisterous world of mirth and reading about it silently in a room of one's own hundreds of years later. This spatio-temporal difference matters sensually, formally, economically, and politically.

Theories according primacy to "textual" representations, such as Bakhtin's and more recent postmodern and poststructuralist notions, can be deployed to assert that: (1) there is no world beyond the text; (2) belief that representations relate to the real world constitutes a "category mistake"; (3) identity is a futile concept as we are consigned to repeat performances by which we momentarily confirm who we are. Such theories resonate with and support the dominance of telematics and IT, which, in a variety of ways, as technologies and practices, confirm not only belief that "all the world's a text," but also the apparent cogency of these academic theories, themselves partially extended, as in the case of Baudrillard's understanding of the simulacra, from information theory. Such theories, along with VEs, are premised on communicating agents extending *away* from bodiliness. Instead a disincorporating subjectivity is actively directed toward metaphoric spaces and representational language practices, within which a kind of purified carnivalesque liberatory potential might reign free.

Such inattention to the meaning of form skirts an abandonment of ethics. In Elizabeth Grosz's (1994) project to advance or reintroduce a more holistic understanding of body and mind and thereby reinvigourate contemporary feminist theory, she writes that Descartes' principal achievement was the "exclusion of the soul from nature" – the fabrication of a binary that allowed "evacuation of consciousness from the world" (1994: 6). Now, Walter Ong (1993) has focused on the relationship between orality and technology, paying close attention to the changing medieval European speech practices that would have informed Rabelais' world, which was in an uneven state of transition from an oral to a print culture. Concerned with the impact of print on the development of an interiorized modern consciousness, Ong writes,

> By removing words from the world of sound where they had first had their origin in *active human interchange* and relegating them definitely to visual surface, and by otherwise exploiting visual space for the management of knowledge, print encouraged human beings to think of their own interior conscious and unconscious resources as more and more thing-like. . . . Print encouraged the mind to sense that its possessions were held in some sort of *inert mental space*.
>
> (Ong 1993: 131–2, emphases added)

Reading Ong against Grosz, "inert mental space" first must be readied to receive consciousness as it is evacuated from "active human interchange" with the world and nature. Over time, interiorized consciousness (secured "behind" Elias's wall and other less metaphoric devices) seeks to extend itself "outwards." Desirability of products such as VEs, which allow disembodied communication that seems to penetrate across the wall, will grow. Such products substitute for the "positionality" required of an earlier fleshy collectivity. They act as post-Leibnizian windows admitting vision and light onto a monad-like consciousness, so that each self, alienated from nature and collectivity, might communicate the relative fact of its lonely and conceptual existence to other equally disaffected selves. This disaffection operates in tandem with the very move of consciousness toward inert mental space and the loss of skills implied by the ceding to mediation of embodied human interchanges that permit and advance recognition of differences between people. Such skills depend upon the myriad face-to-face embodied communication and speech practices between strangers in public, and partly require taking place with/in an increasingly fugitive "public square."

As communications technologies, VEs drain the meaning of form and distance in a manner similar to that performed by Bakhtin's theory. Like the theorized voice of the novel *cum* carnival, VEs suggest the technical feasibility and hence cultural acceptability of conceptually relocating the disincorporating self to a reality constructed from a visibilized language set apart from the body. Philosophically speaking, an idealist belief in the possibility of a world constructed entirely from technologies based on language would seem to promise a pure, if unanticipated, victory for social constructionism. This would be a Pyrrhic victory, however, one in which, as Grosz critically observes, "culture in effect takes on all the immutable, fixed characteristics attributed to the natural order" (1994: 21). But VEs achieve something more in their metaphysical aesthetic pursuit of a "perfect copy" of reality (see Bryson 1983). They collapse distinctions between figure and ground, between bodies and space, and thereby also the possibility that body icons in cyberspace might really stand out against or be defined in terms of cyberspace.

One might argue, therefore, that in VEs two scenarios are possible. 1: The ground upon which bodies stand is collapsed into the figure or subjectivity of the viewer, a dynamic suggesting all of the natural world is a cultural product, including the self. 2: Or, one might reverse the argument to say that subjectivity is denied in being made to merge with the ground – a kind of rediscovery of the wider world reformed as information.[3]

Yet both scenarios are entirely representational. They suggest the much ballyhooed cyborg (Haraway 1985; Dery 1993) which in cyberspace takes the form of a quasi-metaphysical and electro-flesh merger between (interiorized) users and an essentialist or absolute conception of (exteriorized) space. This merger operates within a material technology reflecting the technical elite's wish to discard its flesh in an act of body denial disguised behind depoliti-cizing metaphors of transcendence and liberation. This denial fuses with a sense of feeling overburdened by the duty to create self-meaning by a self-consciousness that confuses itself with "mind", along the way confirming to itself the dumb animal status of its own body materiality. Both scenarios also assume that a direct correspondence exists between reality and its models (see Coyne 1994). Applied to VEs, both scenarios assume that all of reality, our bodies included, can be reduced to a representation – in this case a two dimensional one attempting to pass for 3D. Regrettably however, improvisation – the creative ability to respond to contingency and externality on the basis of using "what is at hand" – is difficult if one is trapped "within" the predeterminations of a model or theory.

Language and desirable visions

Conjuring carnival from print depends on a metaphysical blurring of the physical experiencing of one's body with an idea of "the body." In linking this conflation to the opening anecdote concerning the Toronto gentleman's on-line activities, I suggest his actions, and those of the women he communicated with, *represent* the voice of embodied *laughter* Bakhtin identifies; with ever-greater degrees of representation, however, this laughter becomes increasingly detached from its referent. Carnival's sound is transformed into sets of linguistic symbols "outside" the experiences of those described; what once was heard, felt, and spoken is cloaked in visible language-based representations.

Immersive virtual technologies, which rely on symbols, images and text, reduce the actual distance between eye and screen to less than an inch. One *feels* thrust head-first into the screen because it entirely fills the perceptual field of vision. It is *as if* one's body is present in the spatial display – both at the centre of a 360° representational space, yet at the margin of a visual world looking in. Yet, here, one's *person* performs an iconic return to what Williams (1983: 233) defines as its "earlier meaning of a mask used by a player" – an understanding evocative of Roland Barthes's identification of the classical belief in language as "a decoration or instrument . . . donned at a distance . . . according to the needs of subjectivity" (1976: 40). Thrust into this conceptual space, yet still simulating vision's embodied limits which position the viewer "on the side" or "at the margin" of the field of view, the subject, made over into a point-of-view which now may "fly" through a virtual world, cedes aspects of embodiment for the ritualistic promise of iconically possessing, even merging with or surrendering to, the object or its other, made to appear as images within the VE.

In an increasingly visually-oriented culture, virtual technologies achieve more experientially real simulations of sight's physiology than earlier visual communications technologies. VEs cannot actually collapse physical distances between subject and object. If anything, potentially they reify psychic "distances" or alienation in the material world, with users spending increasing amounts of time in all forms of cyberspace to the detriment of maintaining embodied social relations. It is VE's ability to more perfectly mimic the actual spatial relations established by our embodied field of vision that allows users to feel *as if*, bolstered by the technology, they might transcend the psychic and material distance between themselves and the objects of their desire, along the way perforating the "spatial wall" or at least introducing a "window onto space" so as to vault across the monad's visual isolation, yet only virtually. In more faithfully simulating the physiology of embodied sight, VEs, as part of a continual juggernaut of technological refinement and change,[4] foster the irresponsible wish that a further shrinkage of the subject–object dichotomy may well be "on view" . . . just around the next blind-spot . . . almost on the horizon.

Media theorist Scott Bukatman argues that telematics "are invisible, circulating outside of the human experiences of space and time" (1993: 2). The issue of invisibility is important because of the empiricist connection between space and vision. If an object can't be seen perhaps it doesn't exist. By naming this concept "cyber*space*" a potential irony is introduced, as a space now becomes invisible to visual perception (the clash of Idealism and Empiricism), and this has demanded the next layer of technology in the form of VEs to make this conceptual space visible and hence more experientially accessible to our visually inflected contemporary cultural sensibilities.

Using the equation seeing = knowing, VEs spatialize or visibilize the digital language used in virtual technologies. People cannot "see" into, much less physically enter, the conduits and data flows that are now the ironically rhizome-like centres of power for economic and military institutions. Hence the rush to visibilize data flows as VEs, the rush to the money. On-line VEs will allow power seekers who subscribe to seeing = knowing to experience data flows sensually – sensually, that is, in a culture comfortable with arrogating to sight most of the human sensorium's powers. Yet this longed-for consummation between flesh and digital data demands further objectification of power-seekers' bodies that increasingly will be made to adopt fully representational status. If, as research already proposes, multiple sub-identities will be represented within interactive VEs, power-seekers will also multiply their own objectification. They will do so by recourse to a multiplicity of iconic representations extending different aspects of self-interest ever-outward along several spatial axes at once.

Given proliferating and plural subject identities, why predicate the discursive space of future VEs on the strict one-to-one subject–object relationship on display within current models? NASA, in ongoing research, is designing technology that will allow users to represent more than one icon or "puppet" of themselves at the same time within the same environ-ment. In one prototype, a representational puppet of a user may observe, circle or even penetrate a second icon representing a different facet of self-identity (Kroker 1992). I would suggest that NASA's research is predicated on a split subjectivity modeled on a psychologically incoherent understanding of self-perception operating according to a logic of "I know I am here because I see myself there." Stated otherwise, this split subjectivity may be thought of as composed of both the viewer and the visual field or ground in such a way as to collapse distinctions between them. NASA's advanced research nevertheless remains limited to deploying two virtual self-locations restricted to the same field. Other research with more immediate commercial potential proposes on-line virtual environments in which individuals will assign several different icons or semi-autonomous drones of the virtual self to "travel" to different sites at the same time in the performance of their jobs, acting as writers and editors of what I term their own plural schizophrenia. VR promoter Brian Gardner theorizes an aspect of the virtual future for plural self-identity:

> A major problem with existing artificial worlds is that they can only be experienced from one perspective at a time. . . . One . . . solution is the idea of invisible cameramen. . . . [T]hink of these as people who travel through the environment that has been created and try to analyze it based on what the user is looking for.
>
> (1993: 102)

Working similar code-dependent terrain, VR theorist Brenda Laurel names such cyborg agents *characters*, not people. In her schema, people are central controllers, directing "behaviors" of a variety of representational sub-identities in pursuit of multiple goals. Virtual agency "that is responsive to the goals, needs or preferences of a person – and especially an agent that can 'learn' to adapt its behaviors and traits to the person and the unfolding action – could be said to be 'codesigned' by the person and the system" (Laurel 1993: 148).

Bart Nagel's illustration of *The Future of Television*, reproduced below (page 63), captures a sense of this multiple iconography fused to plural VR identity. Nagel's work can be viewed as an implicit critique of the dystopian logic of the cyborg and of the promise of an elite electronic on-line commons in which alienated selves hidden behind real-life "spatial walls" gaze at one another. In this commons, plural identities in flux, feeling overburdened by both the historical and individuated weight of their own subjective freedom, might be *pictured* more as one in their search for a transcendent continuity – as if returned to a prelapsarian state. Hence such outlandish prophesies as Kevin Kelly's "Hive Mind" – a collective, global VE in which the discontinuity of an embodied individuality theorized as a "dumb terminal" or "drone" cedes to a mediated and digital group intelligence modeled after a swarm of bees in transit (Kelly 1994). Kelly's virtual insectarium hums with the sounds of a psychasthenic collapse as postulated by Roger Caillois (1984). The subject who suffers from psychasthenia confuses itself with surrounding space which then comes to seem like a "lure" possessing an independent agency capable of calling the self to join with it, as if to become spatiality disincarnate (see also Olalquiaga 1992; Grosz 1995). Caillois's psychasthenic and mutated individual/space – theorized from the French sociologist's obsession with the insect world – when conceptually transposed to cyberspace – evinces a yearning for "return"

Figure 4.1 "The Future of Television" by Bart Nagel

to something remarkably akin to Plotinus's ancient metaphysics of "world soul."

Though an active subjectivity always in part demands self-extension and motility, the cyborg subject achieves agency and identity only through extension via representations outward from the self. Engagement with the here and now is set aside, along with the meaning of corporeal intelligence.[5] It is worth considering that the demands for extension and motility are both accommodated in Nagel's view of the telematic future, though in ways that might not have been anticipated, for example, by an Enlightenment emphasis on confirming subjectivity through empirical observation or, ironically, by Haraway's concept of the cyborg positioned as the electro-flesh "site" from which new modes of seeing and resistance are possible. Rather, Nagel's image suggests, for example, that Kelly's Hive Mind reflects a world where, echoing *Star Trek*'s alien Borg, "Resistance is Futile."

Space of the monster

As I have been concerned with how a wedding of virtual technologies to a progressively disincorporating visual sense might foster a myopic withering of social concern for, and identification with, the experience of our lived bodies, I want to note Donna Haraway's (1991) consideration of the liberatory possibilities for vision and its engagement with non-human forms of agency. Haraway argues for an embodied reappropriation of our most powerful sense: "[E]yes have been used to . . . distance the knowing subject from everybody and everything in the interests of unfettered power . . . visualizing technologies are without apparent limit" (1991: 188). An extended but still embodied perspective "gives way to infinitely mobile vision, which no longer seems just mythically about the god-trick of seeing everything from nowhere, but to have put the myth into ordinary practice. And like the god-trick, this eye fucks the world to make techno-monsters" (1991: 189). Yet Haraway refuses to cede vision, at least metaphorically, to the forces behind the myth she identifies. Her eye pleases her, and she argues for an embodied objectivity, necessarily partial, as a tool for making common political cause against the god-trick "in order to name where we are and are not" (1991: 190).

In her project to rehabilitate an American cultural approach to science, Haraway's "embodied objectivity" is an effort to move away from subjectivity. Yet the difficulty in relying on vision alone to make any such move lies precisely in the eye's asymmetrical powers limited by the body of which they are part. Hans Jonas (1982: 136) notes that sight is the only sense that establishes a "co-temporaneous manifold . . . which may be at rest." Real space, he asserts, is "a *principle* of co-temporaneous, discrete plurality" (1982: 138, emphasis added). "The *motility of our body* generally . . . is already a factor in the very constitution of seeing and the seen world themselves, *much as this genesis is forgotten in the conscious result*" (1982: 152, second emphasis added).

64

This limit or forgetting (or, adapting a phrase from Walter Benjamin, a kind of "optical unconsciousness") – itself facilitated by according primacy to sight and metaphors of vision – is central to believing that VEs could wholly represent the entirety of our perceived world both as and in "disinterested," *a priori*, and thereby precausal images.

In allowing that embodied objectivity may finally be contained within a high degree of metaphor, Haraway may inadvertently aestheticize the currently overly-subjected feminine object she would rescue from "the gaze." I find this frustrating in its implications that a disembodied "cosmopolitan" future, along with the mutating technopolis that undergirds its structure, is the only way to go, that the most agents can do with the products placed before them by Nintendo, Sega, or the Pentagon is to creatively interact with them. Referring once again to Nagel's dark vision, it is unclear that resistance can remain immune to technical impingement.

I do agree with technology historian Bruce Mazlish who in 1967 already had concluded that refusal to engage creatively with the change to our metaphysical awareness being wrought by the erasure of difference between humans and our machines leaves us with the unpalatable alternatives of frightened rejection or blind belief in machines' ability to solve all problems (1967: 15). As he notes, this Hobson's choice gives Shelley's *Frankenstein* its ongoing appeal. If, like Victor Frankenstein, we spurn our technologies, we also deny our authorship of them. They become "the other" that returns to haunt the fractured self, even to infect the "spaces" or cracks now opening in it – this self that was once more internally unified. Examining how technology inflects our "metaphysical awareness" remains essential, even if such work is labeled reactionary by those who argue this kind of examination depends on a politics of the natural that was always only a cultural artifact anyway.

There are many ways to proceed. As a culture and as individuals, some responsibility for the actions or effects of technologies in use must be acknowledged and not buried under the rubric of a passive and unacknow-ledged belief in technological progress operating as a kind of contemporary providential metaphysics. Hubert Dreyfus (1992) has noted the West's penchant for constructing its dominant philosophies in the form of technologies. Privileged white men have written most of the West's philosophy. Such men set the machine agenda and run the world. Freed from the reformist engagement Haraway proposes, what kind of machines would be designed by contemporary feminists, south Asian subalterns, radical queers, and all others struggling for fuller voice, given that such peoples still struggle to obtain a fuller measure of the subjectivity that privileged Western elites seem so ready to abandon to the cyborg "positionality" of VEs? Is the die of the technical trajectory of virtual technologies already cast? *Seemingly* so. But might it be broken? Yes. It must remain intellectually acceptable to argue for and to philosophize different technologies than the dominant model of the cyborg now under construction – an accommodationist merger that confuses visual metaphors with the body, and which fuses electricity and flesh, number and light, within the disciplinary grid of an already imbalanced equation. I make this argument recognizing that attempts at social responsibility must begin within a context of unequal power relationships influencing whose intentions, and therefore which philosophies, ideologies, or discourses would get transformed into technology and naturalized.

Almost always, it is the intentions of elite groups that get transformed into technologies which direct subsequent human intentions to result in ever-newer technological forms (Rothenberg 1993: 14). Fantasy is used to suggest the universality of elite philosophies and fantastical desires. Consider the following example. Jaron Lanier, pre-eminent guru of the emergent VR establishment, argues that concrete reality constrains the imagination in pernicious ways. Think, he suggests, of how difficult it is to build a house. In VEs this will no longer be the case. The frustrations that attend embodied reality will give way to the imaginative design capacities of "everyman" transformed into a virtual carpenter whose pleasure in building his virtual house will be courtesy of the requisite software. Lanier is intrigued by the "pliancy" of VEs, suggesting that what is striking about such worlds is that distinctions between human bodies and the rest of the world are "slippery" (Lanier and Biocca 1992: 162). Yet he is able to suggest, in virtually the next thought, that in VEs, "the objective world is completely defined . . . therefore, the subjective world is whatever else there is. Suddenly, there's a clear boundary for the first time" (1992: 163). On which side of the distinction that might leave our slippery bodies and any pliant agency remains unclear, the more so given the implicit claims for transcendence from embodied care also being advanced by VE promoters. Lanier also suggests that the making of a house in VEs will be part of a (future) communication without codes.

> If you make a house in virtual reality . . . you have not created a symbol for a house or a code for a house. You've actually made a house. It's that direct creation of reality; that's what I call post-symbolic communication.
>
> (Lanier and Biocca 1992: 161)

Such proposals seem to be a conjuring[6] that denies the interplay between culture, its symbols, and the physiology of sight, as well as the hard won lessons that building has to offer – not the least of which are those learned about the relationship between bodily limits and creative reach. We may rest assured that Lanier's proposals will do little to alleviate housing shortages or homelessness. Socially irresponsible, his vision recalls Marie Antoinette's lifestyle at Versailles and the aristocratic donning of the appearance of workers' labor as leisure. If humans live not by bread alone, even Antoinette's pitiless cake seems preferable to Lanier's pliant and homeless image.

Lanier promotes the liberatory "prospects" of VEs. However, antecedents to his utopian vision are more rooted in exploitation than liberation. Lord Castelreagh's scheme to export nineteenth-century England's urban poor to overseas colonies turned on offering the downtrodden a utopian image of a "new world." The destitute and discarded were invited to imagine a colonial landscape of plenty and ease, almost as if the necessities of life would present themselves as commodities born without the intercession of labor. As David Bunn (1994: 136) remarks, this was "aristocratic control over representation *for* the working class," in which the to-be-expected interplay between labor, land, and social relations was masked behind a description of the fantastic. The use of aesthetics to organize these kinds of utopian appeals is intended to transcend the particular conditions under which life is to be lived (see Bourdieu 1984). The spatial reality of the downtrodden is denied yet appealed to by fantastic, unworldly, and impossible visions of a better life in the spatial "hereafter" always located at a discrete remove. The exportation of Victorian England's underclasses partially depended on an appeal to fantasy designed to deflect emigrants' attention from the conditions of production that would await them on distant shores. A fantasy of a bountiful world of ease avoided issues such as native populations, unexpected climactic extremes, infertile soils, and social dislocations: in short, the to-be-expected intersection of the natural and social worlds. The kind of easy-to-assemble, do-it-yourself landscape that dances before Lanier's eyes is a magical forward vision or advance brigade. The spatial "prospect" – or landscape – opened here is a projection of future development that refashions a long standing and "unbounded "prospect" of endless appropriation and conquest" (Mitchell 1994: 20).

Feeling uneasy with a larger world still situated beyond Elias' wall, the alienated, disincorporating imagination produces new forms of parallel spatialities and imaginary architectures and landscapes more amenable to its self-perceived sense of strangeness (see Vidler 1992). A "psychic home-lessness" informs VEs as the modern carnival turned grotesque – a disconnected and monstrous language game in which machined laughter mocks the environmental evidence of tangible bodies dependent on a "nature" that shifting postmodern identities are encouraged to discard as an outmoded commodity form. Divorced from bodily constraints, "the fully extensible self" floats free in a humanly-authored dataspace, along the way turning its "back" on an intransigent material homelessness embedded in grossly unequal social relations.

In 1916, John Dewey wrote that "[s]ociety exists not only *by* transmission, by communication, but it may fairly be said to exist *in* transmission, *in* communications" (Dewey 1916: 5). His argument connects to his belief that thought is possible only in the presence of language. However, virtual technologies now suggest vision's technical relocation to *within* the *technologies* of transmission. We may therefore expect the "parallel digital world" of communications – based on symbolic movement and speed and to which knowledge reduced to information flow is increasingly relocated – to reflect a technical elite's accelerating impatience with the physical body "parked" on the other side of the interface by the demands of this technology. As is already happening, this privileged group of technocrats increasingly argues the rational efficiency of further minimization of bodily motility and touch through which people make ethical and political sense of the world. As discursive practices, VEs exemplify linkage between such reactionary "efficiencies" and contemporary theory based on unacknowledged desires to confuse bodiliness with *ideas* of the body, and our bodies' movements with *ideas* of bodily movement. The contemporary rush to fuse representation (and culture) with the broader lived world, to "write" the former over the latter so as to suggest "the death of nature" and the "triumph of the cultural and the text," at base depends on communications *technology*. In general, theory has difficulty locating and criticizing its own conceptual and material bases.

NoBody, no space

The limitations of our bodies – seemingly so noxious to those promoting a virtual future – are what help give shape to the judgments upon which decisions and agency depend. Such limits are, therefore, also central to how an ethical understanding of and participation within the world is achieved. If space can be thought of as a matrix for organizing relationships between and among things, it also contributes to regulating human perceptions, in that they appear to occupy the same perceptual field. As Grosz notes, "[t]his perspective has no other location than that given by the body" (1994: 90). We grasp the world through bodily situation. The relationship between virtual technologies and bodies suggests that despite their apparent spatial aestheticization of the lived world, these machines instead move toward anaesthetizing bodily perception.

Cyberspace is an interiorized world within which it is suggested that (assuming the price of admission) we all might come together in the free play of imagination and information. Such a disembodied imagination or "vision" is now also accorded the power of "voice" as arguments organized around freedom of speech in visual cyberspatial environments make clear. Following Dewey, this imagination is being conceptually relocated to *exist in the technologies of transmission*. If the disincorporating eye looks back on the body that was its home and sees only an abandoned shell without vision, what empathetic awareness would inhere to keep this eye from returning as the "pliant agent" of a disembodied imperialism? Who owns, who pre-programmes this technology that would colonize a user's body as a kind of "final natural frontier?"

If an individual is to have the freedom in play or requirement at work to don many different identities, what might history mean, given that these multiple partial identities themselves are idealized hybrids composed of icons, simulations and a merger of others' concepts with one's own sensation? VEs remove the spatial ground of resistance that human bodies have offered against the appropriation of history by those with greatest power. They recreate a simulation of space that is completely open to surveillance, transcription, review, and censorship. The potential for and implications of "electronic shunning" are only now being grasped. Any history on view, and the possibility to root identity in real places, becomes less important than the continuous circulation of hybrid discourses masquerading as space.

A failed American Express ad campaign portrays the AMEX card as a "virtual credit space" holding up a bridge within a natural setting. In suggesting that a society is becoming "an identity commodity . . . trapped by debt" (Schultz 1993: 437–8), the ad previews a reverse embodiment, a telematic wrap-around of the real world by the visible representations of informational space. From the fibre-optic privilege of cyberspace, human bodies become mere things in the nihilistic perspective of data overload. Any temptation to engage with the new virtuality as a form of resistance must append a caveat that calls the imagination back to the place of the human body and the physical world that can never become entirely virtual. An appropriate phrasing for any message recalling the body in a virtual age might read, "Don't leave home without it."

Notes

1 For the purposes of this discussion, telematics is the synthesis of telephony and digital computation, of telephone technology and computers. For example, e-mail and new media such as the world wide web (WWW) are uses of telematics, in that they rely on the combination of telephone lines and personal computing. On-line interactive media such as Web TV or the expanding range of services such as electronic banking and food shopping are also examples, but so too is closed circuit community television (CCTV), based, in part, on centralized data banks and "smart" technologies having the capacity, for example, to recognize "undesirables" and inform state policing mechanisms accordingly.

2 I infer that Hayles makes certain unstated assumptions here about the specific materiality of paper, as well as of flesh.

3 Both scenarios suggest a merger of subjectivity and space. Because cyberspace and VEs are based on language and code, Henri Lefebvre's critique of structuralist semiotics' theorization of space as given with and in language is worth noting. "[Space] is not formed separately from language. Filled with signs and meanings, *an indistinct intersection point of discourses, a container homologous with whatever it contains*, space so conceived is comprised merely of functions, articulations and connections – in which respect it closely resembles discourse" (1991: 136, emphasis added).

4 Though the concept of progress today is widely denigrated, *technological* progress is still widely accepted, even assumed and anticipated by otherwise critical academic

theorists. Hence, social constructionists dismiss examinations of technology's agency as participating in a technological determinism that is the handmaiden of a mystifying diversion of attention away from the primacy of social relations. A naturalized belief in technological progress often assumes communication technologies as conduits through which pass unaltered messages between actors. Too ready a dismissal of non-human agencies partakes of the belief that, since the apparent technical "conquest of nature", nothing beyond the human has affect in this world.

5 I would note the almost complete naturalization of metaphors of "extension," as applied to acquiring human agency. One "extends" oneself to others – a metaphor referring back to the hand's or arm's outreach, or that of human lips and voice. With respect to achieving, say, greater emotional and intellectual understanding and acumen, however, one might equally use the less spatial metaphor of "engagement" with the world and with others to imply the same process.

6 The suggestion that VEs might void the need for carpentry skills partakes of the "blind belief" Mazlish (1967) notes in "machines' abilities to solve all problems." Such a belief dovetails with a rampant optimism that welcomes technological benefits but assumes they come with few or no attached social and material costs. The fiction of cost-free technical innovation (I am reminded of the nuclear power industry's slogan a generation ago – "too cheap to meter") suggests that things will not change, that technology will not magically bestow its benefits while social and material relations somehow remain unaltered.

References

Bakhtin, M. (1984) *Rabelais and His World*, H. Iswolsky (tr.), Bloomington: Indiana University Press.

Barker, F. (1984) *The Tremulous Private Body: Essays in Subjection*, London: Methuen.

Barthes, R. (1976) *Sade Fourier Loyola*, R. Miller (tr.), New York: Hill & Wang.

Bourdieu, P. (1984) *Distinction: A Social Critique on the Judgement of Taste*, Richard Nice (tr.), Cambridge MA: Harvard University Press.

Bryson, N. (1983) *Vision and Painting*, New Haven: Yale University Press.

Bukatman, S. (1993) *Terminal Identity: The Virtual Subject in Post-Modern Science Fiction*, Durham NC: Duke University Press.

Bunn, D. (1994) "Our Wattled Cot: Mercantile and Domestic Space in T. Pringle's African Landscapes," *Landscape and Power*, W.J.T. Mitchell (ed.), Chicago: University of Chicago Press.

Caillois, R. (1984) "Mimicry and Legendary Psychasthenia", J. Shepley (tr.), *October* (31).

Coyne, R. (1994) "Heidegger and Virtual Reality" *Leonardo* 27 (1).

de Kerckhove, D. (1991) "The New Psychotechnologies," in D. Crowley and P. Heyer (eds), *Communication in History*, White Plains NY: Longman.

Dery, M. (1993) "Flame Wars", *South Atlantic Quarterly*, Fall.

Dewey, J. (1916) *Democracy and Education*, New York: Macmillan.

Dreyfus, H. L. (1992) *What Computers Still Can't Do: A Critique of Artificial Reason*, Cambridge MA: MIT Press.

Elias, N. (1968) *The Civilizing Process: The History of Manners*, E. Jephcott (tr.), New York: Urizen Books.

Gardner, B. R. (1993) "The Creator's Toolbox," in A. Wexelblat (ed.), *Virtual Reality: Applications and Explorations*, Boston: Academic Press Professional.

Grosz, E. (1994) *Volatile Bodies: Toward a Corporeal Feminism*, Bloomington IN: Indiana University Press.

Grosz, E. (1995) *Space, Time, and Perversion*, London: Routledge.

Haraway, D. (1985) "A Manifesto for Cyborgs: Science, Technology, and Socialist Feminism in the 1980s", *Socialist Review* 15(2).

Haraway, D. (1991) *Simians, Cyborgs, and Women*, New York: Routledge.

Hayles, N. K. (1993) "Virtual Bodies and Flickering Signifiers" *October* 66.

Heim, M. (1993) *The Metaphysics of Virtual Reality*, New York: Oxford University Press.

Hirschkop, K. (1992) "Is Dialogism for Real?" *Social Text* 30.

Jonas, H. (1982) *The Phenomenon of Life*, Chicago: University of Chicago Press.

Jones, R. S. (1982) *Physics as Metaphor*, Minneapolis: University of Minnesota Press.

Kelly, K. (1994) "Hive Mind" *Whole Earth Review* 82.

Kroker, A. (1992) "Virtual Reality" radio interview, Turning Point, La Bande Magnetique, Montreal, Summer.

Lanier, J. and Biocca, F. (1992) "An Insider's View of the Future of Virtual Reality" *Journal of Communication* 42(4).

Laurel, B. (1993) *Computers as Theatre*, Reading MA: Addison-Wesley.

Lefebvre, H. (1991) *The Production of Space*, D. Nicholson-Smith (tr.), Oxford: Blackwell.

Mazlish, B. (1967) "The Fourth Discontinuity" *Technology and Culture* 8(1).

Mitchell, W. J. T. (1994) "Imperial Landscape", *Landscape and Power*, Mitchell, W. J. T. (ed.), Chicago: University of Chicago Press.

Olalquiaga, C. (1992) *Megalopolis*, Minneapolis MN: University of Minnesota Press.

Ong, W. J. (1993) *Orality and Literacy: The Technologizing of the Word*, New York: Routledge.

Rothenberg, D. (1993) *Hand's End: Technology and the Limit of Nature*, Berkeley: University of California Press.

Schultz, J. (1993) "Virtu-Real Space: Information Technologies and the Politics of Consciousness" *Leonardo* 26(5).

Simpson, L. C. (1995) *Technology Time and the Conversations of Modernity*, New York: Routledge.

Smith, N. and Katz, C. (1993) "Grounding Metaphor: Towards a Spatialized Politics", in M. Keith and S. Pile (eds), *Place and the Politics of Identity*, London: Routledge.

Smith, P. (1988) *Discerning the Subject*, Minneapolis: University of Minnesota Press.

Vidler, A. (1992) *The Architectural Uncanny: Essays in the Modern Unhomely*, Cambridge MA: MIT Press.

Williams, R. (1983) *Keywords*, New York: Oxford University Press.

5

MISSING KITCHENS

Susan Bordo, Binnie Klein and Marilyn K. Silverman

> Topoanalysis, then would be the systematic psychological study
> of the sites of our intimate lives. In the theater of the past that is
> constituted by memory, the stage setting maintains the characters
> in their dominant roles. At times we think we know ourselves in
> time, when all we know is a sequence of fixations in the spaces of the
> being's stability – a being who does not want to melt away.
>
> Bachelard, *The Poetics of Space*

On a February night, three days before the unveiling of our father's stone in a
cemetery on Long Island, we three sisters (Mickey, fifty-eight, a clinical
psychologist; Susan, forty-nine, a university teacher and writer; Binnie,
forty-five, a clinical social worker) assembled around Mickey's dining table
to diagram the apartments we lived in when we were growing up. The chairs
we sat upon were familiar, but that night, as we tilted toward each other,
excitedly scrutinizing each other's memories, they seemed perches more than
seats. Like neighboring monarchs mapping disputed territory, we prepared to
do a gentle battle with the truth that is each of our "truths."

We had our assignment: to write something about bodies, place, and space.
We shared the excitement of a first-time collaboration. Vying claims for the
primacy and veracity of certain memories kept us laughing and crying as our
diagrams of scattered blocks of furniture gave personal orientation to a room,
an alcove, a hub of remembered activity. Mickey drew a "chair where I sat
to do homework," outside her small Oldest Sister's Room. Susan and Binnie
remembered a well-made, yellow leather chair with studded buttons, now
given away to relatives; Mickey did not. We all recalled the one bedroom we
at one time shared, with a crib nearly flush against a double bed. Our
diagrams created open spaces with new angles and vectors onto which the
sentiments and perversions of memory laid their claims. "Wasn't the bedroom
bigger?" "I thought Daddy's chair was over there." "I didn't know you were
still living there then."

Our different birth order in the history of our family and of our culture –
Mickey was born before the Second World War, Susie during the first year of

the postwar baby boom, Binnie at the dawn of the 1950s – ensured that our geographical and familial memories would diverge as well as overlap. Mickey was born in Brooklyn and lived there until she was five, in the "old neighborhood" of Brownsville in an apartment with parents, grandparents, and uncles. Susan was born in Newark virtually nine months to the day after our father's return from the South Pacific, and she did all her growing-up in that city, in the two apartments where all of us lived at one point or another; the first apartment was tenementlike but spacious, in what we now would call the "inner city," the second was just behind the high school, in the racially mixed (Jewish/Black/Italian), lower- to middle-class neighborhood immortalized in Philip Roth's early novels. Binnie, who spent the first fifteen years of her life in Newark, was also the only one of us to experience life in the "suburbs" (a too-bourgeois term for the barrackslike complex adjacent to the New Jersey highway into which our family moved after Susan went away to college).

Once, sometime in the mid-1950s, a local radio show called to tell us that we would win some wonderful prize (probably two tickets to a downtown movie) if only we could identify the "Garden State." None of us could, although we were living in it. Perhaps it was hard for us to imagine New Jersey, which we knew via Newark, in such Eden-like terms. But, more deeply, we lacked the objective markers that give people a sense of place. During our childhood we were the most truly nuclear family we knew. We belonged to no community groups, no synagogue. Our mother, an émigré to this country at age thirteen and now uprooted from her family in Brooklyn, teetered on the edge of agoraphobia, suffering nervousness and various physical symptoms for most of her life. She died at the age of sixty-three from a series of strokes brought on by an intestinal illness that was never diagnosed, after a year of complaints that her doctors largely dismissed. Our father, a traveling salesman, was gone much of the time; when he returned, he seemed restless to "get out of the house" (to a movie, to New York City, even just for a drive on the highway to fill up the tank). They had both experienced the dislocations of the Depression, World War II, and the loss of their original community; first married to other people – a brother and sister – they had fallen in love at a family get-together and left Brooklyn.

Having come together in a passionate affair and then exodus, our parents seemed unable to ever really settle down. They were witty, warm, fearful, easily offended, superstitious, depressed, addicted to crisis. Our father had a fierce intelligence but was prone to brooding, angry moods over a life derailed by the Depression from its rightful course toward college and a career as a writer. Our mother worshiped, protected, and deeply resented him for their growing isolation and for his petty tyrannies. She loved her brothers and sisters and tried to extend our family to include them, but only rarely could she cajole my father into visiting relatives or going to neighborhood parties. Inside the nucleus of our immediate family, however, the intimacy – if not the

73

communication – was intense. And in 1996, as we sisters sketched our maps to fill in the missing years, the missing rooms, we grew increasingly aware that being together was our real place.

In our talk, rooms and emotions surfaced and entwined. Memories of furniture gave way to discussions of how a particular armchair was transformed by whether daddy or mommy was sitting in it and how their presence and absence transformed the emotional climate of the spaces. We all remembered holding hands for reassurance and comfort at bedtime, Binnie reaching through the bars of her crib. In our train of associations we were now into dangerous and safe places. At this juncture we discovered, with an eerie jolt, that in sketching diagrams of the apartments we grew up in, none of us could place and describe the kitchens. We labeled the dilemma "The Missing Kitchens" and immediately realized that we had begun this process, this exploration of bodies, space, place, without acknowledging what, for want of a better term, might be described as our shared "spatial" difficulties, difficulties to which each of us might give different names and associate with different fears but which run through the family like physical resemblances, no two features exactly the same but unmistakably of our line.

Each of us has suffered, each in her own way, from a certain heightened consciousness of space and place and our body's relation to them: spells of anxiety that could involve the feeling of losing one's place in space and time, events that profoundly affected how we existed in our respective worlds and how or whether we moved about in them. Bridges. Tunnels. Elevators. Open spaces. Closed spaces. In an effort to think about "places through the body" in an intimate and collaborative way, we decided to bracket theoretical and clinical questions and move in on these disturbances of self and place by following the associative trails we discovered: missing kitchens, disappearing bodies, presence and absence, safe and unsafe places.

Somewhat arbitrarily, but based on the topics we gravitated toward in our first attempts, we assigned each other one theme to explore more deeply. Binnie, the youngest of us, was still living at home as our mother's world became more and more limited and her body declined; she has written "Disappearing Kitchens and Other Non-Places." Susan, in the middle, had always experienced her life as a battle between the nesting and traveling sides of her self, and she has associated that conflict with the extremes of our parent's very different personalities; her piece is "The Agoraphobic and the Traveling Salesman." Mickey, who grew up before these parental modes had solidified, when our parents were simply trying to find a place to exist, describes the anxieties of space more existentially; the struggle to find location for oneself, to occupy a place, to know where one is and who one is in space: these are the themes of her piece, "Maps and Safe Places."

The missing kitchens, which set us on our exploration, led us to our parent's lives and the intimate imprints their absence and presence left on our psyches and bodies. But the missing kitchens also ultimately led us to consider larger disappearances and dislocations: cultural diaspora, the attenuation and for some – like us – the virtual disappearance of ethnic culture and religious community, postwar dislocations of gender, the fragmentation and increasing isolation of families. We had always thought of our own family as disconnected from the seemingly more communal and integrated lives around us. Writing this piece, we began to see how reflective our family's history was of a certain cultural trajectory of loss and disorientation. Writing this piece has set our little nuclear bubble down in time and place.

BINNIE: DISAPPEARING KITCHENS AND OTHER NON-PLACES

Is home a place?

Home and place are concepts that are frequently idealized and imagined in terms that do not give credence to real life. Academic geographer Tuan says that to know the world is to know one's place; he speaks of "visual pleasures; the sensual delight of physical contact; the fondness for place because it is familiar, because it is home and incarnates the past, because it provokes pride of ownership or creation."[1] I smile, thinking of the word "place," since the Brooklyn candy factory where my father worked on the days not spent on the road was referred to by him as "the place," never anything else. He never said he was going to the office, or the factory, just "the place." Yet for my father this "place" bore no pleasure, mostly drudgery and defeat, at the hands of the more successful relatives who ran the company. Of course, many people do in fact experience their "place" in life negatively.

Philosopher Gaston Bachelard writes that "when we dream of the house we were born in, in the utmost depths of reverie, we participate in [an] original warmth, the well-tempered matter of the material paradise."[2] But a house is only truly safe and idealizable when the people who inhabit it live in a context of the outside world. The fire at the ski lodge feels the best after a day on the mountain. Even in paintings in caves (the earliest homes), the walls of the cave are dotted with depictions of what goes on outside, the hunt, interaction with the larger sphere. These symbolic totems refer to the outside as viewed from within the protected lair. For me, with a mother who was disconnected from the outside world, her kitchen was a place like an elevator, where you could get stuck, or like a great oceanic void, where you could get lost.

Yet even if cave/home/kitchen is the site of despair or fear, in the midst of a panic attack the idea "I must get home" becomes the only possibility for escaping from the mind-numbing terror and confusion generated by a flood of chemicals in the brain. Then home is imaged as the safe harbor, where one will calm down, reconstitute, regain composure, remember and therefore potentially reexperience oneself as a competent entity traveling through space. Home becomes a floating anchor. Memories can soothe; the infant's thumb recalls the mother's breast. But is home a place? What if the sense of safety does not refer to a place at all but to a collection of objects, feelings, bodies? I stare at a certain clock and the familiar brightness of the hands recalls my mother's face, as if she, like an object seen every day for many years, ticks on through the object. Throughout my childhood I took daytime naps with my mother on one twin bed in my parents' bedroom. With one small part of my body in contact with one small part of hers, I could rest. My eldest sister recalls best being able to study on the living room couch, her two sisters on either side of her, leaning casually against her, keeping her anchored in place.

Simulations of the peace given by these moments of contact are projected onto the floating anchor of the home, the site of remembered juxtapositions. Yet the same sites of comfort may also recall places of pain where certain negative exchanges, heightened moments, and their emotional residues have resonated in time and space. These images shimmer with dread, the air charged with ions of anxiety. I remember living rooms in various apartments where we sat, my mother's eyes searching my father's face and gestures for evidence of the start of a dark mood. We were warned not to do anything to bother him. "Upsetting daddy" became a place, as agonized silences, mean-ingful glances, and tensed muscles imprinted themselves onto the scene. In our living room daddy occupied a certain chair when he was not traveling, and when he sat there the air was heavy both with his cigar smoke and the invisible but palpable tension of his unhappiness. If daddy is home, then mommy feels a certain way. Tense, maybe, but less lonely. If daddy is away and mommy is in the daddy chair, her cigarette smoke is less noxious than his cigar's, but her loneliness is a junior version of his despair. Many feeling states in me were encoded during moments like these while people were sitting in their places, doing very little. The atmosphere left a residue, and the imprint of these fossilized moments remains on my body. The pain of this inheritance becomes a place that is a bodily feeling; pain itself is a place.

Time can create a private space that becomes a place. At 4 AM, in the living room, the chairs are empty of their people; my parents are asleep in their back bedroom. I am lying on the couch trying to reach sleep by watching a series of late movies until the screen goes blank the way it used to. It is 1968 and I have returned from the larger world called Boston, where I have dropped out of college, and I am home to rest my head upon the shore of my mother's lap. I have created a place for myself defined by time, in which the vectors link, as personal as a signature, the juxtapositions are sad and claustrophobic, but they are my creation and I am attached to them. My solitude is immense, it fills the room, it is a place of secret thoughts. If someone stirs in the back room, threatening to emerge to get a drink of water in the kitchen, I lose my place.

Entering a missing kitchen

We have mentioned that we couldn't locate the kitchens in our sketched diagrams. In one sense, perhaps, this absence is not surprising. Viewed as the province of our mother, whose powerlessness was often shown through her dependency and passivity, her inability to drive, her not working outside the home, the kitchen did not offer much comfort. Duncan Hines cakes might sit atop the refrigerator, with names like "Cherry Supreme," occasional interesting meals from the good recipes passed down from my orthodox grandmother might delight us, having my mother wash my long hair in the kitchen sink had a reassuring constancy, long talks at kitchen tables surely must have happened, although I cannot remember any. Perhaps the kitchens are lost because, as feminists have pointed out, such rooms have long served as mere background to the more important places men have traditionally occupied. Yet that background space is the place where family life often centers, where preparation and consumption of food and therefore life itself resides.

Through television we have been imprinted with ubiquitous images of the wife and mother of the 1950s, but they are drained of specificity and particularizing texture. The archetypal housewife/mother (pre-*Roseanne*) spins through our psyches with good-natured grace, in a cotton shirtwaist dress, giving lilting instructions to the rest of the family: "Come on now, dinner's ready!" "Sally, would you pass your brother some potatoes, please?" "Not now, Hon, I've got to do the dishes." These images, bleached of ethnicity, do not reflect the reality of most people's lives.

The kitchen is where crucial aspects of culture and ethnicity are maintained, but it may also be the site where assimilation occurs. Among the features of the religious community of observant Jews are the potentially meaningful moments created by holiday gatherings and rituals. But our family was insulated, isolated from that community. My only contact with a kosher kitchen was through the orthodox practices of my one surviving grandparent, my maternal grandmother. In the kitchens of religious Jews the arrangement of dishes inside cabinets choreographs the "koshering" of that space: certain dishes for dairy, others for meat, used and washed separately, even the dishtowels. These practices were viewed as extreme in our family; as nonreligious Jews our kitchens were mostly stripped of ethnic lessons from the past. I would perform ordinary yet sacrilegious tasks (cutting with scissors, using electricity) on the Sabbath while my grandmother glared at me. I glared back, suspicious of her foreign tongue, her strange practices, her rigid notions of right and wrong. I shared a tendency that members of oppressed groups are vulnerable to, a view of our own identifying characteristics as unacceptable, "too much," because they mark us for curiosity, judgment, and exclusion by others.[3]

77

While the Jewish kitchen was "too much" for me, the missing kitchens of our diagrams, once we had discovered them, felt "not enough," hovering like sad balloons whose structure could not be grasped or appreciated. If mother is of the kitchen, and the kitchens are missing, then mother is missing. But our mother, although limited by her anxiety, was not missing for us. Simple absence causes wounds of neglect, a non-space, but ours are not exactly diagrams of neglect, they are diagrams of the pull toward accomplishment, movement, and embodiment and the simultaneous regressive pull backward to stasis, ill-health, fear, disorientation and isolation. On my first day of school I clutched my mother, the teacher pulling my arm while I screamed, "No, I don't want to!" In my mental picture both teacher and mother want a part of me. I didn't consciously know then that teacher represented the active, external world and mother a homebound passivity. I only knew that in the middle I was in agony, in no place at all.

The missing kitchens, as the rooms of confusion and loss, reflect such tension in spatial terms. When we sisters gazed at our diagrams, the kitchens were difficult to place. They were not completely gone, after all, but they were clouded by a mind-stopping rush of contrary images and emotions: too much, too little, oppressively there and not really anywhere. A space is not filled, something is missing. Is something wrong then? Perhaps what is wrong is that ambivalence makes it difficult to reconcile loss. Anger is a clarifying emotion, one that helps us differentiate and separate. If only the kitchens were . . . angry!! Instead they appear to be lost. "I don't know where I am," we say when we are lost. "I don't know if I am," I say in the boundaryless expanse of nature. I can't feel my borders – the rooms have all merged, there are no maps that apply.

And if the kitchen is the body/belly of the mother, I approach my relationship to my mother's body through the squinty protected vision of my fingers, as if I am watching a slow, plodding foreign film with disturbing images that occasionally poke through. I cannot take it all in at once. My mother's physical presence is a chunky, impenetrable block of tension. Standing, she looked like she was sitting. Sitting, she could be standing; you couldn't bend her. I have no image or memory of my mother assertively angry, and that leaves a hole inside me. The special block my mother's body inhabits in my psyche is how my head feels when I experience a preverbal sensation of my head being too heavy, too thick to bear, or perhaps simply to lift, as if I am an infant negotiating a new stage of physical mastery.

My mother's cooking was lazy. She opened cans, mostly peas, carrots. I didn't eat a mushroom until I was over twenty, and I didn't know what artichokes or avocados were until then either. I felt like I had spent my life eating in a fallout shelter. What was my mother's cooking about? It was partially about her mother, who was orthodox Jewish, born in Poland, and had lost her husband to a vague tale of the times; a soldier, he had gone to America before her, to pave the way, but he was unfaithful and didn't send for her at first. When she did arrive in America, it was with her five children: Regina, Esther, Leon, Arthur, and Bobby. Many years later, widowed, depressed, pining, my grandmother was finally installed in the Brooklyn ghetto in her small apartment. She had never learned English out of that crazy defiance and passion for Yiddish, as if to say, "I refuse to engage with you, although I am here in my body in this country I am not really with you." Grandfather, dead the day after I was born, sat in a framed photograph atop a doily on her television console. In my grandmother's kitchen, meat purchased from the neighborhood kosher butcher and fashioned into elaborate recipes simmered on the stove for hours, fatty, impenetrable, inscrutable.

Tales are told by my sisters of special cooking done by my mother, but by my time she was getting tired and the simmering meat was just too much of an investment in an increasingly irrelevant process unnecessary to modern life. In general, my mother's kitchen efforts were seen as mundane daily events, neither special nor deficient, her cleaning of ashtrays an expected duty. I don't recall any words of thanks from my father. My father's relation to the values of the kitchen was both paradoxical and typical. Although he had been a chief baker aboard ship in the navy during World War II, he never made one meal, prepared one hot drink that I can remember, but always alluded proudly to the apple pie he had baked for 1,200 men. When he suffered in certain years from the discomfort of new teeth made to replace long neglected ones, he would sullenly push away the plate of food my mother had prepared, muttering, "I can't eat this . . . it's too tough." "What can I make you instead?" she would plead, but by then he was in the rapturous arms of bitterness and martyrdom. Nothing would do now.

My mother's deep and growing fatigue was clearest to me after her death. We had gone to my parents' last apartment in New Jersey to go through her things and select those items we each wanted. When we examined mother's cabinets, the most heart-wrenching vision for me was of her inexpensive casserole dishes, which on inspection were dirty and stained with marks that suggested her distraction or preoccupations. I took some of the dishes, imagining her in her later years (she lived to be sixty-three) vaguely nauseated, headachy, and bloated. She was a hypochondriac for years, but like the boy who cried wolf, her last complaints were real – and not taken very seriously. She died of a mysterious illness whose only clues, found during exploratory surgery, were a foot's length of gangrenous intestine and a sigmoid adhered to a colon; my father forbade an autopsy. I am now left with the imprecise image of her body as a congested house.

In that last apartment in New Jersey the kitchen had just about disappeared. The last kitchen was a kitchenette without room for a table, without a door, spilling into a small dining "area" without a door, which spilled into a small living "area" without a door. So now from every direction my father could be seen, pushing his meat away at the dining area table or putting his head into his hands after work, while from every direction I could see my mother, seated temporarily in his club chair (she was in the living "area" while he ate because we hardly ever ate together as a family anymore), tensed, inhaling a shallow breath, watching him.

Now I find disquieting the realization that the small carriage house in which I have lived for many years has no distinct rooms on the first level; the open floor plan is like a loft: kitchen, dining, and living areas merge. If there are no accidents in life, as interpreters of behavior would say, then I have architected a repetition of the last setting in which I observed my parents together. I think there is another, more compelling source: I do not have to deal with a separate kitchen. I do not have to deal visually with the loss of my mother or the isolation of her life. And I do not cook, the man in my life does.

SUSAN: THE AGORAPHOBIC AND THE TRAVELING SALESMAN

Panic

"Agoraphobia," I wrote in 1987, "which often develops shortly after marriage, clearly functions in many cases as a way to cement dependency and attachment in the face of unacceptable stirrings of dissatisfaction and restlessness.[4] "What I did not write was that my own agoraphobia had developed shortly after my own marriage, into which I had drifted – at age twenty-one and a college dropout – like a sleepwalker, giving up bits and pieces of myself in such small increments (an ambition here, a fantasy there) that I was startled awake one day to discover that I was almost entirely gone. What woke me was my body, whose very being in the world suddenly shifted and changed everything. It was as though I had been in a fog, bobbing in a familiar sea between icebergs unknown to me, and then all at once I was stranded on an enormous one, rising high out of the sea, perched, precarious, desperate for walls to plant my hands against. The physical sensations of panic were so new and frightening – and so seemingly arbitrary – that my only anchoring thought was to get myself someplace where they would stop, where I would be safe. That place seemed to be the dreary, gray Hyde Park apartment in which I lived, without much pleasure, with my husband.

It happened on the Illinois Central, which had been taking me and my younger sister Binnie into the Loop. She was visiting me and we were planning to see a movie. I became faint, but instead of putting my head between my knees as a normal person would do, seeking to restore equilibrium through the trusted processes of one's own body, I responded like a drowning person, my only thought to find air. *I must get off this train. Get me off this train!* In the logic of the panic attack, it makes absolute, unanswerable sense, as places and bodies collude, become one: "If I don't get off this train, I'll faint." "This tunnel is suffocating me." "If I can only make it home, I won't die."

But what gives certain places the power to make panic? Freud theorized that his own train phobia was the result of having seen his mother naked during a train trip from the family village in Moravia to Vienna. But Freud was only one of many late Victorians who suffered from "train neurosis," a condition most physicians of the era attributed to traumas suffered in accidents, abrupt stops, loud noises. In the late 1960s the psychiatrists and therapists that I went to, desperate for understanding and help for the baffling and seemingly inexorable thing that was happening to me, were uninterested in the places of my panic; true to their time, they were more interested in the "home" front, arguing that my agoraphobia was the result of my failure to accept my femininity and accommodate myself to marriage.

Only when I began to savor train trips — traveling to colleges later in my life, as a visiting speaker — did I understand the fine line that separates panic and excitement. The charge of leaving home, knowing that your body has been cut loose from the cycling habits of the domestic domain and is now moving unrooted across time and space, always to something new, alert to the defining gaze of strangers. . . . What made this terrifying to me at one point in my life and invigorating at another? Victorians, utterly unused to the massive, indifferent, steel power that could bear them from the rural village of home to the noisy chaos of the city⁵ suffered from railroad anxieties and phobias virtually unheard of today. (Do we really need a naked mother to account for Freud's panic?) They had to learn to tolerate the stimulation, the "nervousness," as Freud put it, of modern life. Could it be that my numbed and muffled self, on that train in Chicago, had become a Victorian, allergic to excitement, experiencing any opening of limits, any fluttering of heart, any intimidation that the world was far, far bigger than my home, as panic?

The road

The father that Binnie and I knew was a traveling salesman. Everything changed when he came home. Fog would lift, the air became oxygenated as doors opened and the brisk outside rushed in, waking us up. We would have waited all day, wondering what our presents would be. There would be tiny hotel soaps, plastic dolls dressed in buckskin, pecan pralines, restaurant matchbooks to add to my collection. The thrill of suitcases opening. Whisking away for Chinese food, to an air-conditioned movie. The tenuous delight of the huge, good mood that he would be in for at least an evening as he recounted stories of the brokers who would deal only with him, the fabulous hotels and restaurants they ate in as they did their business, assorted oddballs he had run across. The drama was always high, no matter what the story. We often got giddy. And, as if to verify my parent's favorite superstitious warning – "If you laugh too hard, you're going to cry" – a crash was inevitable. Usually it began with a petty squabble between the kids. Always it escalated to something global, between our parents, and then metaphysical, between my father and God. "Why can't you ever . . . ?" "Why do I have to put up with . . . ?" "What did I do to deserve . . . ?" The downward spiral almost always happened on the way home.

Ever since I can remember, my father's comings and goings structured and colored my sense of time and place, masculinity and femininity, dream and reality. Other kid's fathers were always dependably but boringly around. To me, used to a father who descended only periodically, like Santa, they seemed almost extensions of the other kid's mothers, vaguely and unheroically domestic. They could stop faucets from dripping (which our father definitely could not), but only our father could do the *New York Times* Sunday Crossword – in ink – in a morning. I was proud that my father condescended to join the mundane world as a visitor rather than a regular inhabitant. He loved telling the story of how I, at some impossibly young age, still an infant, amazed everyone at Camp Whitelake by uttering a complete, perfectly grammatical sentence: "My daddy is coming home Saturday!" He would break up in laughter as he underscored what a "midget" I was and how agog the other bungalow dwellers were to see words coming out of my midget's mouth. (There was always a whiff of the freakshow even in my father's highest praise of us, as though our best talents were signs that we had dropped from another planet.)

What I cannot remember, however, is just how much time he spent "on the road." If I had to do it arithmetically, I"d say he was home one week out of every four. But then when did I get to see all those Broadway shows? Learn to play cutthroat scrabble? Memorize the meaning and portent of his every gesture? Perhaps he was home more than I remember. Or maybe real time only started when he walked through the door. My mother seemed always to be waiting, smoking or dozing in my father's armchair, putting canned goods away, feeding the cat, until called upon to spring into action on his return. She rarely left the house alone, accept to buy groceries or occasionally to visit a nearby girlfriend; I have barely a memory of going anyplace with her without my father. School was real enough in its way, but anticipatory, preparatory – for summer, for the next term, for the transformation that would allow my life to really begin. But my father was undeniably present. And he made the external world present for me; he offered it as a bracing tonic against some domestic stupor into which I was continually being lured, a preventive measure against my own disappearance. The Midwest brokers, the hotels, the highway that led from Newark to Manhattan, trains and cars that took you from one state to another; it was as though they had a sign on them: "This way lies life."

And the other door? It's not that easy to say where that door leads. Unlike my sister Binnie, I love cooking, and my images of my mother's kitchen, snapped at an earlier, more communal time in her life, are warmer. In *Unbearable Weight* I wrote of

> the pride and pleasure that radiated from my mother when her famous stuffed cabbage was devoured enthusiastically and in huge quantities by all her family – husband, children, brothers, sister, and their children. As a little girl, I love watching her roll each piece, enclosing just the right amount of filling, skillfully avoiding tearing the tender cabbage leaves as she folded them around the meat. She was visibly pleased when I asked her to teach me exactly how to make the dish and thrilled when I even went so far as to write the quantities and instructions down as she tried to formulate them into a recipe (it had been passed through demonstration until then, and my mother considered that in writing it down I was conferring high status on it). Those periods in my life when I have found myself too busy writing, teaching, and traveling to find the time and energy to prepare special meals for people that I love have been periods when a deep aspects of my self has felt deprived, depressed.[6]

Yet for many years, in recovery from the debilitating agoraphobia that kept me housebound for the first part of my twenties, I had to go out for a while every morning before I could settle in to do my studying or writing. I needed that inoculation, that first contact with the outside world before I could feel safe at home. Before I could feel safe, that is, from the lure and illusion of safety, from my inclination to get stuck inside, habituated to a contracted world. Being outside, which when I was agoraphobic had left me feeling substanceless, a medium through which body, breath, and world would rush, squeezing my heart and dotting my vision, now gave me definition, body, focused my gaze. Armored, assured that I was vigorous, of the world, I could return home without fear of dissolving. For a while, my safest places were trains and hotel rooms; I love being en route, and it being someone else's professional business to care for me, bring me my coffee.

But for most of my life, whenever I strayed too far from that domestic world in which I, like my mother, was the professional caretaker, I would reel myself in, calling on my phobias to help. (Before my Ph.D. orals my panic disorder revisited me, and then again after a year unsuccessfully trying to become pregnant.) At those times it seemed that the most delicious thing in the world would be to be sitting again next to my mother, watching daytime soaps with her. But she was gone, and I had to console myself by reproducing her world, by bringing her body back through a more recessive invocation of my own. During one of those retreats, I wrote in my journal (only realizing later that I was writing about my mother as much as myself): "There is something touching to me about my diminished state. Puttering around, cooking soup, putting together packages for people I love, my mind gentle and nondemanding, the strict compulsive self gone, I feel a compassion and care for myself as I would for another person but never have before for myself. Who is this person? As I feel myself conquered, accepting my own diminished state and its requirements, the sweetness and dignity of the little tasks I do, desiring nothing more than simple renewal of contact with the world, are like those of a baby."

These cycling patterns led me to believe that I had learned how to move around in the big world from my father, while my mother taught me empathy and intimacy but left me prone to panic. Today I'm not so sure. When I was growing up, it was presented to me as the essence of their personalities that our father traveled and our mother stayed home, that he was bold where she was fearful, that he was autonomous while she was dependent. It was virtually impossible for me to put the indisputable accuracy of these definitions together with the much dimmer, historical facts that my mother had supposedly had a "wild" girlhood and also had worked in the office of a factory during the war – and apparently, had loved it. Vaguely I remember arguments about going back to work, which my father always won. "No wife of mine. . . . Never!" (even though he had no life insurance and they couldn't save a penny from his salary toward a house of their own or funds for our education). But I cannot remember who was doing the arguing on the other side. Could it have been my older sister? It's hard for me to imagine it was ever my mother herself. But surely it must have been.

My mother, even when she was most anxious and depressed, would smile at strangers, strike up conversations, flirt with shopkeepers. Once, in a crowded supermarket, I felt a panic coming on and tried a technique my therapist had taught me. I imagined that everything that was frightening to me – the noise, the crying children, the pushing and shoving – was a warm, colorful blanket like the one my therapist had me imagine in his office, a kind of visual mantra of comfort. The blanket I had imagined was all green and gold and burnt sienna, colors that I recognized as those of a throw that my mother had knit me, rustic colors that still bring my mother to me unexpectedly as I pass by sunlit trees in autumn. In the supermarket I put that blanket around me with a great effort of imagination – and I amazed myself. I really did feel enveloped and calmed by the chattering people in the store who had seemed so alien to me just a moment ago. At the same moment I felt my mother's presence strongly, and I recognized that there was a terrible flaw in my picture of her. Her capacity for human connection, her warmth, was not some compensating factor, developed only to make a small and limited world bearable and less frightening. It was big, it was strong, it was powerful, and it was highly unusual, a remarkable gift. She wasn't afraid of people! She was an adventuress!

Today, looking through a folder of old letters, I found two postcards. Of all the cards and letters I received from my parents, somehow just these two survived. One, sent to me by my father while I was a freshman living in a dorm in Chicago, was from the "Fabulous White Way" of the "World Famous Las Vegas Strip." Inside, via a foldout display of casinos, hotels, and showgirls, it told the "story" of the Strip. My father had written, beside the photo of the Dunes, that he was staying there, and he sent "greetings from the land of lost wages, daughter!" Over the years I had received scores of cards like this from him, exuberant with the romance of travel. But the other postcard startled me. It was from my mother, written to me while I was living in Canada, struggling out of my agoraphobia and splitting up with my husband. It was from Florida, where she had apparently gone – on a plane! – with her best girlfriend at the time. My mother had not been afraid of planes, I now remembered, that was me and my sisters. But a vacation? Without my father? I couldn't remember it, couldn't imagine it, yet the evidence was in my hand.

The cartoon on the front showed a shapely, tanned woman on the beach, beside which my mother had written "Me," and, separated by a line, a little girl bundled up beside a snowman, beside which my mother had written "You." "I'll shovel sand for you in Florida . . . if you'll shovel snow for me up north," it read, with a space for the temperature, which my mother had filled in at 86 degrees. On the back was her more personal message to me: "Hiya Darling: The flight was smooth as silk. Got a little high and didn't feel anything but happy. Wish you were here. Love from Florence. Say hello to your roommates. Love you, Mom. P.S. Later Dad will come."

MICKEY: MAPS AND SAFE PLACES

Waking up

In the middle of an obscure and heavy dialogue on "geography" as reflected in places, bodies, and madness, one of my sisters tactfully inserted a simplifying question: If they asked you where you were from, what would you say?

I would say: I am from the land of the three sisters – from Never-Never Land – from the near/far planet that sent out small bands of thin-skinned envoys to test their survival skills in the ghetto neighborhood of Newark. I was never, in fact, from anywhere that I actually lived.

I lived inside my family and even then in a private place far from its borders. I have since learned that family is often the conduit of the culture of a place, the specifier of the boundaries of its geography. When family *is* your place – encapsulated in its own climate – you gain a measure of independence from the arbitrariness of local places in exchange for the claustrophobic logic of a tenement apartment suspended above the stores and the street.

I had my maps:

- the map of lost treasures, like the house in Brooklyn that burned down with the Persian rugs, piano, and an immigrant family's newly acquired wealth . . .

85

- the map of the town and house we would have had if the depression hadn't come and my father's father hadn't been deceived by his nephew, and the war hadn't come . . .
- the map of the lost village in the Old Country where my mother was born, that no one could ever find or pronounce . . .
- the map of the peopled/storied neighborhood in Brooklyn from which my parents fled in exile to live out their "illicit" romance in New Jersey.

No wonder then that they never showed me the way to school, never warned me that Clinton Avenue was a dangerous place, never told me that New Jersey was the Garden State. The most important parts of my mother and father never existed in the places we were living.

I lived inside my mother's body, inside my father's dreams and nightmares. Lived on the shelf next to my books, lived in my books, and it never mattered that we had the best car on the block to take us to the Jersey shore. When I began the project of leaving, I took a bus to the downtown terminal, the Hudson Tubes to Port Authority. I walked across 42nd Street and took the Fifth Avenue bus to Washington Square. I did this five days a week for three years. I never knew that this was hard or unusual or impractical or risky. I calculated the trip as the distance between the bookshelves in my room and the bookshelves in the college library, the distance further foreshortened by the books I carried with me.

When I was attacked and tied to a tree, walking along Riverside Drive at midnight, the police told me I should have known that this was a dangerous place. I knew – frozen by sudden moments of discontinuity and wordless dread – that I sometimes didn't know who I was. I never realized that I didn't know where I was and that this was something I should have – could have – known: the cradled anchoring in the reality of place, the grounding in the present, the topography of daily living in real/time and real/space.

My books had come with their own maps of safe and dangerous places. My parents, in an effort to protect me which proved to be both misguided and inspired, had left me to my books, believing as I did that my immersion in my books gave me special status, like a time-traveler in a 1940s version of virtual reality. Unaware of real dangers but constantly on the alert for the romantic and tragic possibilities of life as described in my books, I navigated the blocks near our apartment with an odd combination of courage and fear and trembling. What were the dangers of an occasional mugging compared to the prospect of being snatched from the London streets by Fagin? Why should I be afraid of the American Indian on the street who was tormenting cats when I was dreaming of sharing a tent with the last of the Mohicans? When I panicked at the sound of air-raid sirens, I was thinking of tidal waves in Japan I had read about in a novel that morning or a piercing call to arms against an alien invasion as recounted in an H. G. Wells story. I traveled with "Lad, a dog" to protect me and a book-inspired notion that if I meant no harm to anyone I would not be harmed, and in fact I never was harmed during my childhood.

As part of the same effort to protect (and prevent me from ever leaving them?), my parents rarely explained or put into context or helped me to anticipate anything contemporaneous with my own time and existence. Mostly I was given, as a very special confidence and charge, the stories of their past life, which I carried inside me like a great weighty epic novel of the Brooklyn streets which had yet to be written. It was a novel about exodus, success, catastrophic financial reversals, passion, scandal. The British may have needed two hundred camels and as many servants to transport the accoutrements of their culture and comfort to the colonies. My father's family arrived from England with a suitcase and their wits, my mother's father, preceding his wife and children, with a knapsack and a sewing machine. I thought of them all as carrying within them invisible treasures, like a caravan of Russian nesting boxes filled with stories, Jewish lullabies, mandates for humanistic living that were thousands of years old.

I was proud to have European roots. This was a romantic image that tied me to cherished intellectual and cultural values and linked me to the authors of my books. I never really knew until I saw faces like my mother's in a pictorial history of Eastern Europe that in fact my mother had come from another place. She hadn't brought her memories with her or chose not to share them and I had never questioned her silence. When my grandmother tried to tell me about village life in a language I had forgotten, my mother did not translate. When uncles referred in passing to Grandpa's army service she never elaborated. It was in our kitchens that these aborted dialogues took place. It was the kitchen in which I felt my mother's silences most palpably. When I first began to lose the Yiddish in which the earliest experiences of my life were encoded, she never helped me to retrieve it, just as she had never spoken of the first twelve years of her own history in the Old Country. She stopped dressing me in lace blouses and flowered skirts. She stopped wrapping my braids around my head. She stopped baking apple cakes. She never laid claim to the mementos and photographs of her parents' early life. I wonder now if she thought that she had little of real value to contribute to the lives of her American daughters.

When I stepped out of my books and daydreams into real time, I understood finally the nature and scope of the real dangers and losses my parents and extended family had faced. I understood that the air-raid sirens had signaled a real war. I understood that my father, in his romantic naval uniform, had fought that real war in a very remote, very dangerous place in the world. I understood that my mother and other women left at home like a colony of the disenfranchised, negotiating dramas of reappearance and disappearance, were in fact dealing with real danger and real absence, listening to Gabriel Heater,[7] rationing food, diverting each other with stories of childbirth in their kitchens. When over the subsequent years my mother became increasingly withdrawn and her life constricted, I knew that part of what she had struggled with were actual losses, deaths, separations from family, and I wondered whether any of the psychiatrists whom she consulted had ever dealt with where she had come from, where she was going, and why she got stuck.

I have been tempted at times to see my parents lives as a contemporary political parable: woman unfulfilled, housebound, man bitter and neglectful, daughters emancipated through education, psychotherapy, and the cultural endorsement of women's work. I had been tempted to think of my mother as a prisoner of her house, with my father, braver and more vital, able to escape into the wider world. Our mandate then seemed to be to get women out of their houses, out of their kitchens, winning for them the mobility that men enjoy. In my childhood I had been exempted from kitchen duties, given the status of a scholar who shouldn't have to be concerned about such things. In this view kitchens are demeaned the way some musicians relegate lyrics to a lesser place than melody and chords. I shared both biases. I saw home and hearth as a bird's nest with its fragments of stuff, used to support life but subordinated to the more important job of flying off and getting on with the real business of living. Now I understand that kitchens matter as the lyrics of melodies matter.

I realize now that both my parents were traumatized and diminished by the dislocations of their lives: immigration, the Depression, war and wartime, personal upheavals of the diaspora, uprootings from their families. My father, although working with persistence and success in the "outside world," traveled with his sample case of stories that were to him possessions salvaged from the destruction of a richer life. Our mother's tragedy was not confinement to her kitchen and women's work but her growing incapacity to occupy her house, to claim it as her own, to inhabit it fully. Her early years in Brooklyn – and mine – had been rooted in family and in neigh-bourhoods. Friendships were always a part of her life even in the worst period of depression and anxiety; the therapeutic group she joined became her friends. She had deep capacities for sympathetic, loving, nonjudgmental human attachment. But she could not hold onto her entitlement to her own history or the mandates of her own self-development in the face of cultural and emotional dislocations, a possessive husband, and, finally, the threats to the vitality and continuity of memory posed by aging. Our mother didn't disappear into her house, she disappeared within it.

Connections and dislocations

Dramas of spatial meaning and safety from the perspective of self and identity are continuously played out around us. Like a play within a play, people move from point to point following the patterns and routines of apparently practical lives. But they carry with them maps of safe places and safe distances, maps of their internal landscapes, its topography built up of the history of past and present human relationships. These maps designate the way stations where one can be refueled, loved, reminded of one's identity.

Places have always been inseparable from people for me. Syracuse exists on my internal landscape because my sister Susie lived there and when she left, its continuing existence was guaranteed by my purchase of two paintings I loved from people I knew there. New Haven emerges in bold type on my map (Yale University noted in the legend) because my sister Binnie lives there now. Poland will always exist for me no matter who appropriates it because my mother was born there in a town I am still determined to locate on a pre-Second World War map. England first came into being as the birthplace of Virginia Woolf, whom I admired. Though my childhood relationship to places was through my books and remote from the real places I lived in, it was still always personal. As a student I found only geography impossible to learn. I could not keep straight places that were not linked in some way to people with whom I had a connection.

Like the famous Steinberg cartoon of the New Yorker's view of the world, with New York City presiding huge and detailed over three-quarters of the map, the rest of the world misplaced and reduced, we all draw the places on our meaning-maps, large or small, depending on our emotional investments and worldview. But for me places hardly existed at all unless animated by memories of people. More intimate spaces take on the same dynamic. I am most aware of the different architectural features of the house my husband and I built as reflections of the preferences and biases of each person who contributed to the design. Objects in my house are links to specific relationships and are therefore always difficult to discard. In stark contrast, on a bureau inherited from my mother-in-law, arranged ever so artfully, visually compelling, sits a collection of shells from oceans all over the world which I discovered at a consignment shop. They seem to belong to no one and their anonymity never fails to be intrusive and disquieting.

I picture each individual's history of human connections as a flow of conscious, unconscious, and embodied memories that are the substrate of the continuity of mental life and the basis of the most fundamental sense of identity. Throughout life, concepts of home base and the safety of other places remain intimately linked to this sense of identity. Knowing where one is, knowing one's place, understanding the nature of real places in themselves requires first a centered self, grounded, embodied in deeply imprinted maps that record the memories of our history of human connections. Like tethers to a dock or a lifeline to shore, this living, internal history protects us from forgetting who we are, helps us to reconstitute when lost. Without that secure housing for self, places can be seen only as a threat to the stability and cohesiveness of the person. Places become metaphors for states of mind.

Spatial metaphors are well suited to capture the phenomenology of the particular form of panic that occurs when the most fundamental sense of existence and connection is at stake. You are on a well-lit stage. The scenery stands as background and support. You know your lines. Suddenly the stage disappears. The floor drops out. The players and set vanish or persist as unfamiliar figures in another script in which you have no part. You are in a stalled elevator, a traffic jam, on a becalmed sailboat, in confined or open spaces. Suddenly something inside stops. Is it that flow of images of past and present human connections that has been temporarily disrupted, eclipsed, leaving you in a freeze-frame moment of heightened awareness of discontinuity of self and the dread that there doesn't seem to be any way to get back to those way stations that remind you of who you are? Places, to the diminished, ahistorical self, may hardly seem to exist at all.

I try to imagine what my mother's map might have been like in the last decade of her life: a large, featureless area marked with an "x" and a "no reentry" zone and a faceless population, the Old Country; a small place in New Jersey disconnected from the rest of the states; an endless road needing repair leading to a handful of residents on a few named or numbered streets in Brooklyn. The most prominent feature would be the bridges to a few friends and friendly shopkeepers, bridges depicted as spanning hundreds of miles across uninhabited stretches. I can imagine arrows marking the directions her daughters would take in moving away and her husband would follow in his travels. During my visits to my mother that last decade, I always felt anxiety and tremendous sadness at her increasingly constricted life. The road there felt like limbo, an out-of-time and -place stretch of highway, forever foggy. But I never left without feeling oddly revived, reminded of who I was, what I wanted to do with my life, who and what I valued and loved.

Home, as a truly safe place, the container and springboard for integrated living, is the foundation of an ontologically secure existence. While my parents had not been able to provide that, they did give me some very special provisions for establishing roots in my own life in another place. Despite the dislocations, eclipse of tradition, loss of wealth in the Depression, upheavals of wartime, disillusionments of daily living, my parents' humanity, humor, intelligence, and idealism are after all what I invariably draw on to "place" myself in the universe and what fuels my capacity to survive and prevail, wherever I am. Where am I from? Through the process of writing this piece, I realized that I am from England, Poland, New York, Connecticut; I am from all the places I have lived in and that my relatives have lived in because I bring myself, my family, and my history with me. Periods of crisis in my life, spells of anxiety, have almost invariably been associated with a temporary eclipse of that connection.

*****

Now we are sitting in Binnie's real kitchen, surrounded by her totems and linking objects, grounded in the real work of our present, real lives. In tracking the missing kitchens together, we have also retrieved something vital that was passed down to us by our parents. Our father's continuing passion for words and stories, his mythic castings of everyday occurrences; our mother's deep perceptions into other people's feelings; fierce family arguments about the interpretation of past events – what really happened, who was to blame – imprinted on all of us a powerful desire to understand things, to give shape to them, to communicate them. Working together, we actualized these values once again as a family and so reconstructed another part of the missing kitchens. Momentary loss of boundaries, intensities of emotion, anxieties of "influence," all those dissolving and resolving challenges and redefining confrontations have not compromised the safety of the place we are for each other. We are holding hands again through the bars of cribs, leaning against each other on flowered couches, this time not to assure our existence but to celebrate it.

Notes

1 Y.-F.Tuan, *Topophilia: A Study of Environment, Perception, Attitudes, and Values* (Englewood Cliffs, N.J.: Prentice-Hall, 1974), p. 247.

2 Gaston Bachelard, *The Poetics of Space* (Boston: Beacon Press, 1969) p. 7.

3 For Henry Louis Gates, Jr. the kitchen was not only for cooking and bathing but was the place where his mother would "do" peoples' heads – shampooing, curling, straightening. The kitchen was transformed into a hair salon via the presence of one important object, the shears. This place created by an object became then a larger community of people and a site of conversation and the dissemination of value. The word *kitchen* thus acquired another meaning . . . it was used quite literally to describe "the very kinky bit of hair at the back of the head, where the neck meets the shirt collar . . . which you trimmed off as best you could" (*Colored People: A Memoir* (New York: Vintage Books, 1995, p. 42). That bodily marker of black identity signaled what was desirable and what was not, arguably defined by white aesthetics. It was "too much," and that unwanted piece was cut off in the kitchen.

4 Susan Bordo, *Unbearable Weight: Feminism, Western Culture, and the Body* (Berkeley: University of California Press, 1993), p. 176.

5 In *The Birth of Neurosis* (New York: Simon and Schuster, 1984) George Drinka describes one aspect of the changes in life and landscape in the Victorian era.

Between 1850 and 1910 the railway systems of America and Europe became webs binding together the landscapes into skeins that could be traversed in hours rather than days. Between 1850 and 1860 the French railways system expanded from 3,000 to more than 9,000 kilometers of track. By 1880 the French railways had almost tripled in size again and by 1910 redoubled so that many of the provincial sections of France could be reached from Paris in hours. Similar statistics and comparable mobility were also the case in

Germany, Britain, Italy, especially America, and even Russia. The railway – like the telegraph and later the telephone – stands out as a technological wonder tying the national landscapes together and telescoping time and distance in a manner unimaginable to earlier generations.

(p. 110)

6 Quote from *Unbearable Weight*, pp. 123–4.
7 Gabriel Heater was a radio commentator known for his deep and ominous voice, reporting nightly on the war abroad.

6

THE BODY AS "PLACE"

Reflexivity and fieldwork in Kano, Nigeria

Heidi J. Nast

1 **REFLEX**, an involuntary response to a stimulus which passes through the lower spinal and nerve centers without conscious action by the brain. The most common automatic reflexes are *swallowing* chewed food, *laughing* in response to tickling, *jumping* out of the path of danger, *sneezing* when the nose is irritated, *coughing* on inhalation of tobacco smoke, *scratching* in reponse to itching; *erection* of the penis and *ejaculation* in sexual excitement and intercourse; and the *startle reflex* on hearing a sudden loud noise . . .
Fishbein's Illustrated Medical and Health Encyclopedia 1977. Westport,
CT: H.S. Stuttman Co. Inc (their emphasis)

2 **re-flex** (ree'flecks) n. [L *reflectere, reflexus*, to turn back, bend back]. A stereotyped involuntary movement or other response of a peripheral organ to an appropriate stimulus, the action occurring immediately, without the aid of the will or without even entering consciousness. —re-flex-ly, *adv.*
Blakiston's Gould Medical Dictionary Fourth Edition 1979. New
York: McGraw-Hill Book Company

3 **reflex** [L *reflexus* (past part. of *reflectere* to bend or turn back, from RE- + L *flectere* to bend, bow, turn, akin to *plectere* to plait, twist, turn, *plicare* to fold, and to Gk *plekein* to twine, twist) bent or turned back]. 1 The reflection of light from a curved smooth surface, such as the cornea, fovea centralis, or retinal arterioles. 2 An involuntary or stereotyped movement induced by a peripheral stimulus . . .
International Dictionary of Medicine and Biology in Three Volumes 1986.
New York: Churchill Livingstone[1]

Preface

This chapter reworks notions of reflexivity, exploring how it can be considered as a process through which bodies and places, always mutually constitutive, are continuously and relationally negotiated in ways that materially and spatially decenter subjectivity. Negotiations are out of anyone's control, the material making of bodies and places being partial, ephemeral, unsynchronized, and incomplete; bodily and spatial (sub)regions are only ever momentarily stabilized through highly contingent spatial and historical processes. These processes involve being and becoming a subjective *place* of and for chaotic, multiply directed and multi-scalar, demands, expectations, and practices. I argue accordingly that reflexivity in the context of cross-cultural fieldwork involves becoming skilled in epistemologically "falling into" the many material and social processes that derive or impact us from the outside. It involves learning how to accept, reject, and respond physically to external forces and subjectivities in ways that allow us more possibilities *materially* for entertaining difference. In this sense, reflexivity is less about self-introspection, self-reflection, "self-conscious practices . . . in thinking and writing" or self-emanating contemplation of how one "positions and includes oneself in relation to a subject of study" (Marcus 1992: 489) – as it is characteristically defined in the social sciences – than about learning to recognize *others'* constructions of us through *their* initiatives, spaces, bodies, judgment, prescriptions, proscriptions, and so on. In metaphorical and material terms, it is about allowing our bodies to become places which "field" difference.

Introduction

In 1993 I perused one of the CD-ROM data banks (InfoTrac) and the subject index of our University library[2] to find out what kinds of work had been done on reflexivity across a number of social science, humanities, and "hard" science disciplines. As a 1987 convert to human geography,[3] I had heard the term reflexivity referred to largely in the context of anthropological fieldwork, wherein being reflexive was ultimately about comparatively reflecting upon one's cultural practices in relation to others', so as to recognize and come to some understanding of "difference." The way the term was framed empirically and theoretically sounded radically different from what I remembered from the doctor's office: there, reflexivity was checked with sharp, rubber-tipped taps to knees and elbows, resulting in jerky, uncontrollable movements. While in the library, then, I decided to look up medical definitions of "reflexes," an endeavor which ultimately provided me with an embodied counterpoint to definitions otherwise (in non-medical contexts) framed in terms of reflectivity and the "mirror." The stark contrast between the two made me think about why and how the social sciences – when thinking through reflexivity – dwell on the mirror rather than on the body. More importantly, it made me rethink fieldwork and how our bodies might be considered places where we field or draw in difference, or as sites where difference is placed. In effect, bodies are physical field sites upon which the world inscribes itself – *places* to which others come and mark their difference, places like any other place – localized and with continuously negotiated boundaries and subregions.

In the first part of this chapter, then, I offer a very schematic archaeology of the term reflexivity, cursorily reviewing some disciplinary ways in which non-medical discourses frame the term through the metaphor of the mirror. I do this partly to point to the variegated ways in which understanding "difference" through mirroring depends upon a dichotomy that reduces otherness to a singular distinction: me (here)/not-me (there).[4] The me-centered process tames "otherness," making it something one ably decides upon and crafts and something one *chooses* to learn about or reject. Spatially, academic renderings of difference are made mostly in the comfort of an office, study cubicle, home, or library. Our objects of inquiry are deployed passively: books, field notes, and journals, films and photos, guest speakers or bus tours, and a pen writing "about" something/someone on paper.

In the end, the library exercise showed how reflexivity-as-reflectivity is something relatively comfortable and, ultimately, manageable. At the same time the exercise showed that we are rarely taught to locate, creatively work with, and *recognize* reflexivity in terms of what is out-of-(our)-control. We are not taught the skill of engaging with processes and materialities that draw us bodily into other worlds and that require that we "let go" of carefully crafted objectives, agendas, and models and give our bodies/spaces over to other bodies and places. The "giving over" is not always voluntary, nor is it necessarily a gesture or sign of powerlessness. Rather it allows bodies materially and strategically to field or capture, and to be a locatable material field of, difference.

Analytically recognizing that others often *do* set our agendas, *do* show us what is important, and *do* place us in our bodies and assign us to spaces not of our choosing, allows for reflexivity to be re-cast as an embodiment skill, a means for enabling bodily, spatial difference to register in creatively decentering, fragmentary ways. Reflexivity as such is not anarchic, though the language used here may make it appear so. Perhaps the anarchic sense derives from the fact that in academic traditions there are few ways of valorizing, speaking, and inhabiting worlds that are decentering, unknow-able, that mark us. Instead, we[5] often frame our agendas and lives in terms of being in control (good) or not (bad). Loss of control spells depression, confusion, feelings of loss, a lack of physical/emotional equilibrium, alienation, un-centered-ness.[6]

The second part of the paper anecdotally relates examples of my "involun-tary" responses to experiences I had while working on my dissertation in the royal secluded domain of the massive palace of Kano city, a large Islamic urban center in northern Nigeria. The anecdotes foreground what were invaluable formational experiences that led me both to greater subjective decentering and fragmentation *and* to an enlarged experiential field of difference. As such, the stories provide a means of partially re-thinking reflexivity in terms of an embodied "fielding of difference." Such fielding requires that we develop a positional awareness of, and appreciation for, how bodies and places mark themselves upon our culturally owned body/place in nonvoluntaristic, circumstantial ways. I have chosen Kano because it is a cross-cultural context very different from my own. Consequently, the kinds of corporeal, practice aspects of reflexivity I intend to foreground are thrown into exaggerated relief. As cultural outsiders with numerous material unknowns and demands to negotiate, fieldworkers in foreign settings are typically not in control, not subjectively centered, and *not* culturally mirrored.

Unfortunately, mis-fitting and the resulting discomfiture are commonly effaced from fieldwork narratives. Researchers typically present a flawless story – one effect of which is that they become inflated, powerful figures (in control and therefore considered "good" researchers) who are able because they are in control to exert authority, successfully achieving objectives, even though these may change along the way; "bad" fieldworkers are unable to assert control – and fail. The main way a social scientist's assumed powers are mitigated, in this scenario, is through reflecting voluntarily and objectively upon others' "differences," a reflection process which in the end largely reinforces rather than disturbs notions of subjective integrity and control. An anthropologist, for example, goes through the most important disciplinary rite of passage by doing fieldwork, preferably in an exotic place. S/he reflects upon the data, writes up the results, and emerges out of the "work-data analysis-writing" ritual to become a "seasoned" professional. I would argue, however, that the voluntary reflection process is exactly that – reflection – and as such should not be deemed "reflexivity." Moreover, such "reflection" is only a minor part of mitigating assumed power assymetries. What is needed is a means for recovering, re-membering, and re-claiming the uncontrollable – those emotion-laden bits we typically, in our writing at least, ignore. To do this, we need methological and analytical means for analyzing the bodily *as a place* where material differences and subjectivities are registered from without and which speak to and of embodied fragmentation and decentering, something I tentatively and partially broach in the second part of this chapter.[7] Finally, I conclude by discussing how bodily "fieldings" of difference, particularly evident in cross-cultural field work, speak to everyday contexts.

Reflexivity/self-reflexivity as reflection/self-reflection

The number of InfoTrac and other subject-index-located files I worked with in 1993 numbered over fifty, all of them collectively making a strong distinction between *reflexivity* and *self-reflexivity*. While there are clear differences in how the terms are theoretically deployed, at a smaller scale both terms are explained, used or defined explicitly, implicitly, or metaphorically in terms of a mirror. In general, the term reflexivity is taken to imply some sort of structural patterning replicated at some distance from the speaking, reading, viewing, or performing subject – as in the repetition of rhetorical patterns in a text, reiterations of similar gender constructions in different cultural contexts, or the reproduction of the same thematic or narrative structures in a film.

In a psychoanalytic–anthropological piece by Caton (1993), for example, a Yemeni male tribal wedding ritual is construed by the author to be a normative mimetic (reflexive) performance that mirrors ("out there") ideal images/practices of Yemeni manhood. Ritualized identification is immediate – outside the mediation of language: it is performed not to make players aware of its fictional quality, but to provide an absolute image of how masculinity should be narcissistically chosen and lived out. Caton's understanding of the ritual-mirror explicitly draws upon Lacan's theoretical formulation of the *imago*, a mirror image with which an infant initially identifies and internalizes to form a fictional "self," an internalization that is the precondition of its subjectivity.

In a literary criticism piece at some disciplinary remove from Caton, reflexivity is again discussed in terms of mirroring, in this case through two texts. In particular, Hoon-Sung Hwang (1993) in "One Mirror Is 'Not Enough' . . . " locates three forms of resonance or reflexivity in two of Samuel Beckett's plays (author–text relationship, the character–audience relationship, and the relationship between subtexts), each of which he explicitly relates to notions of echoing and mirroring. In similar fashion Bailey (1993) talks about how a poem by Arthur Clough is reflexive in that it reflects both on itself (as a poem) and on the act of writing.

Writings on *self*-reflexivity, in contrast, are more often about finding ways to compel a subject to locate itself and/or to engage in a certain degree of conscious and critical *self*-reflection. Self-reflexivity is therefore commonly portrayed as an introspective act that solidifies as it identifies a subject. Such usage is evident in Swigonski's (1993) work which argues that feminist standpoint theory can be used to make social work researchers more self-reflective about their positioning *vis-à-vis* their clients. Here, clients are envisaged as registering their standpoints voluntarily onto a social worker/ screen, in the process effecting a kind of two-way mirror which both worker and client can view. Viewing through the worker-mirror allows for subjective centering and is a controllable process.

Similarly, family studies teachers are enjoined by Allen (1993) to have students read texts about the life experiences and "personal value systems" of a diverse range of people. Students are expected to keep journals in which they (quietly?) reflect on their own experiences and values *vis-à-vis* the others they have read about (or listened to – Allen suggests that guest speakers be integrated into the curriculum). "Other" practices are thus used as unthreatening, passive background (texts, talking heads) to throw students' images of themselves (i.e., as they script themselves in their journals) into higher relief. This, says Allen, makes students more self-reflexively aware of their values and social positioning, consequently making them more conscious and appreciative of difference. In a similar vein, Bochner and Ellis (1992) tells of an "experiment" involving a heterosexual couple asked to reflect upon, and write their reflections of, the woman's abortion. The couple circulated these written feelings (mirrors) among friends who then formally commented on the writings (a juxtaposing of mirrors), making (according to Bochner) the couple more reflexively "enlightened" about their experience. Lastly, both Man (1993) and Recchia (1991) argue that exaggerated camera techniques in the films *The Third Man* and *Psycho*, respectively, draw attention to the films' filmicness, in the process making viewers ("self-reflexively") aware that they are watching a movie. Again, becoming aware of a film's filmicness is deemed good because it centers-as-it-separates-away-from the viewing subject: the films serve as a kind of cracked mirror.

The terms reflexivity and self-reflexivity, and the metaphorical mirror, are also drawn upon for making *political* arguments, such as those presented by several writers whose works counterpose modernity and postmodernity. Escobar (1993), for example, claims that contemporary (read postmodernist) identity politics in Latin America paradoxically *require* a modernist bi-polar distinction between native and other, subject and object. Their distinctiveness, he argues, needs to be inculcated reflexively, by which he means that "natives" (subjects) must produce "native" art images (objects) which they can claim and identify as their own, in the process materially (and narcissistically) shoring up (reifying) their difference. Such reflexive mirror-art not only requires an absolutist identity (e.g. Native); it requires that appearance (Native Art as mirror) and reality (Native) be accepted as ontologically distinct, both factors negating postmodernist assertions. A very similar epistemological argument is made by Perloff (1993) with respect to reflexivity and poetry. In particular, she claims that the rules of much 1990s (Western) poetry are not postmodernist because they require that authors reflexively imprint (reflect) their own absolutist, legitimating identity (Chicano or Lesbian, for example) onto their works: the unifying identity "'Poet' will not do." Lastly, the reflexivity–modernity association is drawn upon by Jackson (1991) in his discussion of turn of the twentieth century poetry by Japanese Symbolists. According to Jackson, these poets were in part attempting to make sense of modernity, replacing constructions of language and meaning as transcendent and immediate with ones which emphasized dyadic (subject–object) relations mediated and structured through the ego and the mirror.

Even works that attempt to break down the association of reflexivity with modernity end up reproducing the passivity of the mirror. Mellor (1993), for example, argues that Giddens' assertion that reflexivity is to modernity as normativity (or, hierarchies crafted through social laws) is to premodernity is overly rigid. Instead of questioning Giddens' epistemic bases for differentiating between normativity and reflexivity, however, Mellor analyzes Pentecostalism to argue that reflexivity and normativity have always coexisted.

Campell and Freed (1993), in contrast, calls attention to two American television shows (*The Simpsons* and *Northern Exposure*) to posit that reflexivity is potentially a *postmodern* tool that promotes social reflection. For him, the television shows parodically mirror contemporary social norms, drawing attention to their artifice and contributing to critical social awareness. A similar claim that reflexivity is postmodern is made by Van Reijan (1992) who in part favorably compares Leibniz's and Lyotard's notions of mirroring. Instead of searching for an originary subject of the mirror and bemoaning the endless and antagonistic aspects of reflexivity-as-reflectivity that the mirror promotes, Van Reijan argues for a postmodern stance in which the mirror is used only as a means of becoming "familiar" with (and thus accepting) constellations of mirrored identities, in this way refusing dualist modernist notions of image and reality.

In all the renderings above, what seems to be important is how social relations (however those are defined – through books, poems, journal entries, interpersonal communications, ritualized cultural practices, or television shows) are structured and made recognizable through processes of mirroring. Processes include replication, duplication, identification, or bringing an image of oneself (through the mirror-relief of the other) into better focus. Because all of the authors and texts describe dyadic encounters between subjects and mirrors, reflexivity and self-reflexivity work to center and locate subjects. Depending upon one's theoretical position, what is located might be cast as a developmental self, a subject, or a unique cultural or social identity. What is important for the argument here is that the reflecting surface is passive; it is "out there" in texts, mirrors, paintings, and films – to be taken up and reflected upon inside human heads. The relative passivity of the reflection process leads Morton (1993) to conclude that reflexivity in itself cannot produce social change.

Despite the definitional variety of reflexivity, then, the dyadic, antagonistic (there is a me *or* a not-me), and rigidifying structural capabilities of the mirror are reinforced. Reflexivity is rendered a closed system, one mirror or a hall of mirrors, not unlike those theorized in McAllester's (1993) article on the relation of reflexive set theory to automated reasoning systems.

According to set theory, reflexivity is only possible when, first, a numeric set is symmetrically ordered (or transitive) through a regulating formula or rule set and, second, when set limits are placed on variable values. Again, both closure and reflection (symmetry) define reflexivity, attributes that underpin closed systems theories – whether these be biological, physical, psychological, or political economic.

In all cases (and they have their significant differences), mirroring works continually to reinforce the primacy of the subject. The "out there" of the mirror is taken as given, while bodily, intersubjective, and mirror-transcendent *placed* struggles are largely ignored: we watch the television shows and either "get it" (and become more critically aware of social norms) or we don't; literary structures either make the writing process apparent or they don't; alternatively, they shore up particular mirroring constellations of relations (between reader, writer, and text) or they don't; Yemeni men presumably accept the ritualized (mirror) image of themselves to become masculine; and we have guest speakers or we read different perspectives, thereafter quietly retreating to reflect on who "we" are through it all. This is not to suggest that such mirrors are materially or socially ineffectual. But their effect is highly limited and limiting.

What is lost are the embodied and spatialized political relations that produce, erode, support, subtend, and surround the mirror. What remains is an overly narcissistic, voluntaristic, and materially stable view of subjectivity and social relations. Most importantly, "readers" remain in control. As such, the theoretically built-in inescapability of locating and identifying through the mirror places us at some material remove from difference.

Just like Lacan's mirror-dependent theory of subject formation, theorizing reflexivity through the mirror assumes complete identification with visual images or structural wholes "out there" such that corporeal, *placed* relations with the out there and transcendence of the mirror's reified ordering is refused (see Blum and Nast 1996). Such theorizing additionally depends upon a very particular viewing context, one that speaks of privilege and bodily/spatial control. One needs a theatre, a social worker's office, an art gallery, a TV room, a bed (and bedtime), an anthropologist's notebook, pen, and post-fieldwork cubicle. But what happens when we do not have the privilege of this particular kind of bodily/spatial artifice? How does reflexivity work "on the run"?

Fielding difference

Introduction

Between 1988 and 1990 I worked in Kano, Nigeria as part of my dissertation training. Initially, I had hoped to assess the feasibility of a national brickworks program. As a "scientist" (I was a geologist who had returned to school – this time for a PhD in Geography) I was almost completely, culturally unprepared. I had a mental map of the postcolonial political geography in my head and had done a reasonable amount of reading on colonialism, brickworks, industrial minerals, and the links between import-substitution and "development" in contexts around the world. Through letter-writing, I had established a cooperative research agenda with geologists at Ahmadu Bello University in Zaria, a city close to Kano. My rational objectives would be dramatically re-worked through experiences over the next six months.

Eventually, through a number of twists and turn, I wound up working inside the Kano palace, a mini-city within the large and ancient walled city of Kano (Nast 1992). By then (1989) I had jettisoned "hard" scientific pretensions about economic development or geology, although for grant-related reasons I remained administratively attached to a re-worked and well-funded brickworks project. I entered the palace as a stranger and guest of the Emir/King of Kano (Alhaji Ado Bayero) who, until 1967, was post-colonial Native Ruler over what had been, prior to British colonization in 1903, the most economically successful Islamic state (emirate) within the large caliphate of Sokoto.[8] Within a short while of my first visit to the palace, his third wife, Hajiyya Abba Ado Bayero, had not only become my official patroness, but a much-needed confidante and friend.

The palace is massive in size (over 200m × 500m) and contains over 1000 persons organized spatially according to what is still a very strict socio-political hierarchy. Gender and master-"slave"[9] relations inform the spatial ordering of the landscape: two peripheral male "slave" areas, within which reside powerful "slave" families headed by male servants of slave descent, border a separate, secluded female "inside" or *ciki*. The latter term derives etymologically from the word "stomach." It is in the stomach that royal wives (three) and concubines (about fifteen) reside, attended to by eunuchs – now hermaphrodites or impotent men – and women of slave descent. All of the emir's youngest children also live there, about fifty in number (cf. Rufa'i 1987: 112). The architecture is monumental and strikingly beautiful, maintained by a special group of royal builders.

One of the most important social distinctions made in the palace is that between nobility and royal servants. The latter call themselves "slaves" or *bayi*, resisting colonial abolitionist decrees and allowing them access to a wide array of pre-colonial slave entitlements (Nast 1994); their faces are scarred to indicate their status, although the practice is gradually being abandoned; they live in highly circumscribed palace places; they wear particular kinds of clothing; and they carry out very specific kinds of roles – hierarchies of a kind expected in any royal setting. Royal servants and commoners inside the palace prostrate themselves at the feet of Kano nobility, while massive arched entranceways speak of grandeur and the one-time priviliging ability of the emir and his men to arrive or leave on horseback. While only the aristocracy and most powerful "slaves" wear turbans, only males in the emir's direct lineage can tie them up so as to have two pointed "ears" rising out of the top. The bodily and spatial means of marking out different social positions is, as in all contexts, continuous and everywhere. Within the context of this paper, however, I am concerned to show how I came to *field* differences in the palace "stomach," the secluded domain of royal women.

How, for example, did I recognize and negotiate social hierarchies? How did I figure out what I thought I was expected to do, and did the figuring out process change over time? How was I, as a bodily field, continually marked and emplaced into a foreign order of things? How "in control" was I? These questions are addressed anecdotally below using examples of what I consider to be moments when I was forcefully made aware of difference. The examples foreground how I was more or less drawn out of my own presumably self-controlled thoughts about palace life – a relatively disengaged reflective process – into positionings and identities not of my own choosing. I use the "stomach" as the analytical site because it was here that my crosscultural engagements were the most sustained and intense. The stomach is also a site which foregrounds differences amongst "women," a social category typically considered to be relatively homogeneous; moreover, it is a site where *Muslim* women reside, an Orientalized social grouping whose assumed homogeneity is even more pernicious.

Women and difference

One might assume that within an Islamic "fundamentalist" place such as Kano city, women's spaces are all hidden and secluded, tightly packed off into nest areas in households, especially so for the royal family in the palace. But this is not the case. As in all cultures, significant variations exist, depending upon a woman's social and marital status and age. In the palace, non-royal young girls and female *bayi* perform tasks that take them into outside public, "male" domains. Indeed, low status and "outside" labor for women are firmly linked (Nast 1996), just as they are in many cultures. Thus, female *bayi* carry out arduous labor in open, low-status places within the palace stomach, while young girls and elderly women sell goods in what are otherwise public "male" palace parts, places taboo to female royalty. The bodies of female *bayi* are additionally encoded through a hierarchization of stomach-based sleeping quarters: female *bayi* sleep on the floors of hallways or other places considered interstitial or marginal to royal wifely and concubine spaces.

Wives are a leisure class, each one (Islam allows four) having her own palatial quarters called a "place." It is expected that a wife will not leave her place, even to visit other women in the stomach. Concubines live communally in what are referred to as "wards." They function largely as managers and are organized into a complex hierarchy which administers the stomach and female *bayi*. Overall, the stomach is a place for reproductive and some trade activities. Children are raised, annual royal wedding galas and feasts are prepared, the emir's monumental quarters are maintained, food for the royal family is cooked and formal, daily audiences with the emir are conducted to air family news.

Anecdote 1: Royal concubines

After an initial six month field stay in Nigeria in 1988, I decided to change radically the thematic focus of my dissertation. For a number of reasons, I left Nigeria twice – in mid and late 1988 – thereafter deciding to begin Hausa language training before my third and final 14-month return. In 1989 I returned to the palace following a month-long intensive Hausa language course in Holland. Given that I was no longer working on the federal brickworks project (the federal lingua franca is English) and that future work would be in a local, non-English context, some knowledge of the Hausa language was required. Upon this final return, I chose to wear traditional clothing as a sign of respect and out of fear of being excluded (dress is an extremely important marker of Hausa identity). One of the first social visits I made was to the Kano palace. Besides wanting to greet Hajiyya Abba, I had come to ask the emir if he could help me find living quarters in the old city. The new language training, my new clothing, and my anticipation of living in the old city made me feel like a different person.

My enthusiasm and confidence dimmed, however, once I entered the stomach and passed a group of concubines seated on a bench near the main concubine ward with a guardswoman. Upon passing them I heard one of them say in a relatively loud voice, "*Ga ita can, ita sha-sha-sha ce*" (Look at her there, she is an idiot!). Ordinarily, I would not have understood what was being said, given my linguistic rudiments. But "idiot" (*sha-sha-sha*) is one of the first words a new speaker learns because it is easy to remember. So, too, the sentence structure, "s/he is . . . " (*ita . . . ce*) and locative words like "here" and "there" (*nan* and *can*). Frozen with embarassement, but determined to maintain control and composure, I turned to them and asked in Hausa, "*Who* did you say is an idiot?," knowing they in turn would be stunned by what might seem to be my Hausa language skills.[10] Soft murmurings ensued and I walked off dejectedly to Hajiyya Abbas' quarters, relaying to her with feelings of disappointment and shame what had happened.

Later, she told me that one of the reasons they said they had singled me out was that they felt I dressed in a ridiculous way. My head gear, for example, was all wrong. Hausa women's head gear is made up of two parts: a kerchief secured over the forehead and generally tied at the back of the head, and a larger shawl-like covering. If one is married, the shawl covers the head; if unmarried, the shawl is placed around the shoulders. I did not know about the kerchief part(!), however, and merely placed a shawl over my head, which at best made it look like I was trying to pass myself off as married. Worse, married or single, according to palace protocol, all women's heads must be bared in the inner secluded domain of the emir, a protocol I had breached. Hajiyya Abba presented me with a kerchief, her very young daughter Salamatu later teaching me how to tie it into the latest, elaborate fashions. By inhabiting their assessment of me I learned much about local dress codes, leading to further "falling into" of local practices. Nevertheless, I never again felt comfortable passing that particular concubine ward.

There was more. I heard through the grapevine that concubines, mostly from *that* ward, had criticized me for making *yanga*, a seductive and sexually boastful swaying of the hips. To my embarassment and chagrin, Salamatu (Hajiyya Abbas' youngest daughter who was about seven years old at the time) was called upon to imitate me in front of a roomful of palace women visiting her mother – producing shame-inducing (for me) cackles of amusement! *That*, I argued, was *not* the way I walked! Feeling singled out and humiliated, I defensively pointed out that I did not have slim hips like Hausa women *and* that one of my legs was shorter than the other *plus* my legs from time to time got caught up in my wrapper (the traditional wrap-around skirt), making me walk awkwardly. The humiliation I felt seemed compounded at the time because my body movements were re-enacted, re-coded, and appropriated by a small girl for the amusement and pleasure of adult strangers. My engagement with their gossiping impact took at least two forms, both largely defensive. First, I disengaged from concubines, for the most part studiously avoiding those particular concubines *and* their spaces, despite the fact that important historical details about palace history might have been gained from discussions with them. Second, I scrupulously walked with minimal hip movements, in the process paradoxically inhabiting their scripting of me. My body, heterosexualized as a "seductress," was uncontrollably not my own.

Only later did I realize that their scrutiny of me may have been culturally expected, stemming from the Hausa female tradition of *kishi*. *Kishi* translates into English as jealousy, and is characteristically expected from a Hausa wife towards unmarried women or co-wives who, through their very presence, threaten her sexualized position *vis-à-vis* her husband (Muslim Hausa men can have four wives and divorce at will). *Kishiya*, a word derived from *kishi*, in fact translates as "co-wife" and "the opposite of something."[11] Thus, I later came to understand concubine actions in terms of a jealousy expected from them towards someone like myself who had no spouse and was consistently in the emir's presence or household; I was eminently marriageable or at least open to sexual entanglement.

Sexualized jealousy is especially endemic to concubinage: concubines are former "slaves" who rise through the gendered ranks to become a "slave-wife" by being given as a gift to the king *or* through their own sexualized attracting-efforts, that is, by capturing the king's attentions. The latter process is a precarious one. Moreover, concubines control the secluded domain; they are its premier managers, continually filtering through as they administer and survey. To them, I was an outsider with no apparent reason for being in the most private of royal domains. Some persons felt my questions about palace history (for example, numbers of heads of cattle at the turn of the century) were ridiculous or, worse, a sign that I was a spy. Royal wives, in contrast, are less sexualized and more socially powerful than concubines, their power and position deriving in large part from their pedigree. Wives derive from high ranking families and serve as important political patronesses to women outside the palace, a position not occupiable by concubines. I diagrammatically depict my bodily and spatial positioning relative to concubines in Figure 6.1, an oppositional positioning into which I was interpellated, which I did not choose, and within which I felt captured and identified.

Figure 6.1 Royal concubines' interpellation of me as a heterosexualized outsider and social equal

Anecdote 2: Royal female bayi

Some time after my arrival, I was informed that some female *bayi* considered me a miser; otherwise, they said, I would give them gifts. They complained that I did not give gifts to palace children either, despite the fact that children showed considerable loyalty towards me, this being their encoding of the fact that children often accompanied or played with me when I visited or worked in the palace. As someone explained, the children could have chosen anyone as a patroness, but they had chosen me. Why did I not recognize and award their allegiance?

My initial reflex was disgust: how could I pay children to like me (or so I encoded and rationalized their request)? I also felt anger: "Of course children like me" I thought smugly, "I regularly play with them, and don't chase them away with a stick!" I also refused to understand why I was expected to "dash" (give gifts to) the women-servants. Were they not already rewarded – royally? But their disapproval and distancing made me timidly, begrudgingly comply and "fall into" their world. I began to bring sweets for the children and occasionally to give cash to female *bayi*. It was only later, while occupying this uncomfortable place, that I realized how much their identification of me as master was informed by the historical fact that these women had always occupied *slave* posts. They do not, nor have they ever (unlike male *bayi*) received regular, formal pay.[12] They, more so than men of slave descent, depend upon a master-"slave" system. By heeding their assertion of *bayi* identity, I was compelled to become patroness and provider, something additionally brought home by the fact that some of them prostrated before me. Again, I could not have imagined prostration (before me!), nor would I have chosen to occupy such a (to my mind) disturbing, non-pc position, a position I caricature in Figure 6.2. Paradoxically, my "letting go" to take up

Figure 6.2 Female *Bayi*'s positioning and identification of me as master and provider.

"slave" women's expectations of me, involved enabling them to assert their lesser status and, in Hegelian fashion, the consequent right to demand gifts from a master.

Anecdote 3: Royal wives

The superiority of royal wifely embodiment and place was not immediately apparent to me, not only because of my almost complete ignorance of Hausa culture prior to my arrival, but because of a Western arrogance (the world is the cultural geographers' oyster), and because Hajiyya and I were good friends. While I noticed how concubines came and went to her quarters, delivering to her special greetings on their knees, their eyes averted, *I* felt special. But a mini-fall from grace helped me realize her superior powers and place. It happened one day when, during a conversation Hajiyya became angry with me (the details of which I have unfortunately forgotten), abruptly and forcefully asking me to remove my body off the surface of *her* mat; I had never even noticed that her mat was, in fact, a kind of sacred surface, upon which no one, except for me, ever sat! Likewise, I had never noticed that no one except for me dared to sit with her on her sofa. I just assumed that her mat fit two and that sofas were for sitting.

How it is that I never realized that other palace persons never approached these places, I really can't say. Nonetheless, her moment of anger and displacement made me realize with striking clarity the privileged position next to her that I had heretofore taken for granted. After that, I was never as sure of myself in her presence; I was always aware of my lesser body and place, an inferior status with which Hajiyya felt comfortable. From then on she insisted that I learn how to formulate and give greetings befitting someone of noble stature and, occasionally, assigned me labor tasks which only her *bayi* or children ordinarily carried out.[13] Allowing myself to feel the impact of Hajiyya's anger, pushed me not only off her mat but into a world of difference, the rules of which I only very partially ever learned. My inferior social positioning *vis-à-vis* royal wives is shown in Figure 6.3.

Fielding difference

The anecdotes provide an exaggerated cultural and material context for understanding reflexivity not as a voluntaristic and leisured process of the mirror, initiated and/or controlled by a subject through mental exertion, reflection or contemplation. Rather, they show reflexivity as an embodied process wherein the body is itself a *field* for registering and negotiating difference. In all of the cited situations I could have refused to "fall in." I could have opted to wear Western clothes, work only with translators, bring and eat my own food, express extreme cultural indignation upon meeting opposition or ridicule, and/or leave the palace to find a different study site. Responding

107

Figure 6.3 Royal wifely disciplining of me as social inferior.

to how others called upon me and how others defined the terms of engagement, my body became reflexive and a place where others' subjectivity and cultural inscriptions could be made, mapped out, and remembered. In this sense, reflexivity was less about "self"-reflecting upon difference within safe spaces and more about about continually negotiating a "mis-recognizing" of my"self" through others' expectations, scriptings, power networks, and material and physical demands. Becoming a field for others' subjectivities often involved bodily jolts rather than mental musings. Through interpellations from the outside, I became master, maker-of-*yanga*, and client, positionings I did not recognize as my own and which occurred despite my intentions and reflections. Physically identified and marked as seductress, master *and* client, I had to go beyond the mirror to negotiate and live out identities I had previously never imagined could be my own.

The reflexivity described here was not a cleanly dichotomous process involving two completely distinct subjectivities, me being passive and others active. Nor was it a process evenly distributed across time and space. Rather, there were disparate, competing demands registering unevenly across time, place, and person, bringing different bodily parts into significance and play. Prostrations before me, accusations of me making *yanga*, and my learning how to prostrate fragmented me bodily into bits (feet, hips, and knees), different bits meaning something different according to the social situation and status group with which I interacted. I am not, moreover, advocating a putting-to-sleep of researcher subjectivity, but rather, developing a keener appreciation of how to respond and elicit subjectivities in a way which allows for bodily fieldings of difference; obviously, all persons are changed in encounters. Who has more "power" at any one time depends upon existing structural hierarchies, material conditions, and historical contingencies.

In the case of concubines, the reflexive process was initiated abruptly by them – from "outside." I could have ignored their comments, pretending to myself, at least, that they were minor and that I was in control. For a number of reasons, I did not, their calls marking me and drawing me in. Yet concubines were not completely or fully in control, either; my queries to Hajiyya, for example, led her later to use her authority to call upon them for an explanation. Similarly, concubine surveillance of me was neither complete nor debilitating. Rather, their surveillance formed part of their everyday duties and the everyday fabric of palace life – which my body and place only ever partly fielded. Feeling their surveillance and listening to their sexualized comments helped me recognize *them* as much as they helped me recognize myself in terms other than my own.

Re-casting reflexivity as an embodied process of engagement and place limits the extent to which those of us in the "field" are able to create fictions of (at best, guilt-ridden) omnipotence (England 1994). The anecdotes additionally show that bodies and spaces are not discretely bounded, plump objects or plenitudes. Instead, corporeality and place partly produce the meaning and physicality of one another, making it difficult to ascertain where a body ends and a place begins. That is, Hajiyya Abba was a royal wife *through* her ability to sit on a sofa, have her own royal mat, inhabit and embellish upon her "place," and be a site of prostration. The reflexivity of our engagement was just as much about me learning about her place and having those things mark me, as it was about me interacting with her "personally."

Who is to say that any of these anecdotes are "true?" What if they are just one field researcher's paranoid ramblings? What if the stories I have told are ones based upon personal encounters which I have then (unfairly) re-scripted in terms of generalities of social rank (wife, concubine, and "slave")? What if someone reads this who lived in the palace and knows these encounters to be much more complicated than I have described them? Obviously, these are all important (and nagging) questions beyond my analytical reach. My framings of events are an attempt to *make* sense out of at times very confusing circumstances. Nevertheless, I would argue that determining whether or not the stories are really "true" is less important than understanding and valorizing the integrity of the nonvoluntaristic, socially assymetrical, embodied, and material aspects of reflexivity.

It might also appear that I have told the stories with the assumption that I and the various palace women with whom I interacted were fully *conscious* of the relative powers we held and of the dynamics underpinning our actions. Accordingly, this would make us ignorant of the ways in which individual actions are enmeshed in larger political relations that structure everyday meaning (about which we are only ever partially aware) and indicate that I have ignored *unconscious* motives and meanings (see Rose 1994: 509). Yet I would argue that my stories are not meant to convey all meanings or all truths. I have used them in very partial, tentative ways to explore how we become (and can to some extent choose to be) captured and enmeshed within particular political constellations, the fact that such constellations exist being the ground upon which the stories are built.

Conclusions

In the first part of this chapter, a very schematic archeology of the term reflexivity as it is used in the humanities and social sciences was laid out and rhetorically questioned. I did this to draw out key definitional characteristics of the term that, when juxtaposed against the more corporeal medical definitions cited at the beginning, create a productive tension. The medical definitions stress the involuntary-ness of what results when either two objects collide or when an organism involuntarily responds to external stimuli. In either case, the term connotes an openess and a grounding in the physics of movement, collisions and involuntary impacts; in the case of light, for example, what is studied is the actual bending of the rays, *not* the image left lying on a reflecting surface, social scientists mistakenly equating the mirror-surface effect of reflection with the material process of reflexion (see note 1). In the context of cross-cultural fieldwork, what is "involuntary" is the fact that we are often forced to assume and move into social, cultural, and political positionings of which we either know nothing or which we (given our druthers) might not choose.

Although the list of disciplines covered by the InfoTrac and library search is far from exhaustive, the examples clearly point to important political–epistemological assumptions underpinning current constructions of reflexivity. The InfoTrac examples are *not* empirical "dots" which can (if connected) trace out particular discursive lineages along the lines of a Foucauldian archeology. Instead, my analysis should be understood as an attempt to question the problematic way in which reflexivity is characteristically deployed as *reflectivity*, reflection relying upon 1) a dichotomous and antagonistic separation of subject and object and, 2) an effacement of the materiality of movement, body and place. What is particularly problematic is that this deployment is used almost irrespective of the theoretical framing. Behaviorists, set theorists, postmodernists, and poststructuralists ironically call upon the same closed, structuralist system of the mirror. Yet, in so doing they scotomize subjectivity and cultural change because the mirror reduces the world to what is seen, making the visual (what is reflected upon) the reified basis for understanding difference; bodies and places – and bodies as places – are effaced.

My re-working of reflexivity alternatively speaks to a co-making of bodies and space and to some extent the collapsibility of one into the other: the body *is* a place. *Theoretically*, the re-working challenges us to refuse equations of reflexivity and reflectivity. Going beyond the mirror means that we need new methodological means for registering situatedness. *Methodologically*, we need to elicit ways for understanding how to register and work with changes and subjectivities registering from the outside. *Ethically and politically*, embodied reflexivity means we need to recognize the instrumental and moral limits of fielding difference: I cared greatly about what certain Nigerian immigration officials thought of me, for example, and tried to craft myself accordingly. I did not care as much when the cook in one of the guesthouses where I stayed claimed I was a thief who had engineered the theft of the guesthouse VCR, though his claims did cause me to move involuntarily in certain ways with respect to him. The politics of why, in what way, and for whom I fielded and cared is, however, another question, beyond the realms of this chapter.[14]

In some ways, the re-working of the term reflexivity is related to Katz's (1994) re-working of the geographical term, "field." She challenges us to think of the field not in terms of a discretely bounded area, but as something less reified and definitive. For her, the field is at once defined through our research questions, through the particular politics of a place *and* through the socio-spatial channels available to us for interaction – all of which are only ever partially stable. Our spatial interests are, moreover, structured by how and when we define the tentative boundaries of the places we study, the kinds of social relations we enter into, and the changing ways in which we locate and situate ourselves in particular contexts. What I would add to Katz's analysis, however, is that not only is the field potentially everywhere and nowhere; it is every *body*. The body-as-place registers as it maps out othernesses of person and place. Fragmentation of the subject (Figures 6.1, 6.2 and 6.3) is consequently a visceral process requiring physical skill – not in an ableist sense (running, jumping, ducking), but in terms of physically working with, and allowing for, the ambiguities, partialities, and material-ities of difference.

Besides involving a re-analysis of the body-as-place, I would argue that reflexivity requires recognizing the emotional and material qualities of inter-subjectivity. Keith (1992), in his analysis of racism in Britain, talks about the importance for him of bringing his anger about racism into his research agenda and into his writing. I would add that there is abundant room for humor.[15] Laughter allows us to acknowledge how limited we are in "making sense" of fragmentations and contingencies. While laughter may belie deep-felt anxieties, it also spells relief. I can think of many absurd situations in Kano about which all I could do, finally, was laugh.

Acknowledgments

Thanks to the anonymous reviewers and to Angela Martin for her comments on a final copy of the manuscript, for her suggestions on field work "reads" over the years, and for her references to the work of Turner (1992) and Stoller (1984).

Notes

1 Interestingly, the definitions I found for reflexivity dating from the 1960s and 1970s place embodied examples of reflexivity first in their hierarchy of definitions. Definitions from the 1980s and 1990s, in contrast, place discussions of reflection, first. While I think this change is significant, there is no room to discuss it here. It is nevertheless important to note that the ways in which the social sciences and humanities currently use the term reflexivity is more in keeping with a sense of *reflectivity*, as discussed in the text. Even so, there is a crucial difference in the ways in which reflectivity is defined. In the 1980s and 1990s definitions, reflection is seen in terms of light beams bending and turning back – a highly materialist rendition. In contrast, the social sciences and humanities construct reflexivity in terms of the mirror itself, a two dimensional surface at some conceptual remove from the bending beams which create it. It is this latter kind of *displacement of reflexivity onto a mirror surface* which is critiqued in this paper and which should be seen as substantially different from *reflexivity-as-bending*.
2 At the time I was at the University of Kentucky, Lexington.
3 My training and professional experience up to that point had been in the Geological Sciences.
4 In this reductionist scenario, that which is is left over after I have identified what is similar to me is "difference".
5 At least "we" in the "West," though the "we" is neither that geographically universal or specific.
6 While it might be claimed that the "powerful" personally suffer these traumas less, anyone who primarily negotiates difference in two dimensions (through disengagement and looking, a kind of viewing through the mirror) suffers alienation of their own bodies and those of "others." Obversely, immigrants and recipients of racist, heterosexist, religious, age-ist, and class-derived bodily abuse and physical displacement often have no choice but to let others "take control" – not the kind of taking over espoused here. Such abuse and displacement derives its power precisely because the bodily is denied and, through socio-spatial configurations of power, made deniable.
7 Subject-centering and assumed empowerment in the social sciences is exacerbated by the literary, textual sanitizations expected in the reporting of field research. Rarely do we hear of failed, agonistic attempts to communicate or work with groups culturally different from ourselves or how such groups think about,

and effectively, forcefully exert their control. Nor do we hear of how those with whom we work systematically subvert our objectives. At best, particularly difficult cross-cultural negotiations are relegated to fictional or autobiographical accounts not necessarily marketed toward the academic community (Agar 1980; Anderson 1990; Barley 1983; Bennett 1986; Bowen 1964; Briggs 1970; Dettwyler 1994; Kaysen 1990; Russ 1986; Shostak 1983; Spindel 1989; Turnbull 1968. But see: bell *et al.* 1993; England 1994; Gilbert 1994; Keith 1992; Mblinyi 1989).

One effect of textual sanitization is that crises in field work are scripted as theoretical, literary ones: Who "speaks" for whom? Who is an author? How many voices should one include (e.g. Clifford and Marcus 1986; Crang 1992)? All of these questions to some extent evade the relational/political, embodied, and spatial issues of the "field" (Nast 1994: 61, fn 7; but see Pulsipher 1997). One effect of textual sanitation is ironical: researchers feel that they hold more power than they ever had(!), guilt paradoxically assuming and reproducing a centered, all-powerful subject. Such self-centering guilt not only shores up researchers' egos (in a negative way); it also ignores and thus damages local subjectivities and places (Katz 1994).

8 Since his inauguration in 1953, the Emir's powers have been continuously eroded through systematic questioning of his powers following World War I, independence in 1960, and consequent federalization of his powers. Today, supported by the state, he functions largely as a religious and political figurehead, akin to the pope and the Queen of England.

9 I place diacritical marks around the term "slave" to indicate how problematic it is to use the term. On the one hand, the British juridically freed all slaves during the colonial period. On the other hand, the language, practices, divisions of labor, and bodily and place markings of palace "slavery" persist, albeit "master-slave" ties have considerably weakened. What these people "are" is therefore ambiguous, an ambiguity I attempt to retain by either using diacritical marks or by using the term by which many "slaves" identify themselves, that is, *bayi.* The latter term translates into English as slave.

10 It should be noted that we knew each other fairly well by sight from my earlier visits to the palace interior.

11 All definitions are taken from Newman and Newman (1985).

12 Many male *bayi* became wage earners of the Native Authority following the 1909 colonial creation of the Native Treasury, whereupon the aristocracy was also placed on a set salary – the beginning of the attenuated end of slave allegiance to the king (Nast 1992).

13 Hajiyya Abba chose not to have concubine-servants in her quarters. Thus, she, along with her younger children and female *bayi* carried out domestic labors needed to maintain her household.

14 There are indeed many "field" innovations in the area of reflexivity, although they are typically not enuciated or foregrounded in these terms. Nagar (1997), for example, documents the almost comical lengths she went to in negotiating her research agenda with several very different religious and ethnic groups in Tanzania. Her story of changing clothing, mode of travel, and comportment to suit her informants and, conversely, her informants' unexpected interpellations of her are not unusual ones. Edith Turner (1992) in her interpretative ethnography

of the Ndembu of Zambia, for example, consistently draws attention to Ndembu constructions of her (she is made a Ndembu doctor) and how they shaped her field experience. At one point, she draws upon Paul Stoller's work (1984) dealing with sorcery practices of the Songhay in Niger. He states that,

> [h]aving crossed the threshold into the Songhay world of magic, and having felt the texture of fear and the exaltation of repelling the force of a sorcerer, my view of Songhay culture could no longer be one of a structuralist, a symbolist, or a Marxist. Given my intense experience – and all field experiences are intense whether they involve trance, sorcery or kinship – I will need in future works to seek a different mode of expression, a mode in which the event becomes the author of the text and the writer becomes the interpreter of the event who serves as an intermediary between the event (author) and the readers . . . . Anthropological writers should allow the events of the field – be they extraordinary or mundane – to penetrate them.
>
> (In Turner 1992: 162)

15 See Meoño-Picado (1997) for a fascinating look at how a group of Latina lesbians in New York shifted their organizational focus from one based on "dependency theory" (which required a seriousness and singularity of purpose untenable within the organization) to a more flexible one derived from the Brazilian religious form, Santeria, which values playfulness and humor. It is not that the latter form is not serious; rather, it addresses the fractured-ness of everyday life, in the process allowing for different scales or levels of political dilemmas to be broached, other than those frameable in dependency terms. Her work is one of the few that shows the importance of humor in choosing how we frame, and participate in, an analysis.

References

Agar, Michael 1980. *The professional stranger*. New York: Academic Press.
Allen, Katherine R. 1993. Reflexivity in teaching about families (Family Diversity: A Special Issue) *Family Relations* 42(3): 351–7.
Anderson, Barbara Gallatin 1990. *First fieldwork: the misadventures of an anthropologist*. Prospect Heights, Illinois: Waveland Press.
Barley, Nigel 1983. *The innocent anthropologist*. New York: Henry Holt.
Bailey, Suzanne 1993. "A garland of fragments": modes of reflexivity in Clough's "Amours de Voyage" *Victorian Poetry* 31(2): 157–71.
bell, diane, pat caplan and wazir jahan karim (eds) 1993. *gendered fields: women, men and ethnography*. New York: Routledge.
Bennett, Jo Anne Williams 1986. *Downfall people*. Toronto, Ont.: McClelland & Stewart.
Blum, Virginia and Heidi Nast 1996. Where's the difference?: the heterosexualization of alterity in Henri Lefebvre and Jacques Lacan *Environment and Planning D: Society and Space* 14: 559–80.
Bochner, Arthur P. and Carolyn Ellis 1992. Personal narrative as a social approach to interpersonal communication. *Communication Theory* 2(2): 165–73.
Bowen, Elenore Smith 1964. New York: Doubleday.

Briggs, Jean L. 1970. *Never in anger: portrait of an Eskimo family*. Cambridge: Harvard University Press.

Campbell, Richard and Rosanne Freed 1993. "We know it when we see it": postmodernism and television. *Television Quarterly* 26(3): 75–88.

Caton, Steven C. 1993. Icons of the person: Lacan's "imago" in the Yemeni males tribal wedding. *Asian Folklore Studies* 52(2): 359–82.

Clifford, J. and G.E. Marcus 1986. *Writing culture*. Berkeley: University of California Press.

Crang, P. 1992. The politics of polyphony: reconfigurations in geographical authority *Environment and Planning D: Society and Space* 10: 527–49.

Dettwyler, Katherine A. 1994. *Dancing skeletons: life and death in West Africa*. Prospect Heights, Illinois: Waveland Press.

England, Kim 1994. Getting personal: reflexivity, positionality, and feminist research. *The Professional Geographer* 46(1): 80–9.

Escobar, Ticio 1993. The furtive avant-gardes. *The South Atlantic Quarterly* 92(3): 461–72.

Gilbert, Melissa 1994. The politics of location: doing feminist research at "home." *The Professional Geographer* 46(1): 90–5.

Hoon-Sung, Hwang 1993. One mirror is "not enough" in Beckett's *Footfalls* and *Ohio Impromptu*. *Modern Drama* 36(3): 368–83.

Jackson, Earl Jr. 1991. The heresy of meaning: Japanese Symbolist poetry. *Harvard Journal of Asiatic Studies* 51(2): 561–99.

Katz, Cindy 1994. Playing the field: questions of fieldwork in geography. *The Professional Geographer* 46(1): 67–72.

Kaysen, Susanna 1990. *Far afield*. New York: Vintage Books.

Keith, Michael 1992. Angry writing: (re)presenting the unethical world of the ethnographer. *Society and Space* 10(5): 551–69.

Man, Glenn K.S. 1993. "The Third Man": pulp fiction and art film. *Literature-Film Quarterly* 21(3): 171–8.

Marcus, G.E. 1992. "More (critically) reflexive than thou': the current identity politics of representation. *Environment and Planning D: Society and Space* 10(5): 489–95.

Mblinyi, M. 1989. "I'd have been a man": politics and the labor process in producing personal narratives. In *Interpreting women's lives: feminist theory and personal narratives* ed. Personal Narrives Group, 204–27. Bloomington: Indiana University Press.

McAllester, David A. 1993. Automatic recognition of tractibility in inference relations. *Journal of the Association for Computing Machinery* 40(2): 284–304.

Mellor, Philip A. 1993. Reflexive traditions: Anthony Giddens, high modernity and the contours of contemporary religiosity. *Religious Studies* 29(1): 111–28.

Meoño-Picado, Patricia 1997. Redefining the barricades: Latina lesbian politics and the creation of an oppositional public sphere. In *Thresholds in Feminist Geography*, eds John Paul Jones III, Heidi J. Nast, Susan M. Roberts, 319–39. Lanham, Maryland: Rowman & Littlefield.

Morton, Donald 1993. The crisis of narrative in the postnarratological era: Paul Goodman's "The Empire City" as (post)modern intervention. *New Literary History* 24(2): 407–25.

Nagar, Richa 1997. Exploring methodological borderlands through oral narratives.

In *Thresholds in Feminist Geography*, eds John Paul Jones III, Heidi J. Nast, and Susan M. Roberts, 203–25. Lanham, Maryland: Rowman & Littlefield.

Nast, Heidi J. 1992. *Space, history and power: stories of social and spatial change in the palace of Kano, northern Nigeria, circa 1500–1900*. Montreal, Quebec: unpublished PhD dissertation.

Nast, Heidi J. 1994. Women in the field: critical feminist methodologies and theoretical perspectives. *The Professional Geographer* 46(1): 54–66.

Nast, Heidi J. 1996. Islam, gender and slavery in West Africa c 1500: A Spatial archaeology of the Kano Palace, northern Nigeria *Annals of the Association of American Geographers* 86(1): 44–77.

Newman, P. and R. Newman 1985. *Modern Hausa–English Dictionary*. Ibadan, Nigeria: University Press Limited.

Perloff, Marjorie 1993. Postmodernism/fin de siècle: the prospects for openness in a decade of closure. *Criticism* 35(2): 161–92.

Pulsipher, Lydia. For whom shall we write? What voice shall we use? Which story shall we tell? In *Thresholds in Feminist Geography* eds. John Paul Jones III, Heidi J. Nast, Susan M. Roberts, 285–319. Lanham, Maryland: Rowman & Littlefield.

Recchia, Edward 1991. Through a shower curtain darkly: reflexivity as a dramatic component of "Psycho." *Literature-Film Quarterly* 19(4): 259–68.

Rose, Gillian 1994. Engendering and degendering. *Progress in Human Geography* Progress Report 18(4): 507–15.

Rufa'i, Mrs. Ruqayyatu Ahmed 1987. Gidan Rumfa: The socio-political history of the place of the Emir of Kano with particular reference to the twentieth century. Unpublished Masters Degree in History, Bayero University, Kano, Nigeria.

Russ, Joanna 1986. *The female man*. Boston: Beacon Press.

Shostak, Marjorie 1983. *Nisa: the life and works of a !Kung woman*. New York: Vintage Books.

Spindel, Carol 1989. *In the shadow of the sacred grove*. New York: Vintage Books.

Stoller, Paul 1984. Eye, mind, and word in anthropology. *L'homme* 24: 91–114.

Swigonski, Mary E. 1993. Feminist standpoint theory and the question of social work research. *Affilia Journal of Women and Social Work* 8(2): 171–84

Turnbull, Colin M. 1968. *The forest people*. New York: Simon & Schuster.

Turnball, Colin M. 1972. *The mountain people*. New York: Simon & Schuster.

Turner, Edith 1992. *Experiencing ritual*. Philadelphia: University of Pennsylvania Press.

Van Reijan, Willem 1992. The crisis of the subject: from Baroque to postmodern. *Philosophy Today* 36(4): 310–24.

2

ConfiningPlacesBodies

7

HAREM

Colonial fiction and architectural fantasm in turn-of-the-century France

Emily Apter

If, in *fin de siècle* European culture, metropolitan topography could be aligned in the human sciences with sociology and social psychology (Gustave Le Bon, Gabriel de Tarde, Emile Durkheim), then, by contrast, home, particularly the Viennese drawing room, was the place of psychoanalysis – a theater of changing décors supporting the then highly popular *drame de famille*. Architecture, per se, had no integral or obvious connection to either of these burgeonoing disciplines. Freudian psychoanalysis implied that psychic states were independent of specific architectural topoi, though spatial and architectonic metaphors could be marshaled in the service of the talking cure. Freud's *The Interpretation of Dreams* offered a panoply of loaded symbols: the house metonymized the self; smooth walls, the facades of houses, and inner courtyards were tabulated to the insides of women's thighs or womb-cavities; "bed and board" meant marriage, and so on. One such metaphor emerged strikingly in *Totem and Taboo*. Freud evoked a primal horde dominated by a patriarch who had secured for himself exclusive rights over all the women in the socius. Though the story proceeds with disgruntled younger men banding together, committing parricide, and sharing the women among themselves, it is interesting for our purposes to note that what Freud placed at the psychic origin of civilization was nothing other than a *harem*. The old man and his seraglio, according to Freud's mythic scenario, functioned as a spur to the world-historical enactment of the Oedipal drama. In the beginning, was a harem . . .

Blurring distinctions between inside (psychic) and outside (the body, the social realm), the harem loomed large in Europe's phantasms as an archaic erotic idea – one man to many women, unchecked sexual domination over totally submissive members of the weaker sex. In full-dress pictures of the harem throughout the nineteenth century (Delacroix, Ingres, Gérôme, Alma-Tadema), it was as if the unconscious had secreted itself into spatial and

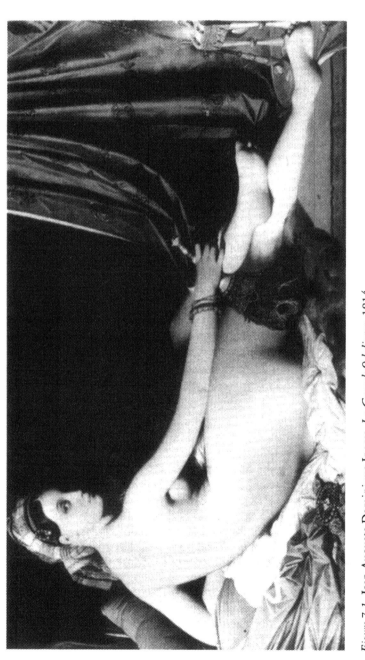

Figure 7.1 Jean Auguste Dominique Ingres, *La Grande Odalisque*, 1814

material form. Like its prototypical counterpart, the Parisian *maison close*, the harem was and continues to be a haunting Occidental figure of pure *jouissance* made into architecture. A harem, of course, is not *stricto sensu* an architectural form, for it refers to a social rather than a physical configuration; it is a space made out of an assembly of women rather than a composition of historically coded orders, elevations or building types. But in its culturally reiterated domestic plan and gendered distribution of spatial drives, the harem has been for centuries easily recognizable as a distinctive *locus sexualis*. A synonym for "transgression" and "taboo," the early Arabic word *harem*, as Malek Chebel has pointed out, refers to the gyneceum, to the "sexualization of interior space," and to the restriction of male access to the community of women.

In contradiction to the *salem'lik* or seraglio (broadly, the sultan's palace, a versatile place designating the threshold where male visitors are welcomed as well as the lodgings of the pacha's personal female museum), the full term *harem-'alik*, according to Chebel, refers more specifically to gendered zones of interdiction within intimate dwelling spaces.[1] Similar to the veil (*hidjâb*, "curtain," "that which hides"), a shield or carapace ensuring that a layer of domestic interiority remains stuck on or wrapped around the female body even when it is exposed to public view, the harem, like the metaphorical "skin" of a building, is a kind of inside on the outside; a concentrate of interiorized interiority, a miniaturized version of the Casbah self-enclosed and circumscribed within the boundaries of the colonial city.

The harem has always been a staple of North Africanist discourse in French, serving as a site of psychosexual obsession that was particularly potent during the heyday of empire-building between 1880 and 1931 (date of that most spectacular stage-production of imperialism, the *Exposition coloniale*). From romantic orientalism (Chateaubriand, Nerval, Balzac, Flaubert, Gautier, Fromentin) to symbolist and post-symbolist travelogue poetics (Baudelaire, Rimbaud, Segalen, Loti) to the little-studied genres of pseudo-ethnographic *littérature coloniale* (the Tharaud brothers, Louis Bertrand, Guy de Téramond, Charles Géniaux, Michel Vieuchange) through to exoticist *écriture féminine* (Jane Dieulafoy, Isabelle Eberhardt, Elissa Rhaïs), the harem trope was obligatory in colonial French fiction. As Marc Hélys, alias Leyla, alias Marie Léra, alias Djénane (the heroine of Loti's novel of a Turkish harem, *Les désenchantées*, 1906) wrote in her confidential memoir "Une Parisienne dans les Harems de Constantinople:" "there was no word more flattering and poetic to Occidental ears than *harem*." Like the world *odalisque*, "an entire asiatic poetics resounded in the two syllables of this word": "Hot colors, strange countryside, penumbras forming a halo over searing images: perfume boxes, amorous indolence, feverish passions, tragic pleasures, silent, heavy wall hangings splashed with blood. . . ."[2]

Sophie Monnerot's recent *L'Orient des Peintres*, though a highly selective survey, shows the extent to which harem themes in European nineteenth and twentieth century painting emerged as an almost pathological *idée fixe*,

Figure 7.2 Frederick Goodall, *A New Light in the Harem*, 1884.

Figure 7.3 Jean Léon Gérôme, *The Guardian of the Harem*, 1859.

Figure 7.4 Eugène Delacroix, *Women of Algiers in their Apartment*, 1834.

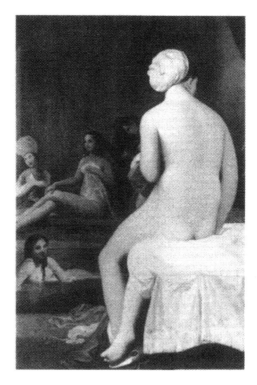

Figure 7.5 Jean Auguste Dominique Ingres, *La Petite Baigneuse*, 1882.

generating a veritable industry.[3] Delacroix's *Femmes d'Alger* (1834), Ingres' *La Petite Baigneuse ou Intérieur de Harem* (1828), *Odalisque à L'esclave* (1842), and *Le Bain Turc* (1862) are only the most well-known.

Academic painters and pornographers (whose work was often indistinguishable) exploited the harem for every possible venue of erotic fantasy. Abduction, exhibitionism, voyeurism (between eunuch and captives), bestiality (as when monkeys intervene in lieu of the sultan), sapphism, onanism, masochism (female bondage), and visions of the "dark side" of femininity itself comprised a cornucopia of "aberrant" sexual preferences to be grouped under the rubric "harem". One could even stretch this last formulation to argue that the word evolved at some level into a western synonym for femininity *as perversion*. In this regard, the coincidence of veiled female bodies and veiled interiors becomes particularly significant as a visual reiteration of one of the master tropes of philosophy and psychoanalysis: from Nietzsche to Joan Riviere and Derrida the idea of "womanliness as a masquerade" has been theorized through the metaphor of the veil.[4]

Femininity, which "pretends" itself into existence according to Joan Riviere, in order to assuage male trepidation towards the spectre of masculated women, has historically been described in terms of masks and screens. In Balzac's *Le Fille aux yeux d'or*, for example, a profusion of semi-transparent fabrics, fringed curtains, and throw-rugs theatricalize the associations between false appearances and the sartorially illusionistic female body (carpets are thrown over the shoulders of furniture just like shawls).[5] The same romantic dressing up of the interior appears in Pierre Loti's *Les Désenchantées* in which the trappings of a modernized, pseudo-orientalist interior form hymenal screens: "An April sun [. . .] filtering through the blinds and muslins into the bedroom of a young girl asleep. A morning sun, bringing, through those curtains, screens and bars, that ephemeral joy and eternal illusion of renewal on earth, which takes everyone in, since the beginning of time, from the most complicated to the most simple of souls."[6]

Loti presents an odalisque in layered space in the manner of tableaus by Ingres. Adopting the flattened and foreshortened perspective of Persian miniatures, Ingres had collated bed covers, interior panels, walled courtyards, and guarded doors, superimposing the harem woman in the round, so to speak, like a classical, free-standing sculptural western nude. Loti may be seen as performing a similar operation in prose: walls dissolve into curtains (the French word "mousseline" derives from *Mosul*, the name of a village on the Tiger river in India where this material was produced) and Venetian blinds (*persiennes*), thus highlighting, with the help of a pun, the Persian woman (*persienne*) who lies at the heart of the westernized harem. With its cargo of cultural ambivalence, the harem as veiled space reveals the place of orientalism within the traditional "enigma" of femininity.

Perhaps more than any other orientalist construction, the harem survived successive literary and pictorial changes of fashion, bridging broad gaps between nineteenth-century realism, neo-primitivism, and psychoanalysis. Though French surrealism and avant-garde modernism, from Apollinaire to Michel Leiris, privileged black African and Oceanic cultures over the Maghreb, it nevertheless preserved the harem in residual, abstracted form in the pictures of Picasso, Matisse, Duchamp, and Le Corbusier. Picasso painted a fauvist *Harem* in 1906 in which a naked eunuch lies sprawled in a bare room in front of several anguished looking nudes. Pressed flatly into the walls, their bodies fused with the minimalist interior, the occupants of this gyneceum seem to have lost their status as human subjects. Picasso returned to the theme in the 1950s and 1970s with erotic parodies of Ingres (*Le Bain Turc*), Delacroix (*Femmes d'Alger*) and Degas (*Degas dans la Maison Tellier*).[7] Consistently, the old man in the harem, surrogate of the painter himself, is identifiable as a Peeping Tom in old fashioned European attire, while the women, sapphically entwined and exotically costumed, compete in showing this percipient intruder "la chose". This transgressive, deictic geometry of sexual showing (prefigured in the early painting by the smudged vaginal line formed by adjoining walls), seems in one way or other to be symptomatic of the harem genre as a whole, most likely because, at least in architectural terms, the harem is so obviously the place that can never be shown. Maurice Barrès' *Le Jardin sur l'Oronte* (1922) – a novel in large part about the showing of the unshowable harem – provides a literary corollary to Picasso's late engravings in its use of rhetorical pointers. The phrases "Ecoute!" ("Listen!") and "Ah! si tu la voyais!" ("Oh! if you could only see her!"), uttered by a Syrian Emir as he prepares to introduce a European dignitary to the secrets of his harem, enhance the visual delectation of occluded spectacle.[8] In the crudest terms, the harem is to the architectural anatomy what the female sex is to the woman's body; both "signify" in unison, spatial zones of pleasure that have no ultimate indicative object.

Picasso's harem scenes, as has often been pointed out, contain an allegory of impotence as it afflicts an ageing modernist and an aging modernism. By contrast, in Matisse's œuvre, one could speak of a "haremization effect," in which, depending on how one reads it, orientalist motifs prevail either as a means of working out a modernist decorativism, or, as a way of reimposing fabric remnants of a disappearing bourgeois interior onto modernist spatiality. An early example of this haremization effect may be seen in the transition from *Les Tapis rouges* (1906), a still-life, to *La Desserte rouge* (1908), a domestic scene. The pattern and brilliant red of a Persian rug in the former tableau seem to seep out into the backdrop of the latter. Here, a maid austerely arrayed in black and white, is engulfed by an orientalist sea of redness, as if to suggest a double claustration of the European domestic through constraints of gender and social station.[9] When in the 1920s, Matisse painted faintly orientalized odalisques in poses of reverie mixed with sexual abandon – as in *Odalisque au Pantalon Rouge* (1924–5) or *Odalisques Jouant aux Dames* (1928) – the confusion of mural arabesques and scalloped harem trouser-legs (flung

Figure 7.6 Giulio Rosati, *Inspecting the New Arrivals*, undated.

Figure 7.7 Jean-Léon Gerôme, *The Slave-Market*, undated.

open at the crotch) can be read as a modernist play on an older orientalism in which female body and interior wall are fused as pure decoration, pure surface.

In its retrievals and renewals of the harem *mise-en-scène* modernism showed itself unwilling to let orientalism fade into art history. In addition to providing inspiration for pictorial experimentalism, such orientalist fixtures constituted an erotic sign language that brought imperialist and sexual questions into collusion. As Anne Fernihough, commenting on the impact of Leo Frobenius's 1913 *The Voice of Africa* on D. H. Lawrence, has suggested, an intuitive grasp of the connection between two symbolic orders of the unknown – the territorial and the psychoanalytic – characterized this period of early modernism.

> Frobenius entitles his preface to *The Voice of Africa* "Fiat Lux": he sets out to shed light, he tells us, over Stanley's "dark continent". Freud's parallel project is that of reclaiming land from the primordial chaos of the unconscious mind, and again the opening of Genesis provides the analogue: "Where id was, there ego shall be. It is a work of culture – not unlike the draining of the Zuider Zee". [. . .] While Frobenius and his co-imperialists scar the African with their "spades" and "picks", Freud, the self-styled "conquistador", engages in his colonization of the human psyche.[10]

From romanticism to modernism the harem was a dual sign, for while on the one hand it prompted forays into those recesses of the unconscious from which violent sado-masochistic fantasies hailed, on the other, it reinforced the picture of a subjugated, domesticated female sexuality. This is precisely what makes its architectural transposition particularly intriguing for the Western interpreter; it is a site that is already a site of the unconscious, hence a materialized example of the mythic rootedness of tropes of otherness within that most quintesssentially eurocentric human science – psycho-analysis. Though the historic familiarity and formulaic conventionality of harem iconography could make the European viewer feel, paradoxically enough, at ease with the spectacle of radical cultural alterity, there was also invariably a moment when colonial discourse became transparent to itself, revealed as a sickness involving the violent, sado-erotic management of cultural territory and human subjects. Albert Memmi (*Portrait du colonisé, Portrait du colonisateur*), Octave Mannoni (*Prospéro et Caliban: psychologie de la colonisation*), Frantz Fanon (*A Dying Colonialism*) and more recently, Malek Alloula (*The Colonial Harem*) have each, from the outbreak of the Algerian Revolution to the present, theorized this moment of existential transparency. The veil and the harem, though symbols of authoritarianism and male chauvinism within Islamic culture itself, are suddenly perceived as bicultural, ambivalent signs that reveal the camouflaged ideological strategies of colonial

Figure 7.8 Frank Dillon, *An Apartment in the Harem of Sheik Sâdât in Cairo*, 1873

repression. In this scheme, Algeria "herself" is placed in the position of the cloistered Muslim woman, while France, in the role of the pacha, looks on with malaise at a feminine dominion teeming with *ressentiment* toward its metropolitan "master." A space of two-way *scopomanie*, discomfiting in its visual hierarchies and tyrannies (and for the western female viewer, scenes of sequestered women will never, I think, come to feel naturalized as banalities of Arab life or fixtures of Islamic mores that merit endorsement as markers of difference), the harem thus emerges as a synecdoche of what the West most desires and fears in the Other as well as what it most wants and detests in itself. As a mode of European voyeurism, as a tangible expression of the (largely repressed) psychosis of coloniality, the harem desublimates the asymmetry of relations between Europe and North Africa even as it points to problems of culture and gender essentialism in architecture: should spaces and architectural styles subscribe to traditional codifications of cultural difference? Should they dispense with or adhere to the conventions of sexual difference? How might these visual fixations be dislodged by a critical spatial ethnology?

Notes

1 Malek Chabel, 1988, *L'Esprit de Sérail*, Lieu Commun: Paris, pp. 118–19.

2 Marc Hélys, 1908, *Le Jardin Fermé*, Plon: Paris, 4. Marc Hélys was putatively the woman who acted as the model for Loti's Turkish heroine Djénane in *Les Désenchantées*. The was some scandal when it became known that the woman on whose "authentic" report of harem life the novel was based, was a *Française* posing as Turkish with two women friends (they in fact were genuine members of the Istanbul bourgeoisie stifled by the social strictures of the harem). When Hélys's book was published shortly after Loti's death it was assumed by some that Loti had seen through the hoax but had played along for fun. Others justified the adventure, as did Hélys herself, with the argument that the play-acted canular of *Azyiadé* "Part II" had furnished Loti with wonderful material for a novel. A detailed appraisal of the affair may be found in Pierre Briquet's monumental *thèse d'état*, 1945, *Pierre Loti et L'Orient*, La Presse Française et Etrangère: Paris.

3 Sophie Monnerot, 1989, *L'Orient des Peintres*, Nathan: Paris.

4 See Joan Riviere, 1929, "Womanliness as masquerade" in V. Burgin, J. Donald and C. Kaplan, eds, 1986, *Formations of Fantasy*, London: Methuen, pp. 35–44. See also Mary Ann Doane, 1982, "Film and Masquerade: theorising the female spectator," in M. A. Doane, 1991, *Femmes Fatales: feminism, film theory, psychoanalysis*, London: Routledge, pp. 17–32.

5 Honoré de Balzac, 1988, "La Fille aux yeux d'or" in *Histoire des Treize*, Garnier Flammarion: Paris, p. 91.

6 Pierre Loti, 1989, *Les Désenchantées*, Presse de la Cité: Parigi, p. 877.

7 See Picasso in Centre d'Art Contemporain in Château de Tanlay, 1990, particularly the section on "Gravures 1970–1972," *Degas dans la Maison Tellier*. A useful essay by G. Picon entitled "Picasso et l'érotisme" (1974) is republished

in this catalogue. For more on Picasso's play on Degas's brothel monotypes, see Carol Armstrong's monograph on Degas, *Odd Man Out*, Chicago University Press: Chicago, 1990.

8 Maurice Barrès, 1922, *Le Jardin sur l'Oronte*, Libraire Plon: Paris, pp. 23, 32.

9 I should mention that this interpretation goes against the grain of standard art historical appraisals of the painting. Most commentators refer to the maid as a sign of peace and security reigning in the domestic interior.

10 Anne Fernihough, 1990, "The tyranny of the text: Lawrence, Freud and the modernist aesthetic," in P Collier and J Davies, eds, *Modernism and the European Unconscious*, Polity Press: Oxford, p. 47. For further analysis of the relationship between psychoanalysis and psychogeographical colonization, see Christopher Miller, 1985, *Blank Darkness: Africanist Discourse in French*, Chicago University Press: Chicago and William Pietz, 1987, "The pornograph in Africa: international phonocentrism from Stanley to Sarnoff" in D. Attridge, G. Bennington and R. Young, eds, *Post-Structuralism and the Question of History*, Cambridge University Press: Cambridge, pp. 263–85.

8

DANCING IN THE DARK

The inscription of blackness in Le Corbusier's Radiant City

Mabel O. Wilson

This picture is not symbolic. It is a large painting composed in 1930, at the same time as the plates for the Radiant City were being worked on in our studio. It is possible that there is a relationship between these works despite their wholly different intentions. The human creative work stands midway between the two poles of the objective and the subjective, a fusion of matter and spirit.

(Le Corbusier 1967: frontispiece)

Black slavery enriched the country's creative possibilities. For in that construction of blackness and enslavement could be found not only the not-free but also, with the dramatic polarity created by skin color, the projection of the not-me. The result was a playground for the imagination.

(Morrison 1992: 38)

Architects imagine and create buildings through the scribing of drawings, treatises, manifestos, and theoretical texts. Given these disciplinary and professional practices, architecture is a discourse and thereby situated within social spheres informed by racial categories, institutions, and beliefs known as racial formations. The articulation of racial formations within architectural discourse, however, is often difficult to discern. In methodologies similar to the objectification of the text in modern literary criticism, architects, historians, and critics, for example, often limit their analysis to intrinsic qualities of architectural form – typologic relationships, qualities of light, and internal organizational concepts such as axialities, figure/ground relationships, and so forth.[1] These methods of evaluation privilege the building as the locus of critical inquiry and divorce it from key generative processes: theorization, techniques of representation and production, and those social, political, and economic factors which also contribute to the making of architecture. To expose how racial formations operate within architecture, we must sift through a variety of ways in which architects realize ideas and forms, as well as how people receive and live through these architectural creations.

In *Playing in the Dark: Whiteness and the Literary Imagination*, novelist Toni Morrison offers a useful critique of how racial formations, in this instance articulated through literature, come to serve the formation of America's socio-cultural beliefs, practices, and national identity. In particular, Morrison posits that colonial Euro-Americans meditated upon the modalities of enslaved Africans in order to imagine their democratic nation, a supposedly enlightened socio-political body whose founding principles of individual freedom and liberty would guide the destiny of their new civilization. In this process the cultural hegemony of this white identity, one steeped in individualism, is established by writers who strategically situate within their prose and poetry an "Africanist" presence – a character, an event, or locale – metaphorically connoted by blackness to represent black identity. American novelists Edgar Allan Poe, Willa Cather, Mark Twain, and Ernest Hemingway conjure up in their narratives images of whiteness when by itself as "empty," "mute," and "vacuous." These flat, singular images often representing an individual's search for affirmation and wholeness, acquire complexity and depth, signifying self realization, through an encounter with blackness. The presence of blackness unleashes "self-contradictory concepts of the self" that are complexly "evil *and* provocative, rebellious *and* forgiving, fearful *and* desirable" (Morrison 1992: 59). These now canonic literary works carefully deploy metaphors of blackness, "Africanisms," that shuttle between dialectical pairings of images and concepts to construct and stabilize white identity. Significantly, it was through the fervid imagination of both author and audience, according to Morrison, that the writing and reading of these novels constituted a white American identity.

Morrison's understanding of whiteness on the level of literary representation proves extremely useful in examining how these same socio-cultural forces of identity formation operate spatially and are thus underpinned by architecture and architectural discourse. In order to discern in architecture an Africanist presence, connoted by metaphors of "blackness," we must sift through a variety of ways in which architectural ideas and forms are conceptualized and circulated through writing and drawing, as well as through building.[2]

I shall focus on the work of modern architect Le Corbusier whose prolific career spans over fifty years. As the Father of the international style, his abstract, white forms represent the core of modernism's utopian impulse. His *œuvre* includes numerous commissions, speculative designs, polemical books, and a significant body of paintings and sculptures. But while historians and architects often foreground the design of his buildings, ostensibly a vast and rich source for inquiry, it is crucial to consider how these buildings are positioned within a web of creative and conceptual endeavors that include writings, paintings, and sculpture. Whether he is critiquing the ostentatiousness and impracticality of domestic wares in *Decorative Arts of Today* (1925), or designing a modern domicile such as the Villa Savoye (1929–31), Le Corbusier experiments in each genre with new theories, forms, or techniques of construction. In fact, Le Corbusier was keenly aware of the interrelation between these various modes of theoretical and aesthetic experimentation; he wrote in the frontispiece of the *Radiant City*: "it is possible that there is a relationship between these works [urban theory and painting] despite their wholly different intentions" (Le Corbusier 1967 [1933]: frontispiece).

Le Corbusier's experimentation with urbanism commences with the reformulation of garden city planning tenets in his design of the Ville Contemporaine of 1922 and reaches its apex with the monumental master plan and designs for India's new administrative capitol in the Punjab at Chandigarh, begun in 1951. I focus on the period of the late 1920s and 1930s, because it is at this moment Le Corbusier, observing of the effects of industrialization on European cities and populations, theorizes a new city – a Taylorized urban form where industrial production organizes social and political life. His theories on contemporary urban design evolve accordingly through a plethora of written works: journal articles, pamphlets, books, and visionary urban schemes.[3] In 1933 many of these writings were compiled into a complex opus entitled *The Radiant City* featuring designs for an ideal city of the same name.

In order to disseminate his ideas and rally financial backing to realize his vision of a modern metropolis, Swiss born Le Corbusier left Paris and traveled to New York, Buenos Aries, Rio de Janeiro, Sao Paulo, Moscow, Geneva, Antwerp, Stockholm, Rome, Barcelona, and several cities in French colonial Algeria. During these forays into Western and colonial capitals, exuberant followers who had been won over by his widely circulated utopian manifestoes invited Le Corbusier to lecture. At venues crowded with eager architects, urban planners, and students, Le Corbusier gave animated performances highlighted by the telling of anecdotes and the constant jotting of sketches and diagrams. Typically, these lectures culminated in the application of Radiant City planning techniques and forms to an area of the host city. Often during these visits, Le Corbusier met with local authorities – mayors, municipal officials, and venture capitalists – and challenged them to realize his vision of a modern metropolis. He implored them to implement his plans for modernizing their cities to avert devastating physical and social collapse caused by the poor condition of the urban infrastructure: dilapidated housing stock, streets ill-suited for high-speed vehicular traffic, and sparse green spaces and parkland.

An account of a 1935 US junket, a lecture tour sponsored by the Museum of Modern Art, is recorded in the travelogue *When the Cathedrals Were White*

(1936).[4] Over a four month period Le Corbusier visited universities and museums in the northeastern and midwestern United States. In the book Le Corbusier recounts in a sometimes jocular tone, a myriad of adventures, from delivering a radio broadcast from deep inside the towers of Rockefeller Center to a delightful automobile excursion through the suburban parkways of Connecticut. He describes a colorful but restrained masquerade ball in Manhattan and relates the social nuances of a haughty business lunch at the Plaza Hotel in Manhattan. These events and others confirm his opinion that Manhattan's culture, nurtured by America's international economic prowess, exudes a raw physical energy that should interest the culturally stagnant French. Yet the brilliance of American economics pales in comparison to the "timidity" and "puerility" he observes pervading American society. In America Le Corbusier detects an ominous blackness permeating the cavernous winding streets of New York City, he senses its presence it in the "roaring cadence" of Negro jazz, and discerns it in his encounters with "Amazon-like" society matrons who roam Manhattan's cocktail party circuit. To the European eyes of Le Corbusier, Manhattan is in its infancy, merely twenty years old. Manhattan was born in the first decade of the twentieth century with the erection of its imperial architecture – its skyscrapers. This young, metaphorically black and primitive America awaits a renaissance in which its culture will come into fruition. In contradistinction to America, Le Corbusier reminds his French readers that France has had skyscrapers for centuries, grand gothic cathedrals, majestic monuments to rational building techniques. France, according to Le Corbusier, teeters on the brink of a second Renaissance in which French arts, music, and architecture, cultivated and refined over many centuries will lead the vanguard in the salvation of Western culture. In this tumultuous period of rebirth, Le Corbusier's urban plan to erect a new metropolis composed of crystalline, "white" cathedrals of glass and steel will rescue French cities from imminent destruction brought on by decades of poor planning and neglect.

Imagining racial patriarchies

Significantly, three key elements: trees, mannequins, and blackness encodes racialized metaphors of American life into the narrative of *When the Cathedrals Were White*. These metaphors underwrite a social order which architecturally and spatially structures the Radiant City. Crucial to Le Corbusier's analysis of America is that an overarching metaphoric "blackness" incites complex dialectical pairs sometimes described as demonic *and* playful, desirable *and* fearful, or spiritual *and* material. In his narrative blackness registers upon the bodies of the unfettered American woman and the Negro, both of whom fulfill a requisite position in America's social and economic order as laboring bodies. The work of these bodies not only reproduces the racial stock and produces capital, but also accords freedom from physical work and power to those who assume roles of intellectual labor. When underwriting a social hierarchy defined by racial categories, this metaphysical distinction between mind and body becomes a racial patriarchy.

As a social order, racial patriarchy privileges and accords power to dominant racial groups who monopolize education and intellectual labor, while leaving subordinate groups to carry out physical labor (Doyle 1994: 6). In *Bordering on the Body*, literary theorist Laura Doyle writes that racial patriarchy is "an inherently metaphysical social formation – one that rests on the metaphysical distinction between a ruling 'head' and a laboring 'body' and one that genders and racializes this distinction" (my emphasis, Doyle 1994: 20).

Concepts of racial patriarchy are in rooted in the nineteenth century. During this period science ascribed to certain peoples, particularly those biologically linked through spurious racial classifications, various physical attributes: diminutive bodies, hulking physiques, and other deformities. Substantiated by data accrued through cranial measurements, for instance, it was widely accepted that certain racial groups had a limited intellectual capacity and were therefore in need of supervision and governance. A cadre of scientists and those who avidly ascribed to Social Darwinism rationalized that these racial groups were biologically equipped for physical labor. Many of these notions, disseminated in part through eugenics, gained currency in America's burgeoning industrial society and continued in popularity well into the twentieth century. Scientific hierarchies of biological difference were neatly overlaid onto the ideology of capitalism's striated work force of owners, managers, and laborers. The result: a new American social order in which those in possession of valued intellectual faculties, now corroborated by scientific research, assumed positions of power and wealth and those with less legitimate abilities became laboring bodies – blacks, women, ethnic immigrants and poor whites.

Returning to Le Corbusier's *When the Cathedrals Were White*, we find intertwined in his urban adventures a racial–patriarchal schema inscribed through a dynamic, invigorating, yet threatening blackness. These concepts underwrite a critique of the area surrounding Manhattan's Empire State building. Le Corbusier proclaims:

> And once again this: *that black* and *those mannequins*: Aeschylus. Once more this: there are *no trees* in the city!
>
> (my emphasis, Le Corbusier 1947: 188)

Since the implications of these metaphors will be more fully elaborated upon later in my argument, I will briefly mention here that the phrase "that black" alludes to both the ebony marble interiors of the Empire State Building and jazz, the "frenzied" music and dance of Harlem, whose dangerous explosive energy must be harnessed to erect Le Corbusier's transcendent white cathedrals. "Those mannequins" refers to both the statuesque mannequins poised in the shop windows around 34th Street and the domineering bourgeios white women who threaten to depose, in Le Corbusier's eyes, the patriarchal order of America society. And finally "trees" whose ubiquitous absence in Manhattan is duly noted by most visitors, in Le Corbusier's narrative their scarcity reveals an imbalance in the natural order, whose stability is key to nurturing the proliferation of the white race.

While seeking an American society founded upon and organized by a racial patriarchy, Le Corbusier instead discovers a culture churning with racialized bodies and unbridled feminine figures. As the cure, he presents his Radiant City as an urban and architectural mechanism that socio-spatially enforces and guarantees racial patriarchal order. Le Corbusier's skyscrapers, contemporary "white cathedrals," symbolize the restoration of a Western culture that transcends and masters filth, the infiltration of "blackness," and the materiality of the body. Aspiring to stem off an ideological upheaval and revolution ignited by the destructive forces of a heretofore mismanaged industrialism, Le Corbusier made plans, architectural plans for a Radiant City. His new metropolis would be a panacea for the current ills of modern urban life. Ironically, Le Corbusier's theories and architecture for a gleaming white metropolis and ideal society depends upon a controlled blackness; a blackness transposed onto bodies not only to racialize them, but to articulate normative heterosexual ideals of family and motherhood.

No trees in the city

There are no trees in the city! That is the way it is.

Trees are the friends of man, symbols of every organic creation; a tree is an image of a complete construction.

Sun, Space, and trees are the fundamental materials of city planning, the bearers of the "essential joys." Considering them thus, I wish to restore urban man to the very heart of his natural setting, to his fundamental emotion.

(Le Corbusier 1947: 71).

When theorizing the geographic context for his city in the *Radiant City*, Le Corbusier stipulates that it should be located within what he terms "natural regions," areas possessing "permanent elements that dominate the machine-age adventure: climate, topography, geography, race" (Le Corbusier 1967: 193). In other words these raw elements: soil, topography, and climate of a "natural region" nurture biologically specific races or groups which share physiognomic features. Additionally, a natural region fosters a social order that begins at the micro-scale of the family, the "cell of society," next spirals outward to the tribe or the race, and finally moves outward to the macro-scale of the region. Besides the specific racial character of a natural region, its topography also forms a "natural frontier." Since these natural frontiers or boundaries are underwritten by racial difference, they would supplant what Le Corbusier asserts are the arbitrary borders of nation-states. By planning new cities, Radiant Cities, within these specified topographic natural regions, conflict and war, the outcomes of different racial groups cohabiting in the "unnatural" regions of nation-states, could be thwarted (Le Corbusier 1967: 193).

Notions such as natural regions and frontiers, illustrates the degree to which Le Corbusier's new architectural metropolis is predicated upon reified conceptions of nature. Le Corbusier believed that a dense richly planted landscape would bring modern man, overwhelmed by the grim-laden, disorderly industrial city, physically and psychologically back into alignment with nature's order. The triad "sun, space and trees," emblematic of nature in the text of *The Radiant City*, are base elements with which to build the green city. His seventeen illustrations of the Radiant City depict a lush verdant utopia. Broad carpets of greenery stretch below elevated blocks of housing units, skyscrapers, and highways. Trees are plentiful. Wide swathes of grass are dotted intermittently with gardens and athletic facilities. Rooftop gardens and artificial beaches for sun bathing and recreation cap residential units, housing 1000 inhabitants per hectare. Each building is also equipped with "exact respiration" or conditioned air, a novelty in the 1930's, providing for clean, mechanically controlled breathing for inhabitants. Collectively, these planned amenities not only aimed to improve the physique of the residents, but also attempted to restore their "fundamental emotion" and psychological well-being.

Beliefs such as Le Corbusier's in the necessity of physical vitality for intellectual acuity, reflect ideals first propounded in the nineteenth century. The doctrines of scientists Comte de Gobineau, Karl Pearson, and Francis Galton often invoked soil/tree metaphors and

> take on more than local significance in the light of an emerging, if still infant, science of national or racial character. Although thinkers and scientists . . . were beginning to explore the genetic transmission of racial characteristics, the predominant secular explanation in this period was still the climatic theory, namely, that racial features were shaped by the soil and climate of a country.
>
> (Doyle 1994: 43)

These tracts of the founders of eugenics inferred that racial superiority was discernible in certain physiognomic characteristics. Interpreting Darwin's theories of natural selection toward their own ends – justifying a racially segregated capitalist society – Eugenicists postulated that the highest intellectual capabilities would certainly be an attribute of the dominant racial group. They theorized that European and Euro-Americans were intellectually superior and thus the guardians of history and the forebears of progress.[5]

At the other end of the eugenics scale were racial groups positioned lower on the Social-Darwinian chart of biological, social, and cultural evolution. These groups – ethnic immigrants, working classes, peasants, and newly emancipated slaves – were thought to be biologically best suited for physical labor and predestined to ill-health. As Doyle concurs:

> Dominant kin groups associate themselves with mind or spirit and associate subordinate groups with body or matter. The conflated kin and metaphysical distinctions in turn justify a division of power and labor by which hand workers serve brainworkers.
>
> (Doyle 1994: 28)

These conceptions of racial difference had profound effects upon social divisions and urban development. Surfacing among middle- and upper-class white urban populations during this period, for instance, was an alarming fear of miscegenation in which "good blood" when mixed with "bad blood" would eventuate in the debilitating traits of feeblemindedness and degeneracy. Biologistically racialized and classed bodies were thereby sorted out and maintained under the strictest of controls. Under the sway of these ideas, racially inferior groups or those seen as socially deviant and dangerous, were forced to live at some distance from bourgeois whites. In the service of elite interests, municipal authorities eventually zoned and planned cities so that bourgeois and upper class whites, safe within suburban enclaves abounding with sun and verdure, were far from the threatening unsanitary slums of racially inferior groups. Under this misconception, those factors such as inadequate wages and squalid housing conditions: consequences of social inequalities affecting these marginalized groups, could be and were for a long time ignored.

Returning to the narrative unfolding in *When the Cathedrals Were White*, Le Corbusier asserts in New York City there are "no trees." Trees, a key indicator of a thriving "natural region," are noticeably absent. By pointing out their scarcity a vigilant Le Corbusier wants to alert his reader (remember this book was originally addressed to a French audience), that something in the natural order of America has gone amiss.

Those mannequins

The wax mannequins in the windows of the smart dress shops on Fifth Avenue make women masters, with conquering smiles. Square shoulders, incisive features, sharp coiffure – red hair and green dress, metallic blond hair and ultramarine blue dress, black hair and red dress.

The mannequins in the windows have the heads of Delphic goddesses. Green, lamp-black, red hair. . . . Polychromy. When polychromy appears it means that life is breaking out.

Next door I note the funereal entrances of the Empire State Building.

And once again this; that black and those mannequins: Aeschylus. Once more this: there are no trees in the city!

(Le Corbusier 1967: 165)

While the natural order of society may be predicated on geographic and climatic specificity, stabilized through racial homogeneity, it is women, precisely because they reproduce and nurture the family, who become the guardians of the racial and natural order. In *When the Cathedrals Were White*, Le Corbusier puts forth a "dangerous hypothesis:" the death of the American family, and it is the white bourgeios woman, as Mother, who is held accountable for its murder.

In a section entitled "Searching and Manifestations of the Spirit," white American women, typically middle and upper class, appear to Le Corbusier in one of two guises: either desirous or threatening. While visiting the Connecticut countryside to lecture at the women's college of Vassar, a delighted Le Corbusier tells us of its learned, privileged, and desirable coeds: "they are in overalls or in bathing suits. I enjoy looking at these beautiful bodies, made healthy and trim by physical training" (Le Corbusier 1967: 136). A few passages later, a harried Le Corbusier informs us that following his lecture these attractive coeds metamorphosed into aggressive "Amazons," greedily grasping to retrieve a shred of his sketches as a souvenir. Both alluring *and* ominous these women are an indication to the discerning eye of Le Corbusier that America's natural order, in this instance a biological one, is in disarray.

A similar assessment of American women can be ascertained from Le Corbusier's kaleidoscopic description of the fashion displays in the shop windows near the Empire State Building. The unnatural wax figurines outfitted in resplendent, shapely Delphic sheaths, remind him of Clytemnestra, Aeschylus' infamous Greek murderess. In brief Aeschylus' classic *Oresteian Trilogy* tells a tragedy in which adultery, patricide, and matricide destroy the royal family of King Agamemnon. A firestorm of destruction is unleashed by the machinations of the all-too-powerful Mother, Queen Clytemnestra. This strong-willed Matriarch, whose "words are like a man's," murders her husband King Agamemnon to usurp the throne (Aeschylus 1986: 55). Clytemnestra frantically racing to escape detection of her deed, enwraps the King's lifeless body in a luxuriant purple silk robe soaked in crimson blood. Aeschylus' vivid prose captures the tumultuous events, reckless deeds, and uncontrolled passions of this domestic tragedy.

141

Along Manhattan's 34th Street a modern Oresteian tragedy is evoked by the omnipotent posture of its mannequins and the polychromatic setting. Le Corbusier spies in the shop windows, similar phantasmagoric hues of the "lamp-black," "green," "red," and "metallic blond" hair and exaggerated artifice of the "Delphic dresses" sheathing "those mannequins." This urban drama culminates in its own funereal spirit, set against a vertical shaft of blackness: the Empire State Building. If "those mannequins," like Clytemnestra, is a metaphor for defiant American women, then how do they imperil the natural order of America's racial patriarchy?

Historically ascribed as the primary biological reproducers of the race, women's bodies have been regulated through social practices and codes of morality in order to guard against the infiltration of the genetic traits of inferior races. Beginning in the mid nineteenth century, the editor and writers of women's magazines and leaders of bourgeios women's associations circulated notions of feminine propriety and motherhood. Mothers were advised as to how to provide for the moral education of their children. Wives were instructed on methods of maintaining a fastidious, healthy, domestic environment. Many women were led to believe that they should not undertake professional employment. Nor should they profess an interest in politics, economics, theater, or any subject matter that might be the purview of their spouses. On the contrary, women were persuaded to direct their energies into their reproductive duties (Merchant 1989: 163). Such disciplining practices associated with motherhood not only focused women's labor toward replenishing the racial stock, but also aided in the maintenance of its purity. The vaunted white Mother became the guardian of the racial patriarchy. Doyle compellingly argues

> This metaphysical division [between matter/mind and spirit/body] further determines the function of the dominant group mother: she sorts out bodies not only into kin and non-kin but also into brainworkers and handworkers.
>
> (Doyle 1994: 28)

In other words, the bourgeois white maternal body sat precariously at the border between the dominant race group who were the intellectual elites and all others who were the physical laborers. Following this logic of racial division, if a daughter or wife gave birth to child by a man of a lesser racial group, she ruptured and polluted the family's bloodline. Not accorded status of Mother, the "fallen" woman and her child were cast out of the family and exiled from the Father's house. Therefore, to avoid enticements that would lead to shame and degradation, women were kept busy at home with their wifely duties. Rigorous regimens of housework popularized by ladies magazines and facilitated through the mass production of household equipment, ensconced bourgeois white women in domestic spheres far away from the "dangers" of urban life. Tucked away from the dark, foreboding metropolis teeming with lesser races, the white Mother was secure from possible rape, impregnation, and ruin.

Given that for Le Corbusier American women represent nature's equilibrium gone awry, then what would be the fate of America's natural order, an order in which it is imperative that men determine the discourse of family life and women work to maintain the family and household? To verify the death of the American family, Le Corbusier's "dangerous hypothesis," he tells a parable entitled "The Family Divided." In it he describes the daily regimen of a typical middle class Euro-American household. Leading separate lives the husband works in the city and the wife remains at home in the suburb. The husband commutes via rail to his place of employment in the central city, where he arduously labors, as Le Corbusier quips, to "shower her with attention – money, jewels, furnishings" (Le Corbusier 1947: 154). But while the husband toils, his wife cultivates her intellect through reading, attending lectures, and socializing with her circle of friends. From these events, Le Corbusier deduces: "the husband is intimidated, thwarted. The wife dominates" (Le Corbusier 1947: 154). This reversal in the natural order where man is now equated with the body and woman equated with the mind, precipitates the death of the family.

Le Corbusier concludes that American men, overshadowed by these shrewd Clytemnestra-like women, possess unremarkable intellects. The poignant comedic performances of Charlie Chaplin and Buster Keaton reflect this denigrated, lackluster American male spirit. Le Corbusier speaks of these cinematic characterizations as representing:

> The simple man . . . , a good fellow full of friendly and altruistic thoughts which are often puerile. Around him an overwhelming situation of inhuman dimensions. That disproportion is the rule in the USA: *an abyss opens up* . . . at every step.
> (my emphasis, Le Corbusier 1947: 157)

While searching for "manifestations of the Spirit," Le Corbusier comes upon an "abyss" of unfathomable blackness. An aftermath of the death of the family and the Father caused by the domineering Matriarch, a deep melancholic spirit envelops everyone and everything. A dark abyss swallows the "funereal" cloaked shaft of the Empire State building. An abyss erupts in the geography of the city separating downtown from uptown, white Manhattan from black Harlem. In the dynamic "mongrel metropolis," (to borrow historian Ann Douglas' phrase,) however, mobile white bodies travel across these racial and spatial borders to the phantasmal nightlife of Harlem's clubs, theaters, and dancehalls. However, to be sure, such a journey into Black Manhattan exposes white bodies to the constant threat of corruption and contamination.

That black

On the stage of Armstrong's night club a series of dances follow each other, supported by the music and stimulating the body to frenzied gesticulation. Savagery is constantly present, particularly in the frightful murder scene which leaves you terrified; these naked Negroes, formidable black athletes, seem as if they were imported directly from Africa where there are still tom-toms, massacres, and the complete destruction of villages or tribes. Is it possible that such memories could survive through a century of being uprooted? It would seem that only butchery and agony could call forth such cries, gasps, roars.

(Le Corbusier 1947: 160)

As brilliant stage performers, efficient porters and slum-dwelling social outcasts, African Americans appear in a myriad of incarnations throughout Le Corbusier's travel narrative. Observing African Americans exclusively in servile guises, Le Corbusier overlooks Harlem's diverse social milieu that included various ethnicities and classes, as well as an established community of intellectuals and artists. Those Negroes (I employ his term here,) Le Corbusier does encounter are "good-natured, cordial, and companionable." He believes they pose a thorny question to white Americans, a question that cannot be resolved "in a superficial manner" (Le Corbusier 1947: 87). Beyond their affable character, Le Corbusier observes that many Negroes reside in horrid, squalid communities hidden from more prosperous white New Yorkers who "if they knew the slums, it would make them sick at heart and they would make new city plans." Negroes are social "pariahs," who resettle areas abandoned by whites; and in these "former paradises" they "sow a spirit of death" – again a metaphor of blackness (Le Corbusier 1947: 86). Despite recognizing the unbearable living conditions and racial injustices to Blacks in America, Le Corbusier nevertheless imagines his white city, his Radiant City through tropes of blackness.

Over the course of his stay, Le Corbusier visited some of the popular haunts of Jazz Age Manhattan, including those within and associated with Harlem such as the raucous Savoy Ballroom and the downtown incarnation of Harlem's famous Connie's Inn at Broadway and 49th Street where he saw Louis Armstrong headlining a show called "the Hot Chocolates of 1936." These experiences are retold in the chapter "The Spirit of the Machine and Negroes in the USA" which strategically follows the tale of the death of the family. Louis Armstrong takes center stage. Le Corbusier benights Armstrong:

the black Titan of the cry, the apostrophe, of the burst of laughter, of thunder. He sings he guffaws, he makes his silver trumpet spurt. He is mathematics, equilibrium on a tightrope . . . with Armstrong,

the exactitude leads to an unearthly suavity, broken by a blow like a
flash of lightning.

(Le Corbusier 1947: 159)

Clearly in admiration of Armstrong's brilliance and artistry, Le Corbusier
identifies a regal composure: "he is in turn demonic, playful, massive, from
one second to another, in accordance with an astounding fantasy. The man
is extravagantly skillful; he is king" (Le Corbusier 1947: 159). During the
performance Armstrong incarnates brute physicality – emitting a fierce heat
that combines with the driving rhythm of the drums and blare of the horns.
Imbued with a sublime spirit, Armstrong's "voice is as deep as an *abyss*, it is a
black cave" (my emphasis, Le Corbusier 1947: 159). Paradoxically Armstrong
radiates a spirit of a material sort – a kind of black ore, raw matter, that fuels
the machines of the modern age. In his prose Le Corbusier simultaneously
envies and dreads the "demonic" *and* "playful" Armstrong. In a complex
association, Armstrong's corporeal and metaphoric blackness is emblematic of
the primitive, a modality which in turn engenders the modern.

As a telltale sign of the primitive, acts of savagery are ever-present in the
description of the performers of the floor show at Connie's Inn. In his reverie,
Le Corbusier evokes the typical tropes of blackness – primitive, rhythmic,
frenzied, abysmal, frightful, murderous, and explosive – to characterize the
performance. As he is entertained by a dance review choreographed to the
syncopation of jazz, he senses "savagery is constantly present, particularly in
the frightful murder scene" (Le Corbusier 1947: 160). To Le Corbusier these
black bodies, like those ravenous Vassar coeds, are athletic *and* savage,
incarnating imminent death, and posing a threat to racial patriarchal order.
A similar eruption of primitivism infuses the passages about Lindy Hoppers
at the Savoy Ballroom where "ordinary colored people join each other in very
nearly savage rights" (Le Corbusier 1947: 161). The ethereal environs of the
dance hall envelop Le Corbusier. With the architecture of the hall receding in
the glow of the flickering projector, the kinetic twists and turns of the dancers
apparently evoke memories of his journey to French colonial Africa. He
remembers a similar phenomenon in the fierce eroding winds that batter the
peaks of the Atlas Mountains. Observing from the bird's-eye view of an
airplane, the winds unfold into a primeval scene of nature, a "geological
drama," where one seeks shelter against the "tumult" of the "unfathomable
march of the elements." The eroding mountains elide with the dancing black
bodies to create a sublime performance witnessed by the rational disembodied
eye of the architect.

Ironically, while Le Corbusier fears abysmal blackness, it nonetheless
deliriously intoxicates and spiritually envelops him; therefore, he desires it as
part of his Radiant City. Dancing across the text, dynamic, kinetic black
bodies of African Americans exude the energy and spirit of the machine age:
"new sounds, of everything and from everywhere, perhaps ugly or horrible:

145

the grinding of the streetcars, the unchained madness of the subway, the pounding of machines in factories" (Le Corbusier 1947: 161). He couples the precise rhythm of black tap dancers with the exactitude of machines – sewing machines – producing the raw material to sustain industrial capitalism. These black bodies provide the labor to erect the "foundations of cathedrals of sound which are already rising." As a complex duality indicative of Africanisms, "blackness" is both spirit *and* matter.

In this very primitivism Le Corbusier discovers a rationality and regularity that counters the confusion and unpredictability of the chaotic industrialization of the twentieth century. He aligns the art of the Negro with the art of the engineer, "the old rhythmic instinct of the virgin African forest has learned the lesson of the machine." These performing black bodies are machines: the rhythmic "tap-tap-rap-tap-tap" of dancers are "as mechanical as a sewing machine" and the jazz band's tempo is a "smoothly running turbine." Together they create a symphony of production,

> The Negro orchestra is impeccable, flawless, regular, playing ceaselessly in an ascending rhythm: the trumpet is piercing, strident, screaming over the stamping of feet. It is the equivalent of a beautiful turbine running in the midst of human conversations. Hot jazz.
>
> (Le Corbusier 1947: 161)

Le Corbusier exalts and desires these black bodies because he finds flourishing in them a primal human spirit; it is a essential spirit lacking in the European soul dampened by the chaotic industrialization of the twentieth century. The cacophony of sound and energy emitted by these black bodies and/as pulsing machines are the base elements of a modern world over which nevertheless, Euro-American and European men, must be masters. From the performances of the machines/nature/black bodies Le Corbusier composes an opus, a great symphony for a modern society – a plan for a Radiant City.

This alluring "blackness" however is not confined to Harlem's dance halls and nightclubs, it invades the space of Manhattan where a "vast nocturnal festival . . . spreads out." As with the black entertainers, Le Corbusier admires the raw performative energy of the skyscrapers, noting that "Manhattan is hot jazz in stone and steel." Both jazz and skyscrapers are events, gestures, bursts of activity. For Le Corbusier neither jazz nor a skyscraper is a "deliberately conceived creation," he continues "if architecture were at the point reached by jazz, it would be an incredible spectacle. I repeat: Manhattan is hot jazz in stone and steel" (Le Corbusier 1947: 161). In the end, Le Corbusier informs his reader that black culture, jazz, and the American skyscraper are a folk tune of emotion, not a symphony of rationality.[6] Despite his fascination with their energetic expression, Le Corbusier's sees them as an inefficient use of their verticality, exclaiming in a *New York Times* interview "They are too small!" Although he admires the masterful engineering of Manhattan's high-rises, Le Corbusier observes, in an ominous tone, that the streets of Manhattan are clogged with automobile and pedestrian traffic, there is a proliferation of slums, and the migration of middle-class white populations to garden city suburbs erodes the vitality of the city. The city dies. This spirit of death cloaks the skyscrapers. A funereal spirit connoted, once again, by blackness – "the black polished stones, the walls faced with dark gleaming slabs" enwraps the art deco edifice of the Empire State and overwhelms the city.

White cathedrals and black bodies

Thus far, I've illustrated how Le Corbusier metaphorically scripts "blackness" as an uncontrollable force threatening to depose the order of a racialized patriarchy. But it is important also to consider how these metaphors constitute conceptions of whiteness, which similarly spatialize and structure his desired racialized society. What then are the operative metaphors of whiteness within *When the Cathedrals Were White*?

Following his metaphoric drama of the death of the family, the kinetic black bodies of Harlem, and those phantasmal mannequins, Le Corbusier unveils plans to transform Manhattan into a Radiant city of pristine white skyscrapers. In a sobering chapter entitled the "Necessity of Communal Plans and Enterprises," Le Corbusier's lament ensues: "When the cathedrals were white, spirit was triumphant. But today the cathedrals of France are black and the spirit is bruised." Unsurprisingly, he longs for the era of the glorious Gothic cathedrals in which the world was "white, limpid, joyous, clean, clear." The architecture produced an orderly society whose culture manifested "itself in fresh color, white linen and clean art." Here, whiteness metaphorically evokes purity and cleanliness. I concur with architectural theorist Mark Wigley, who asserts that Le Corbusier's architectural whiteness works as a thin opaque layer of whitewash to master and stabilize the architectural and corporeal body in order to liberate the mind (Wigley 1989: 85). Whiteness as a necessary structural element appears in a variety of modes: as a racialized concept ordering the narrative of *When the Cathedrals Were White*, as a thin coat of white paint ordering the modern facade, as starched white undergarments ordering clothes, and as white skin ordering the surface of the body. Yet, lurking menacingly below the stable surface of whiteness is blackness. Blackness manifested as death, dirt, and lawlessness – posing a persistent threat to the natural order and necessitating containment and control.

This articulation of whiteness ordering a dynamic and unstable blackness also characterizes the socio-cultural relationship between France and America. Towards the end of the chapter on Negro culture, Le Corbusier compares it to his own folkloric cultural heritage. Upon return from America, his wife, Yvonne:

> puts on the record "Fifine", a Parisian java ... here I am in the presence of the real originality of the java; I find in it mathematical France, precise, exact; I find in it the masses of Paris, a society worthy of interest, so measured, precise, and supple in its thought. A controlled sensuality, a severe ethics.
>
> (Le Corbusier 1947: 163)

Reinforcing the racialized maternal ideal, it is his wife – the bearer of his racial heritage – who dutifully brings him this pleasurable snippet of folkloric culture "the java."[7] In Le Corbusier's house in Paris, we are assured that all is in order. In contrast to America's inchoate culture, superior French culture is refined, mathematical, and most importantly "controlled sensuality." Paris, but more importantly France, home of the majestic gothic cathedrals, is the patriarch of fledgling America.

As a tourist, Le Corbusier peruses America's socio-cultural spectrum and imagines through a metaphorically black America what his beloved France is not. From his journey, he concludes that the society, preferably France, that transcends blackness will achieve the order and harmony necessary to erect the "bright whitewashed," "radiant filigreed white cathedrals" of his modern metropolis. The architecture of Le Corbusier's cathedrals will herald a new heroic period, one that echoes a mythic past in which "an international language reigned wherever the white race was, favoring the exchange of ideas and the transfer of culture" (Le Corbusier 1967: 4).

But how will his imaginary society come into being? How is a racial–patriarchal social order reliant on the rift between physical and intellectual labor configured into the design of the Radiant City? And where is Le Corbusier in the scheme of things?

The plans for the Radiant City harbor the promise of an orderly society and city, transcending chaotic infrastructure and grime-laden edifices. To achieve this end, Le Corbusier invents a meticulously functioning urban machine. Enamored by modern production theories such as Taylorism, Le Corbusier designs an urban standard, an "*objet type*," based upon a set of criteria that allow for a variety of possible iterations. As an ideal form, the elements can be modified when applied to a given topography, in this instance, New York City. The layout of the elements of the Radiant City resembles a body: as legs – warehouses, heavy industry, factories; as the torso – housing, cultural institutions, hotels, embassies; as the neck – rail and air terminals; as the head – the business center; as a network of arteries weaving across the entire city – highways and railways. The productive forces of the economic base: warehouses and industries, are the legs that carry the load of the social body: housing, cultural amenities, business, commercial and government buildings. Building vertically by elevating each mass onto "*pilotis*" or columns, achieves the desired population density. And also frees the ground from congestion

148

allowing for ample coverage of park space and recreational facilities. So that mothers (remember they are the custodians of the racial order) would be near their children, the plans include nurseries, kindergartens, and schools within each housing block. A worker would no longer waste valuable productive time traveling by train from garden suburbs to his job in the city, in the Radiant City he could drive his automobile along the extensive network of highways to nearby offices and manufacturing facilities. To maximize exposure to sunlight and air, necessary to sustain the natural order, the entire city would be oriented on a "heliothermic axis" determined by local climatic conditions.

The city's distribution of elements and functions also reflects what Le Corbusier labels in the *Radiant City* as the "pyramid of natural hierarchies." The pyramid, a socio-political order structured according to a resident's occupation, is based upon theories espoused by a French labor movement known as Syndicalism to which Le Corbusier had affiliations in the 1930s.[8] The Syndicalists, in brief, proposed the reorganization of French society and politics away from Republican ideals of citizenry and participatory government and toward a ruling ethos emphasizing industrial production and governance by worker's guilds or "metiers." In this planned economy, at the bottom of the pyramidal order would be workers' groups organized into trades that form "metiers." At the next level are the leaders from these groups making up an inter-union council who would deliberate and resolve inner disputes between trades, as well as implement economic policy stabilizing production and distribution. At the top of this social order is the "extra metier," the grand chiefs, or the supreme authorities who would be "free from all problems stemming from technical insufficiencies. This group of intellectual elites is at liberty to concentrate on the country's higher purposes." The supreme authority – a body of men, not engaged in corporal labor and entrusted with the future of Western civilization would be housed in a grid pattern of "Cartesian skyscrapers" located in the business district at top of the city. The behemoth steel and glass towers, maximizing exposure to light and air, would be a phenomenal feat of ingenuity created by the marriage of engineering and architecture. The Radiant City's political order fits neatly into the schema of a racial patriarchy, since these men would be those with the highest intellectual acuity, biologically predestined to be free from physical work. These Cartesian skyscrapers, translucent white cathedrals "at" and "as" the head the Radiant City, project the gaze of the supreme authority, and ultimately the gaze of Le Corbusier, outward to survey the neatly ordered city of light, sun, space, and trees.

White place, black face

Throughout his narrative Le Corbusier envisions himself outside worldly materiality, transcending the body – a pure mind. Yet, the act of writing itself deceives him by leaving traces of his corporeality. His inscriptions tie him bodily to the text and conversely the political effects of his texts draw them into the world of things. In fact, *When the Cathedrals Were White* teems with passages that reveal Le Corbusier's uneasiness about his own and others' bodily presence. In an amusing story centered around the events of a masquerade ball at the Waldorf Astoria, Le Corbusier remarks of his physique: "Not being a handsome fellow, I keep my anatomy out of sight" (Le Corbusier 1947: 150). During the festivities, outfitted in simple blue and white attire, whilst others are adorned in colorful, brilliantly plumed and brocaded costumes, Le Corbusier relates his disdain at the conspicuousness of his body, "I was neither mad nor clownish, I was a sore thumb. I was out of place" (Le Corbusier 1947: 150). His rational discerning character conveyed through the severity of his dress is lost, displaced by the phantasmal whirlwind of color, costume, and corporeality.

Continually traveling within Le Corbusier's gaze over the course of his journey to America are the bodies of "those mannequins" and "that black." It is their corporeality whether maternal or racialized, which consistently disrupts his much desired patriarchal order. Uncontrolled, women and Blacks impinge upon the smooth architectural and socio-political workings of his Radiant City. Yet, as much as Le Corbusier privileges the mind as the ultimate measure of civilization's progress, the body returns as a specter – to dance with Le Corbusier in his text and in his city – forever haunting his crystalline vision.

My reassessment of Le Corbusier's urban theories and architecture illustrates that architecture although material in nature, nevertheless is conceived, constructed, and lived through a multitude of social formations – practices, institutions, cultural beliefs, political affiliations and so forth. Perhaps it is because of its supposed ideological bent toward universality that Le Corbusier's urban theories, along with those of other modern architects, enjoined quite smoothly with the aspirations of twentieth century colonialism and imperialism. Significantly we find most of the international style of architecture built not in the Western countries, but instead in so-called "third world" nations and enclaves. In former colonial outposts such as Brazil, India, Zimbabwe, and elsewhere, monumental modern architectural projects cater to the economic, political, and social aims of ruling elites, who were formerly backed by powerful colonial empires and are now supported by influential Western nations and transnational corporations.

In America, similar sites of experimentation with the tenets of modernism can be found in segregated and impoverished communities such as Harlem, Newark, Detroit and Chicago – cities that bore the brunt of state funded social housing beginning in the 1930s. Forsaking Le Corbusier's utopian vision of a gleaming white metropolis, large numbers of bourgeois Whites and those various groups of immigrants and blue collar workers who garnered the "wages of whiteness," moved out of central cities (a migration subsidized by government agencies promoting the growth of suburbs) after the Second World War. Those who remained in cities, Blacks and Latinos, the poor and the underclasses – those disempowered in the American socio-political system – were shuttled into towering housing blocks whose designs reflected the then prevailing theories in modern architecture. In the end, these edifices were quite different from those sprouting across the verdant landscape of Radiant City. In Le Corbusier's heroic white towers, the gaze of the supreme authority oriented outward. Ironically in these American reiterations of his skyscrapers, the gaze – a panoptic one of the State and dominant white society – pierced dauntingly inward.

Notes

I am grateful to Mark Wigley, Paul Kariouk, Wallis J. Miller, Ernest Pascucci, and Jerzy Rozenberg who offered helpful comments on various drafts of this chapter. I also want to thank Steve Pile for his thoughtful suggestions and Heidi J. Nast for her inspiration, as well as her patience and tenacity in coaxing this endeavor out of me.

1 These methods of analysis, emerging in the 1960s and 1970s are commonly taught as part the curriculum in many architecture programs. These techniques of formal analysis are found for example in Francis Ching's *Architecture: From Space and Order*, a popular book assigned in many beginning design courses. But one can also find these categories which parallel high modernist conceptions of aesthetic quality utilized in many scholarly articles and books.

2 A recent issue (number 16) of the architecture magazine *Any*, "White Forms, Forms of Whiteness" explores the myriad of the social and formal interpretations of the term "whiteness" and architecture. Drawing out the complex incarnations of whiteness in architectural discourse editor Ernest Pascucci observes in his introduction to the issue: "White is sometimes as dumb as a color, 'an achromatic color of maximum lightness' whose dictionary definition cannot help but enter into a complimentary if antagonistic relationship with black. In a psychoanalytic sense, white envelops the most beautifully dumb substitute objects, onto which fantasies of pure form and good democracy (or is it good form and pure democracy?) are projected and acted out. And sometimes white acts as a visual blocker, concealing a disavowed racial unconscious, especially when opposed to gray, thus leaving black, metaphorically and otherwise, out of the picture."

3 Prior to the publication of the *Radiant City* in 1933, Le Corbusier wrote a number of books outlining his urban theory; these include *The City of Tomorrow, Precisions*, and *Towards a New Architecture*. Architect Rem Koolhaas in *Delirious New York* devotes a chapter to Le Corbusier's critique of Manhattan. For ideological critique of his work see Manfredo Tafuri, *Architecture and Utopia*; for critics of gender and sexuality in Le Corbusier's urbanism and architecture see Beatriz Colomina, *Publicity and Privacy* and Zeynib Celik, "Le Corbusier, Orientalism, Colonialism" in *Assemblage* 17.

4 Le Corbusier arrived in the fall of 1935 and departs in winter of 1936.

5 Recent scholarship on racial science examines the political, economic, and social investment made in the research by those in power. For further reading see Daniel J. Kevles, *In the Name of Eugenics*, Laura Ann Stoler, *Race and the Education of Desire*, William H. Tucker, *The Science and Politics of Racial Research*, and Robin Wiegman *American Anatomies: Theorizing Race and Gender*.

6 Blackness is configured not only as a threat but also as the site of desire. America's blacks and their musical innovation, jazz, become the locus of Le Corbusier's yearnings, remarking "if architecture were at the point reached by jazz, it would be an incredible spectacle." Through their sensuality he can free himself from the chains of the past and conceive, reflexively, his new heroic city of white cathedrals. In *White Walls, Designer Dresses* Mark Wigley thoroughly unpacks the sexual implications of Le Corbusier's remarks, writing "an architecture that releases the sensual potential of the machine age would, like jazz, contain the pre-machine past as well as the present, putting 'dynamism into the whole body' by putting people in touch with the irreducibly sensual origins of humanity" (294).

7 The implications of sexuality underlying this passage were brought to my attention by Mark Wigley. *White Walls, Designer Dresses* examines the banishment of color and ornament in the designs of modern buildings. This examination of repression parallels discussions of clothing, fashion, and the body, now sexualized and racialized, in architectural discourse in the first half of the twentieth century.

8 Mary McLeod, *Urbanism and Utopia: Le Corbusier, From Regional Syndicalism to Vichy*, dissertation (Michigan: UMI 1985).

References

Aeschylus. (1986) [1956] *The Oresteian Trilogy*. (P. Vellacott, trans.) London: Penguin Books.

Doyle, L. (1994) *Bordering on the Body*. New York: Oxford University Press.

Frampton, K. (1985) *Modern Architecture: A Critical History*. London: Thames and Hudson.

Le Corbusier (1925) *The Decorative Arts of Today*. Paris: Editions G. Cres.

Le Corbusier (1967) [1933] *The Radiant City*. (D. Coltman, trans.) New York: Orion Press.

Le Corbusier (1947) *When the Cathedrals Were White*. (F. Hyslop, trans.) New York: McGraw-Hill Book Company.

McLeod, M. (1985) *Urbanism and Utopia: Le Corbusier From Regional Syndicalism to Vichy*. Unpublished dissertation.

Merchant, C. (1989) *Ecological Revolutions*. Chapel Hill: The University of North Carolina Press.

Morrison, T. (1992) *Playing in the Dark: Whiteness and the Literary Imagination*. New York: Vintage Books.

Wigley, M. (1989) "Architecture After Philosophy: Le Corbusier and the Emperor's New Paint." In A. Benjamin (ed.) *Philosophy and Architecture*. London: *Architectural Design*, pp. 84–95.

9

THE SOUTH AFRICAN BODY POLITIC

Space, race and heterosexuality

Glen S. Elder

Introduction

South African apartheid was a geographical process. People moving over space and the invention of place are two processes with which geographers are familiar. For many South Africans who lived under the weight of apartheid, geography translated into devastating effects. Forced removals, evictions from ancestral homes, and the active destruction and re-invention of neighbor-hoods and communities are just some ways that geographical processes were experienced by South Africans living under apartheid. It is estimated that between 1948 and 1985 the white South African police and military forcibly removed close to 3 million South Africans from designated "white areas" or "black spots" and relocated them to "black homelands" sometimes hundreds of miles away (Platzsky and Walker 1985). The apartheid government was thus able to re-map the South African landscape so that it came to reflect their vision of racial separateness. Questionable and historically inaccurate racial categories became indisputable realities once they were written on the landscape in the shape of bounded *racialized* spaces occupied by *racialized* subjects. These later included the black and ethnicized Bantustans (or Homelands), the permanent white suburban sprawls, and the "temporary" black township ghettos precariously tacked onto the peripheries of white cities.

Historically significant, desegregated locales were bulldozed and replaced by segregated (and I would venture heterosexual), stark, modernist archi-tecture. Modernism's architectural forms scarred the South African landscape in the shape of towering "single" apartment blocks and "nuclear family" housing, surrounded by immaculately clipped, frost-bitten lawns in spaces where rambling and dusty multi-racial neighborhoods once stood, where extended families once lived. In one such space ironically renamed Triomph, west of Johannesburg, a stark, high-walled neighborhood arrogantly testifies to the brutal and terrifying destruction of Sophiatown that took place throughout the 1950s (for an evocative geographical account see Hart and Pirie 1984). Like hundreds of other places on the South African landscape, "multi-cultural" and therefore multi-racial Sophiatown evoked a sense of place that challenged the apartheid ideal; accordingly it was eliminated. "Different" neighbors were scattered and relocated in racial terms. At first, racial categories made sense only to apartheid's architects. Racial signs became "real," however, once they were tied to spatial signifiers within the landscape. And so the apartheid map is not a coincidental collection of incidental spatial representations. Rather the map is the product of a complicated and nuanced re-configuring of socio-spatial relations – racial identities invented by inscribing them in space.

While the apartheid landscape is a text through which an explicitly racial policy was mapped and carried, "race" is simply the *endpoint*. Processes lying behind the invention of "race," or the precise racial nature of the "socio-spatial" dialectic of apartheid remains hidden. In this chapter I argue that part of that racial invention through spatial practice also includes the politics of sex and the body. While the geography of race is well documented both in South Africa (for a list of South African work see Rogerson and Parnell 1989), and internationally (for an example of international studies of racism see Jackson 1987), I argue in this chapter that the practice of racism in South Africa (apartheid) and its outcome (racial categories and racialized spaces) were premised upon the re-configuring of sexual–social relations and attendant spatial processes. By sexual–social relations I mean, quite literally, the myriad of sexual activities, sexual prescriptions, and sexual prohibitions that regulated sex lives during apartheid. Moreover, the specific way in which social–sexual relations were controlled by the State was through the micro-spatial regulation of apartheid's subjects. This micro-spatial policing is an important, albeit often overlooked, dimension that underscored much of the racial process that took place in South Africa during the apartheid years (1948–90). Apartheid, I argue, re-mapped social relations (including sexual–social relations) to bring about a racial outcome. In the process, South African bodies were also subjected to exceptionally violent and destructive forces.

Within the context of this book I argue that the re-configuring of sexual–social relations is a spatial exercise that maps out simultaneously the body of apartheid subjects and the nation state. Put another way, racism can be understood as a re-mapping of social relations which traverses public and privates spaces through the body. Starting with the body then, the smallest scale of analysis, we see that the body of the apartheid subject was foundationally coded in heterosexual terms upon which a politics of race was laid by the State. For example all black worker housing built by the apartheid government was built on the ideas that all familes were heterosexual. As familes were forced into these housing arrangements, heterosexuality as understood by the State was imposed. Another example would be the encoding contained in the Sexual Immorality Act. This law held that transgressive sexual acts were heterosexual encounters across the "color bar," but also included sexual encounters between people of the same sex, regardless of color. Little attention has been paid to the preceise way in which heterosexuality and homosexuality were coded under apartheid law and so exploring how spatiality linked racism and heterosexism is the central concern of this chapter.

Race, gender, and sexuality

Geographers have worked to show the role of space in manufacturing racial difference (for example see Jackson 1987; Smith 1989). Feminist scholars have argued and shown that racial hierarchies are premised upon patriarchal gendered divisions of society, and that racialized bodies are premised upon the sexed body (for example see hooks 1991). In response, feminist geographers have sought to explore the spatial dimension of the intersection of race and gender (Sanders 1990; Peake 1993; Kobayashi and Peake 1994; Gilbert forthcoming).

As a result we see that racialization occurs at numerous scales and is better understood as a process supported by networks of meanings (for geographic expansion of this concept see Massey 1993) – inscribed at all levels of analysis, from the macro-scaled nation state, through the city, into the neighborhood, home and, finally, onto the bodies of racialized subjects This conceptualization of identity formation does not "hierachize" scale, but rather points to the interconnected spatial mechanisms which include the body; a site of identity construction.

The production of race through the regulation of sexual relations, I argue, takes place at numerous scales, all of which are interdependent. Because of a history of apartheid, in South Africa active, spatialized inscriptions of race are particularly obvious. To date much of the urban and political geography of South Africa has described a racial landscape, as discussed later. These thick descriptions, however, hide the processes of racialization occuring through other scales and at other sites, including the body. Racial, sexual encodings of the body, I argue, are central to the operation of apartheid. These bodily sites where encoding took place were not limited to black bodies alone. All bodies, under apartheid were subject to different types and degrees of State sexual anxiety.

In my analysis, I draw upon the work of the feminist philosopher Elizabeth Grosz who has theoreized the body extensively. She argues:

> The body is a pliable entity whose determinate form is provided not simply by biology but through the interaction of modes of psychical and physical inscription and the provision of a set of limiting

155

biological codes. The body is constrained by its biological limits. . . .
On the other hand, while there must be some kind of biological
limit or constraint, these constraints are perpetually capable of
being superseded, overcome, through the human body's capacity
to open itself up to prosthetic synthesis, to transform or rewrite
its environment, to continually augment its powers and capacities
through the incorporation into the body's own spaces and modalities
of objects that, while external, are internalized, added to, supplementing and supplemented by the "organic body" . . . surpassing the
body, . . . that represents always the most blatant cultural anxieties
and projections.

(Grosz 1994, pp. 187–8)

Grosz forcefully demonstrates that the pliability of the body makes it
a contested site. In this chapter, I show that apartheid policy was partly
informed by State anxiety about the male body. Recognizing the pliable
nature of the body, the apartheid State sought to organize and control a body's
"capacity to open itself up" when that capacity challenged the State's
racialized and heteronormative assumptions. An important contradiction was
that once encoded as such, white and black bodies (in close proximity)
threaten the State. After all, sexual relations "across the color bar" would
result in miscegenation and thereby challenge the myth of racial purity.

Heterosexual sex between races challenges the myth of racial purity. A
gendered analysis of that threat reveals that men and women's bodies
challenged the myth of racial purity in different ways. Because women's
bodies have the capacity to bear children, women's bodies were gazed upon
heterosexually as biological, reproductive sites. Lesbian sexual activity, in
contrast has no immediately apparent reproductive end and so lesbian women
were seldom a focus for State anxiety. We can conlcude therefore that women's
sexuality under apartheid was controlled *solely* in the way in which it related
to reproduction. Male sexual relations, on the other hand, were officially
defined as transgressive in at least *two* ways. First, sex with women of other
racial groups was outlawed because of the reproductive consequences. Second,
sex between men was outlawed, although this translated into a policy that
regulated homosexual activity between white men only (Elder 1993).

Marking and describing public, racial boundaries of the apartheid city
reveals little about how apartheid policymakers and planners exacted
apartheid policy at other scales. By concentrating on the public urban scale,
the State's racialization process, which took place through the renegoting of
sexual relations in private, remains hidden. Policing that took place in formal
public arenas depended upon spatially regulating bodies. In the South African
case, subjects' bodies were heterosexually and racially encoded and policed
through legislation making it possible to set racial prohibition in place.

Rethinking the apartheid city

Since the mid 1980s, progressive South African geographers have sought to map the apartheid landscape. In so doing they have created a rich and varied body of literature that demonstrates how legislation like the Group Areas Act, the Native Land Act, influx control measures, and the Bantustan policy shaped the map of South Africa (for review see Crush 1991). In post-apartheid South African politics this tendency to map formal public landscapes has continued, delimiting regional boundaries in the "new" South Africa (see Mabin 1995).

The notable and important effort to map the "public" apartheid city must not, however, be seen out of context. The intellectual climate in late apartheid South Africa, particularly at English-speaking liberal institutions, pressured geographers to fill in blank spaces on the map. For almost two decades, English-speaking liberal geographers in South Africa sought to "do relevant" research, with both empirically exciting but theoretically disappointing results. The call to map the "terra incognito" of black space in the early 1980s (Beavon 1982) produced a literature soaked with detail, that helped establish a canon of South African urban geography, characterized by heavily layered and textured accounts of public apartheid spaces.

A limited amount of work has sought to generalize and develop a view-point that explained how the apartheid map resulted in the disempowerment of black South Africans. These theoretical endeavors have not provided a sustained and thorough understanding of the socio-spatial dialectics of apartheid. If the apartheid map is gendered as has been suggested by Robinson (1994), for example, how was domesticity regulated under apartheid, how different were men's and women's experience of apartheid and why were these gendered differences important to the functioning of apartheid and political relationships more broadly? If mining capital in South Africa gleaned aspects of its power from the assymetrical spatial relations of apartheid as suggested by Crush (1994), as a further example, how did pre-existing patriarchal relations shape racial power in South Africa later on? Building on the work of Crush and Robinson we would do well to explore the spatiality of apartheid in all its forms.

I now turn to a theoretical analysis of how two apartheid spaces intersect – the segregated city and the sexed body of apartheid's subjects. To do so, I draw on the amusing and terrifying *Report of the Select Committee on the Immorality Amendment Bill of 1968 (Republic of South Africa*, 1968) to show that the same planners who designed the Bantustan policy and the Group Areas Act (policy that sought to racialize public residential space) sought to control and regulate bodies. Policymakers did not simply aim to control the apartheid body, they sought to name it and construct it in heteronormative ways that served the apartheid end. My aim here, then, is to show how and why the regulation of spaces in the form of bodies and cities was important.

Moreover, I will also show that apartheid was more than an inscription of static identity upon static space but a nuanced and complex interplay of processes fed though a web-like network of policies that were not limited to the public sphere.

Apartheid spaces

The often quoted but only half-understood utterance of Prime Minister Voster in 1973 has been used by geographers for some time (for example see Jackson 1987). The South African Prime Minister is usually cited as saying that if he were to be reclassified as "black," the only difference he would experience would be a geographical relocation. Yet, while making the point that apartheid was geographical, he was also showing that there were intricate links between the bodily encoding of apartheid's subjects and geography. I would argue that policy focusing on the city and the body are part of a set of policies that we can see as a continuum of spatial control mechanisms. When we look at the inordinate amount of attention apartheid architects paid to regulating sexual social relations – who it occurred between, how it occurred, and where it took place – we see State voyeurism was not limited to those having sex across the racial divide; it extended into the bedrooms and parties of gay and lesbian South Africans, amongst others. I now turn to an analysis of the official discourse around urbanism and homosexuality in South Africa.

An inordinate amount of official time was taken up during the late 1960s to name, define, code, describe and explain both the city and the "rise of homosexuality in cities." In the words of the commission of inquiry "it is clear that homosexuality in all its forms constitutes a threat to the Republic." The attention to urbanism and homosexuality and the inter-linkages between them raises a number of inter-related questions – why and to what extent did homosexuality threaten the apartheid architecture, what were the links between urbanism and sexuality, and finally what can we draw from the congruent policing of cities and bodies to better understand the geography of apartheid?

Contextual setting

In the late 1960s, a hitherto unknown urban culture was emerging in South Africa. Part of that urban culture was believed to include vice and so it was only a matter of time before the Nationalist Government officially discovered an urban gay subculture. Chauncey (1994) argues that the linkages between urbanism, vice, and homosexuality in the United States are ongoing and that the targeting of gay men in New York City in the early twentieth century was a response to a wider identity crisis around masculinity in a changing urban economy. While South Africa is a very different context, I believe that Chauncey's point about societal flux challenging the way in which the men in power see themselves is transferable. Put another way, the role of the gay "closet" or "the open secret" as described by Sedgwick (1993) becomes more important than at other times. Government policy in South Africa during the late 1960s can be understood as the delineating of that closet; to assuage the panic felt by many within the government, a vigorous legislative campaign against male homosexuality, in particular, ensued.

A steady tide of impoverished white Afrikaans speakers, feeling the push of rural impoverishment as early as 1930 had reshaped the cultural milieu of major cities like Johannesburg. That voting bloc had guaranteed the National Party victory and it was important for the government to be seen creating and enforcing policy that protected the culture, morality, and economic interest of Afrikanerdom. The white, male, middle-class, English-speaking gay man, once constructed, became a foil for heteropatriarchal Afrikaner culture.[1]

With Afrikanerdom ascendant, anti-British colonial sentiment was also on the rise in government circles (for a gendered analysis of this response see McClintock 1995). In 1967 the British Parliament re-examined and reversed a century of legislation that had sought to criminalize homosexuality and more specifically the act of sodomy between consenting males.[2] The British Parliament argued that the role of the State was to protect the "public"; acts performed in private and with mutual consent fell outside State jurisdiction. As the following quotation demonstrates, the British distinction between public and private was a problematic position for the South African government:

> If we wish to combat the evil, I believe that it will not serve our purposes to close our eyes to what happens in private, because it is precisely in private that young people get waylaid.
>
> (*Republic of South Africa* 1968: 9)[3]

To distance themselves from the British position and out of a paternalistic concern for young (and as I will show) *white* males being "waylaid" in the new urban spaces, the state saw a necessity to protect and regulate activities that occurred in private. A witch hunt ensued into the white affluent suburbs of South Africa. A police raid on a house in one of Johannesburg's wealthier white suburbs was reported as follows:

> . . . a party in progress, the likes of which has ever been seen in the Republic of South Africa. There were approximately 300 male persons present who were obviously homosexuals. . . . Males were dancing with males to the strains of music, kissing and cuddling each other in the most vulgar fashion imaginable. They also paired off and continued their love-making in the garden of the residence and in motor cars in the streets, engaging in the most indecent acts imaginable with each other.
>
> (*Republic of South Africa* 1968: 10)

Thorough, temporalized details about the party suggest that undercover officers stayed at the party for a number of hours, circulating through a number of different private places. Beyond questioning about how the report was written without direct involvment in the "vulgar" acts, the report presumes a normative heterosexual masculinity. The disgusted tone of the report records anxiety over white male bodies in close sexualized proximity. In contrast, a well-documented incidence of homosexual relations between black men in South Africa's hostel system (see Moodie 1994), never enters into the committee report. What the government focus does reveal is the degree to which apartheid officials responded to sexual behavior in heterosexual and racialized ways. That black male migrant workers living under panoptic conditions in urban hostels are tangentially mentioned in the committee report (p. 25), in fact, sheds light on the ways in which policies of apartheid sought to simultaneously map specific bodies and public spaces.

Mentioning the unmentionable . . .

The committee's report focuses on white male homosexuality as practiced in South Africa's growing urban spaces. At one moment we are told that, "In certain parts of the country they are virtually taking over blocks of flats." "They," we read later "occupy flats which they keep very neat and which they furnish fashionably" (p. 32). Throughout the report we read again and again that "homosexuality is deep-rooted in the Republic and it is rapidly gaining ground in the country, particularly in the big cities." The real crime is described as follows:

> Every homosexual makes it his business to recruit as many homosexuals as possible. In other words, their task is to increase the number of homosexuals. We know that most homosexuals are born that way, but it is also know that many of them become that way through indoctrination by other homosexuals.
>
> (*Republic of South Africa* 1968: 25)

Later on a confused chairman asks why homosexuality has gained so much ground recently? He is told by a police officer:

Police Officer: They have no problem identifying one another.
Chairman: How do they do it?
Police Officer: They can identify each other at a distance of 50 yards.
Chairman: How do they do that?
Police Officer: By their carriage, their gait, their manner of speech. Most people regard them as effeminate but they are not.
Chairman: But, how do they know which one will be the wife?
> (*Republic of South Africa* 1968: 19)

From this exchange the committee members hypothesize that high urban densities possibly cause homosexuality. The above passage also shows how white, male, homosexual bodies are coded heteronormatively as either husband (active) or wife (passive).

Through surveillance, government attention constructed homosexuality in urban, "white" and male terms. Interest (once again marked by heteronormative assumptions) in lesbianism and butch/femme relationships is also evident in government documents, demonstrated in one of the more bizarre exchanges about lesbianism and dildo use:

Chairman: How prevalent is this instrument that they use?

Doctor: I should say that it is not very prevalent and not necessary for these women to reach their orgasm.

Chairman: But is this instrument of normal size or of abnormal size?

Doctor: It is rubber and it is definitely bigger than the normal male penis.

Chairman: Is there any relationship between people who use this thing, which is bigger than a male penis, and persons who cannot get satisfaction from a normal male penis?

Doctor: I am afraid I am unable and unqualified to answer that question.

(*Republic of South Africa* 1968: 38)

Although exchanges about lesbians are rare, this exchange and several similar officially recorded ones about male homosexuality reveal how "homo-sexuality" was thought of as a bodily transgression against "natural encodings" of the body. The exchanges reveal the anxiety on the part of the State about "the human body's capacity to open itself up to prosthetic synthesis" (Grosz 1994: 188). In general, women's homosexuality did not threaten the ideas of myth of racial purity. Conversely, it was heterosexual women's reproductive capacity that threatend the myth of racial purity.

Conclusion

Implicit in the *Report of the Select Committee on the Immorality Amendment Bill of 1968* (*Republic of South Africa* 1968) is the need to name and encode homosexuality, particularly white urban males in bodily terms. Unlike apartheid's racial proscriptions, white male homosexuality posed a threat to the Republic because of its "invisibility" and because it was "catching"; a catchiness exacerbated by a rising urbanism. The white apartheid state subsequently sprung into action to regulate and name the activity, the "rising incidence" of which is attributed to rising levels of urbanism.

Homosexuality between South Africa's white men challenged State definitions of race. As hooks (1991) has argued, constructions of race and heterosexuality depend on one another. Within the South African case racial/sexual privilege is undermined when the white male body becomes a locus of desire. In particular, what white male homosexuality threatens is a patriarchal and racial order that shaped interlocking structures that provided many white Afrikaner males access to power in South Africa during apartheid. It is for this reason that the apartheid State regulated sexual activity between men, in particular that between white middle-class urban men. If we agree that the apartheid model was based upon ideas that the mapped body of the apartheid subject was important to the architecture of apartheid, we must inevitably ask what the connection between sexuality and race means for post-apartheid planning.

The new South African constitution provides for a dismantling of sexism and homophobia. Accordingly, a progressive post-apartheid geography must help to excavate the interconnected spatial relations of apartheid as part of the constitutional effort. This chapter has sought to do that by locating the construction of racial identity within a sexual framework. The argument put forward here, that sexual encoding of bodies was part of the larger racial landscape in South Africa during apartheid, is not the only interlocking framework. Questions of ethnicity and religion, for example, also shaped the geography of apartheid. To ignore the interplay of spaces like bodies and cities, however, continues the silence about racism and sexism that shaped political discourse in South Africa but also in geography more generally.

Notes

1 This particular group of men were, of course, not the only ones whose State-constructed identity created a counterpoint for a government-sponsored Afrikaner culture. Black South Africans in particular and women in general experienced similar legislative control, often with far more devastating effect. What makes the case of gay men particularly fascinating is that it marks an instance where white men were disempowered by apartheid legislation.
2 Porter and Weeks (1991) document the lives of fifteen men who lived in England between 1885 and 1967 while homosexuality between men was criminalized. Included in their account is that of a white gay man who emigrated from South Africa to England in the mid 1960s to escape political persecution that included an increasing societal homophobia (pp. 81–96).
3 The quote is a direct translation from Afrikaans: The original reads as follows: "As on teen die euwwel wil optree, glo ek dat dit nie vir on sal loon om ons oë toe te maak vir wat in die privaat geskeid nie, want uit die aard van die saak sou ek aanneem dat hierdie tipe van dae in die privaat geskeid en dit is daar waar jou jong mense velei word."

References

Beavon, K., 1982: "Black townships in South Africa: terra incognita for urban geographers," *South African Geographical Journal* 64, 3–20.
Chauncey, G., 1994: *Gay New York: Gender, Urban Culture, and the Making of the Gay Male World*, 1890–1940, New York: Basic Books.

Crush, J., 1991: "The discourse of progressive human geography," *Progress in Human Geography* 15(4), 395–414.

Crush, J., 1994: "Scripting the compound," *Environment and Planning D: Society and Space* 12(3): 301–24.

Elder, G.S., 1993: "Of moffies, kaffirs, and perverts: the control and regulation of sexual discourse in apartheid South Africa," in David Bell and Gill Valentine (eds) *Mapping Desire*, London: Routledge.

Gilbert, M., forthcoming: "Identity, space, and politics: a critique of the poverty debates, in J.P. Jones, H. Nast and S. Roberts (eds), *Thresholds in Feminist Geography*, Lanham, Rowman and Littlefield.

Grosz, E., 1994: *Volatile Bodies: Toward a Corporeal Feminism*, Bloomington: Indiana University Press.

Hanson, S. and Pratt, G., 1995: *Gender, Work, and Space*, New York: Routledge.

Hart, D.M. and Pirie, G.H., 1984: "The sight and soul of Sophiatown," *Geographical Review*, 74, 38–47.

hooks, b., 1991: *Yearning: Race, Gender, and Cultural Politics*, Boston: South End Press.

Jackson, P., (ed) 1987: *Race and Racism: Essays in Social Geography*, London: Allen & Unwin.

Kobayashi, A. and Peake, L., 1994: "Unnatural discourse: 'race' and gender in geography," *Gender, Place and Culture* 1(2), 225–43.

Mabin, A., 1995: "Negotiating local identities: local government and space in South African cities," paper presented at the Canadian Research Consortium on Southern Africa, Queen University, 27 October.

Massey, D., 1993: "Power geometry and a progressive sense of place," in Jon Bird, Barry Curtis, Tim Putnam, George Robertson and Lisa Tickner (eds), *Mapping the Futures*, London: Routledge.

McClintock, A., 1995: *Imperial Leather: Race, Gender and Sexuality in the Colonial Context*, London: Routledge.

McDowell, L., 1993: "Space, place and gender relations: Part I. Feminist empiricism and the geography of social relations," *Progress in Human Geography* 17(2), 157–79.

Moodie, T. D., 1994: *Going for Gold*, Berkley: University of California Press.

Peake, L., 1993: "Race and sexuality: challenging the patriarchal structuring of urban social space", *Environment and Planning D: Society and Space*, 11, 415–32.

Platzsky, L. and Walker, C., 1985: *The Surplus People: Forced Removals in South Africa*, Johannesburg: Ravan Press.

Porter, K. and Weeks, J., 1991: *Between the Acts: Lives of Homosexual Men 1885–1967*, London: Routledge.

Republic of South Africa (1968) *Report of the Select Committee on the Immorality Amendment Bill*, Cape Town: Government Printers.

Robinson, J., 1990: "A perfect system of control"? State power and native locations in South Africa," *Environment and Planning D: Society and Space* 8, 135–62.

Robinson, J., 1994: "White women researching representing 'Others': from anti-apartheid to postcolonialism," in Alison Blunt and Gillian Rose (eds) *Writing Women and Space: Colonial and Post Colonial Geographies*, London: Guilford Press.

Rogerson, C.M. and Parnell, S.M., 1989: "Fostered by the laarger: apartheid human geography in the 1980's," *Area* 21(1), 13–26.

Rose., G., 1993: *Feminism and Geography: The Limits of Geographical Knowledge*, Minneapolis: University of Minnesota Press.

Sanders, R., 1990: "Integrating race and ethnicity into geographic gender studies," *Professional Geographer* 42, 228–31.

Sedgwick, E.K., 1993: "Epistemology of the Closet," in H. Ablelove, M. Barale and D. Halperin (eds) *The Lesbian and Gay Studies Reader*, London: Routledge.

Smith, S., 1989: *The Politics of Race and Residence*, Cambridge: Polity Press.

10

THE HOUSE BEHIND

Karen Bermann

What is going on behind this door?
A book is shedding its leaves.

Edmond Jabès, The Book of Questions

Down the canal street

Referring to a persistent theme in the dreams of her patients, the psychiatrist Charlotte Beradt, who worked in Germany until 1939, defined the situation of the average German as that of "a life without walls." The metaphor referred to a dream one of her patients, a forty-five-year-old doctor, had in the second year of the Thousand Year Reich.

> It was about nine o'clock in the evening. My consultations were over, and I was just stretching out on the couch to relax with a book on Matthias Grünewald, when suddenly the walls of my room and then my apartment disappeared. I looked around and discovered to my horror that as far as the eye could see no apartment had walls anymore. Then I heard a loudspeaker boom, "According to the decree of the 17th of this month on the Abolition of Walls. . . ."
>
> Robert Jan van Pelt, "Apocalyptic Abjection," in
Architectural Principles in the Age of Historicism

> *Thursday, 5 June 1944 At Whitsun, for instance, when it was so warm, I stayed awake on purpose until half past eleven one evening in order to have a good look at the moon for once by myself. Alas, the sacrifice was all in vain, as the moon gave far too much light and I didn't risk opening a window. Another time, some months ago now, I happened to be upstairs one evening when the window was open. I didn't go downstairs until the window had to be shut. The dark, rainy evening, the gale, the scudding clouds held me entirely in their power; it was the first time in a year and a half that I'd seen the night face to face.*
>
> Anne Frank, Anne Frank: The Diary of a Young Girl[1]

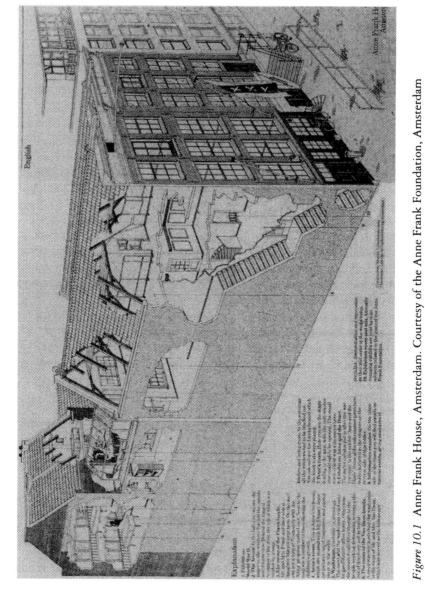

Figure 10.1 Anne Frank House, Amsterdam. Courtesy of the Anne Frank Foundation, Amsterdam

When the envelope from the Anne Frank Foundation in Amsterdam arrives, it contains a drawing of the building in which Anne, her family, and four others were concealed for twenty-five months, along with an accompanying legend of the rooms. The drawing is a cutaway axonometric. In this context it is impossible not to reinterpret the convention: the drawing looks like an aerial view of a building that has been subjected to an act of violence. This attack opens the building to view, revealing for our inspection the rooms, the furniture, the quiet interior. We see behind the bookcase that concealed the entrance to the annex; we see the beds; we see where the sisters stood when their father marked their heights on the wall; we see the space under the window where Anne crouched to sniff the fresh air leaking through the crack.

We see what no one was supposed to see. In the drawing, as in life, it is the compromise of the building envelope that makes viewing possible. But it was this building's resistance to compromise, its intended impermeability and the failure of this intention, that causes us to remember it today. The cutaway axonometric, employed here to optimize understanding, is underlaid with another meaning, because the act of revealing is in this case charged with both the danger it represented to those inside and the pain that we feel upon looking back. The axonometric is a fiction that evokes the omniscient gaze of fascism; the cutaway, with a particularly graphic irony, mimics both the effects of a bombing and the violence of opening-to-view that, in the case of the Anne Frank House, led to the discovery and destruction of the Jews hiding inside. The access to the interior that is the subject of this representation was once a matter of life and death, and this knowledge works its way into our experience of the drawing. We want to look, we don't want to look. The wall tears away, the book falls open, the shawl is parted; we pass inside.

Inside the door and up the stairs

It was like a laboratory. First they weighed us, then they measured and compared. . . . There isn't a piece of body that wasn't measured and compared. . . . We were always sitting together – always nude. . . . We would sit for hours together, and they would measure her, and then measure me, and then again measure me and measure her. . . . You know, the width of, say, our ears or nose or mouth or . . . the structure of our bones. . . . Everything in detail, they wanted to know.

Robert Jay Lifton, *The Nazi Doctors: Medical Killing and the Psychology of Genocide*

Wednesday, 8 July 1942 The first thing I put in was this diary, then hair curlers, handkerchiefs, schoolbooks, a comb, old letters; I put in the craziest things with the idea that we were going into hiding. But I'm not sorry, memories mean more to me than dresses.

Prinsengracht 263, situated on a canal, is like many other buildings in the old part of Amsterdam. Because of the structural and economic difficulties of building on water, it is long and narrow. The structure is divided into two parts, front and rear, with a courtyard through which light enters. Built in 1635, the building at the time of the Nazi occupation contained offices and warehouses in the front and a suite of narrow rooms in the back, connected by a passageway. The rear was known as an *achterhuis*. *Achter* is Dutch for "behind" or "in back of;" *huis* means "house."

In the diary Anne Frank wrote of her plans to publish a book after the war of her life in hiding: *Het Achterhuis* would be its title. When first published in 1974 (in the original Dutch) the diary carried that title. The title refers to the place in which the diary was written, the part of the building that served as a hiding place for the two families who took shelter there between July 1942 and August 1944. *Het achterhuis* has been translated as "the secret annexe."

The *achterhuis* consisted of two upper floors and an attic. The entrance to the hiding place was concealed behind a hinged bookcase; a map on the wall hid the top of the doorway from view. The building housed a wholesale herb and spice business that had belonged to the Franks before German decree forced Jews to give up their holdings. Because the herbs needed to be stored in darkness, the windows at the back were painted over. The obscured glass, the blackout slats ordered by the Germans, and a layer of curtains helped to conceal the inhabitants from view.

Prinsengracht 263 is an ordinary Dutch building, a type appropriate to its location, suggesting a conventional interior. Convention and propriety in appearance were defined by the Nazis: the physiognomy defined as characteristic of the "master race" and all deviations from it were studied and documented obsessively. Nazi anthropologists published pictures of "typical Jewish posture"; physicians investigated facial and cranial physiognomy with the intention of determining racial ancestry through appearance. Yet visual recognition of a deviant physiognomy was unreliable as a method of identification. In Nazi-occupied Europe, homosexuals were marked with pink or black triangles, Gypsies with black or brown triangles. Jews were required to wear yellow cloth stars attached to their outer garments, a marking of appearance that prefigured the deeper marking, the tattoo, that was to come in the concentration camps.

A conventional appearance, a correct physiognomy true to Aryan type, was an asset, a way to "pass." Like proper posture or coloration, architectural convention provided a camouflage, a means of passing behind the mask of propriety. This camouflage created a theater of appearance, a highly visible display that also offered invisibility.

Behind the hinged bookcase

A wall is beautiful, not only because of its plastic form, but because of the impressions it may evoke. It speaks of comfort, it speaks of refinement; it speaks of power and of brutality; it is forbidding or it is hospitable; – it is mysterious. A wall calls forth emotions.

Le Corbusier[2]

Tuesday, 18 May 1943 Although it is fairly warm, we have to light our fires every other day, in order to burn vegetable peelings and refuse. We can't put anything in garbage pails, because we must always think of the warehouse boy. How easily one could be betrayed by being a little careless.

A building is a collection of opacities and transparencies, a theater of appearance and disappearance in which we mask our presence or make it known. Every existing wall contradicts itself with openings, places where the obduracy of matter yields to the necessity of passage: joints and seals, points of rupture, of flow and failure, where water seeps in and air pours through, where materials meet and pull away. These gaps present us with opportunities to be seen and heard. Yet the exchange across the building's porous envelope makes us vulnerable. One's presence may be betrayed by a discarded orange peel, a bit of smoke, the sound of a toilet flushing or a pipe banging as water passes through. We are revealed through these traces, the things that architecture cannot keep, the separation that it cannot provide, its secretions, the excess that leaks through like light.

The expressions of the body, from its secretions to the sounds that "escape" us, are critical evidence that we would hold back if we could. If only these things could be kept inside, as if we were sealed containers; reused, continuously, so that they were always part of us; or worn on the outside, like another layer of clothing, another surface. Anything but a trace, cast out, refuse that we leave behind.

Wednesday, 8 July 1942 We put on heaps of clothes as if we were going to the North Pole, the sole reason being to take clothes with us. No Jew in our situation would have dreamed of going out with a suitcase full of clothing.

A little step, and beyond the plain gray door

I read to the very end. I was surprised by how much had happened in hiding that I'd known nothing about. Immediately, I was thankful that I hadn't read the diary after the arrest, during the final nine months of the occupation, while it had stayed in my desk drawer right beside me every day. Had I read it, I would have had to burn the diary because it would have been too dangerous for people about whom Anne had written.

Miep Gies with Alison Leslie Gold, *Anne Frank Remembered: The Story of the Woman Who Helped to Hide the Anne Frank Family*

Wednesday, 5 January 1944   Each time I have a period – and that has only been three times – I have the feeling that in spite of all the pain, unpleasantness, and nastiness, I have a sweet secret, and that is why, although it is nothing but a nuisance to me in a way, I always long for the time that I shall feel that secret within me again.

Camouflaged within the convention of the building envelope, bound inside a stolen cover, Anne wrote into a compressive, diaristic space, her Secret Annexe, *Het Achterhuis*, the House of the Book.

Saturday, 11 July 1942   The "Secret Annexe" is an ideal hiding place. Although it leans to one side and is damp, you'd never find such a comfortable hiding place anywhere in Amsterdam, no, perhaps not even in the whole of Holland. Our little room looked very bare at first with nothing on the walls; but thanks to Daddy who had brought my film-star collection and picture postcards on beforehand, and with the aid of paste pot and brush, I have transformed the walls into one gigantic picture. This makes it look much more cheerful, and, when the Van Daans come, we'll get some wood from the attic, and make a few little cupboards for the walls and other odds and ends to make it look more lively.

Het Achterhuis, the House Behind, denotes both the part of the building in which the Franks hid and the name Anne chose to give the diary. This shared name suggests the similarities between the hiding place and the diary. Both are defined by the distinction between inside and outside. Both are forms of resistance, variations on the theme of disappearance as another form of appearance, inconspicuous containers with carefully defined interiors, spaces whose borders must be patrolled, whose porosity must be restricted.

In the House Behind, "like being pressed between two pieces of paper,"[3] through a secret door beyond a swinging bookcase, the rituals of living resemble the rituals of secret writing – a bed must be pulled out, a table converted from desk to dining table, a kitchen changed into a bathing room, and then back again. A book with a red-and-white-checked cover, pulled from a drawer and written in quietly while others nap. The diary too constitutes a home, constructs a world in which one is safe, one is known, one can be seen, in which recognition is desirable, a world of visibility and appearance. Diary writing is a near-silent act of inscription in which all that can be heard is the movement of a pen, the rustling of the page, an appearance in which exposure is never dangerous, in which the one who reads is always a friend, until the interior is revealed to strangers. A sanctuary, a prison.

And now inside

When Hans moved in he brought with him a daybed that doubled as a couch. It was a massive boxlike piece built of mahogany, with a lid, hinged on the inside, which could be opened to store bedding. "You know," Marushka said one day as she gazed at the couch, "that would make a perfect hiding place for you if we ever needed one."

"I'd suffocate," Hans said.

"No, you wouldn't. We could drill holes in the bottom."

Using a hand drill, she made the holes and then covered the underside with a loosely woven linen fabric. Inside the top she fastened a latch, so that when Hans climbed inside he could secure the lid; the pad on top would fall back into place. Each morning thereafter, before she went off to her classes at the university – she was working toward a degree in veterinary medicine – Marushka would put a fresh glass of water inside the couch alongside a cough suppressant.

Leonard Gross, *The Last Jews in Berlin*

Wednesday, 5 January 1944   I think what is happening to me is so wonderful, and not only what can be seen on my body, but all that is taking place inside. I never discuss myself or any of these things with anybody; that is why I have to talk to myself about them.

Diminutive spaces too small to be considered useful, voids in the dense weave of inhabited space, pockets of exile. *Crawl space, shawl, interior of couch, hole in the ground.* The darkness of these spaces, their lack of programmed identity, the difficulty in naming them, not *places* but simply *spaces*: this is precisely what makes them potential hiding places. As poché, the space between figural spaces, they are invisible and offer this same cloak to their inhabitants. The visual assimilation within such a space that might make survival possible is matched by a psychological assimilation. It might be enough to be contained inside the couch, but if one becomes the couch, disappears into the couch, the hunter perceives only the couch. The inhabitant disappears *without a trace*, into *thinnest air*. She is gone; she is now *crawl space, shawl, interior of couch, hole in the ground.* Disappearance is the ultimate camouflage.

In Cynthia Ozick's novel, *The Shawl*, a mother forestalls her infant's death in the concentration camp by hiding the child inside her shawl. The space enclosed by shawl and mother's body becomes a home, a dwelling, a domestic envelope that contains the baby's body:

Rosa knew Magda was going to die very soon; she should have been dead already, but she had been buried away deep inside the magic shawl, mistaken there for the shivering mound of Rosa's breasts. Rosa clung to the shawl as if it covered only herself. No one took it away from her.

171

Shawl and breasts become architecture, and the body inside becomes a shawl and breasts; incorporated, it disappears into them. This portable dwelling is the house of the body, an intimate space between.

The pages of the diary unfolded, expanded, laid edge to edge, would cover ground, demand space; its story might cover a street, a wall, the surfaces of a house, a body. Bound, layered, the book, like the shawl, is a portable hiding place. It is an intimate space of pages pressed between covers, a place where geography and emotion, history and invention leave their traces, the long trail of handwriting folded in on itself.

In between floorboards

Up at the hiding place, Anne would take me to the curtained window, the curtains now very soiled. She would point out each new burst of green on the great chestnut tree behind the Annex. What an elegant tree it was, ablaze with rich green buds. Anne studied the progress of the buds each day, explaining to me how much bigger they were and how quickly they were ripening.

Miep Gies

Saturday, 12 February 1944   I believe that it's spring within me, I feel that spring is awakening, I feel it in my whole body and soul. It is an effort to behave normally, I feel utterly confused, don't know what to read, what to write, what to do, I only know that I am longing . . . !

In the strange and cruel paradox that characterizes hiding, physical growth itself – the body's positive movement into the future – endangers the future. In *The Shawl*, the literal expansion of body mass and the mobility that come with growth "give away" the child. When the child runs out into a public space, she is seen; this visibility, and the movement that is a sign of life, is death.

In the room in which the Franks slept, Otto Frank kept track of the Allies' advance on a map of Normandy. On the wall next to the map he kept track of his adolescent daughters' growth; lines on the wall, July 1942–August 1944. In this diminutive space, he left traces that celebrated the optimistic expansion of cells that, paradoxically, further endangered the family. New shoes were needed, goods purchased from salespeople who must not suspect; more food was bought from the vegetable man, who knew but said nothing, whose secret became larger, harder to contain; her footstep became heavier and there was more of herself to conceal when she crouched beneath the window to breathe the air coming through the crack.

The diary, over time, accumulates its dangerous traces, expands inside its cover. Over time, its writing poses an increasingly grave danger to the friends outside who are aiding them in hiding; it reveals more and more secrets about the strategies that make the families' lives possible. It contains, too, secrets of Anne's life on the inside; her changing body, her romance with another teenager in hiding with her, her struggles to conceal herself from and reveal herself to the others within their compressed space. Anne writes about her growing sense of autonomy and psychic distance from her parents; she negotiates that passage of adolescence even without physical distance, within a space of constant intimacy. The diary is the alternate location, and writing an act of making distance, of constructing another place. This writing is a generative act, constituting as well as recording the movements of becoming.

Where the molding and the wall pull apart

Nothing belongs to us anymore; they have taken away our clothes, our shoes, even our hair; if we speak, they will not listen to us, and if they listen, they will not understand. They will even take away our name; and if we want to keep it, we will have to find in ourselves the strength to do so, to manage somehow so that behind the name something of us, of us as we were, still remains.

Primo Levi, *Survival in Auschwitz*

Wednesday, 5 January 1994   After I came here, when I was just fourteen, I began to think about myself sooner than most girls, and to know that I am a "person." . . . Sometimes, when I lie in bed at night, I have a terrible desire to feel my breasts and to listen to the quiet rhythmic beat of my heart.

The childhood terror of hide-and-seek contains the pleasure of being found and returned to the world of appearances. Hiding inverts the crucial relationship of visibility and empathy, recognition and identity. A form of psychic death – disappearance – becomes a survival technique; obliteration holds the hope of existence. You must *give yourself away*: you must get rid of the familiar things by which the self is recognized in order to preserve existence, the necessary foundation of any future self. You get rid of your identity papers or get new papers; you conceal or mutilate your body's distinguishing marks. Not coughing, not crying out, not expressing grief or pleasure, not expressing desire or anger. You comply with the imperative to die, if to die is to disappear, to cease to occupy space, make noise, be known. Compliance and resistance are strangely folded into each other, form a binding opposition. To play dead is to stay alive.

Yet, in trying to play dead you become ever more conscious of being alive. Breathing, trembling, the pounding of the heart, the sweat under the arms: these cannot be withheld.

The diary records and embodies this paradox of self-awareness within self-obliteration. It describes as well the heightened self-consciousness, the acceleration of becoming, that characterizes adolescence. Writing, unlike breathing, is voluntary and consciously generated; but, like breathing, it maintains and evinces your existence. The diary resists the psychic effects of playing dead.

Through the little gap under the door

In post-exilic Judaism, but perhaps earlier, active reading, answerability to the text on both the meditative–interpretative and the behavioural levels, is the central motion of personal and national homecoming. The Torah is met at the place of summons and in the time of calling (which is night and day). The dwelling assigned, ascribed to Israel is the House of the Book.

George Steiner, "Our Homeland, the Text", *Salmagundi*, number 66

Monday, 7 December 1942 Chanuka and St. Nicholas Day came almost together this year – just one day's difference. We didn't make much fuss about Chanuka: we just gave each other a few little presents and then we had the candles. Because of the shortage of candles we only had them alight for ten minutes, but it is all right as long as you have the song. Mr. Van Daan has made a wooden candlestick, so that it was all properly arranged.

The old world was memorized, internalized, and carried into hiding within each person and unfolded through gestures: lighting the candles, making the bed, seeking privacy in which to wash. The world, stored as ritual, is maintained through reenactment. Inside the house behind, clocks were wound, birthdays celebrated, mealtimes scheduled; the families took time into hiding. Momentous events took place within the narrow confines of the rooms and were reenacted within the still narrower confines of the diary through another cultural ritual, the act of writing.

Inside the diary and the hiding place, inside the spaces of playing dead, a new world is invented from the materials at hand; not the world on the other side of the walls, but a ruptured, mutant world both within and without war. The space of playing dead is also a theater of everyday life.

This theater, like the psychic health it signifies, is massively compromised and contaminated by the real conditions of the inhabitants' lives; still, in the diary we read of the extraordinary efforts made by the inhabitants of the house behind and their protectors to maintain it. The hiding place is the container in which this world is generated, and every breach of the seal threatened the integrity of the interior. Danger always existed on the other side of the wall, an infinitely fragile yet significant membrane. Every movement across it threatened to eradicate the critical difference between inside and outside, to tear open the carefully constructed interior and let the world of the war pour in and their secret world pour out. Yet the families and their protectors took this risk to smuggle in birthday presents, books, desserts, flowers – the gifts of an endangered daily life and culture.

> *Thursday, 27 April 1944   What doesn't a schoolgirl get to know in a single day! Take me, for example. First, I translated a piece from Dutch into English, about Nelson's last battle. After that, I went through some more of Peter the Great's war against Norway (1700–1721), Charles XII, Augustus the Strong, Stanislavs Leczinsky, Mazeppa, Von Gorz, Brandenburg, Pomerania and Denmark, plus the usual dates.*
>
> *After that I landed up in Brazil, read about Bahia tobacco, the abundance of coffee and the one and a half million inhabitants of Rio de Janeiro, of Pernambuco and Sao Paulo, not forgetting the river Amazon, about Negroes, Mulattos, Mestizos, Whites, more than fifty percent of the population being illiterate, and the malaria. As there was still some time left, I quickly ran through a family tree. Jan the Elder, Willem Lodewijk, Ernst Casimir I, Hendrik Casimir I, right up to the little Margriet Franciska (born in 1943 in Ottawa).*
>
> *Twelve o'clock: In the attic, I continued my program with the history of the Church – phew! Till one o'clock.*

Like the inconspicuous diary cover, or the ordinary building exterior, routines and conventions provided an enfolding structure for a liminal and asynchronous existence. And like all of the gifts and dangerous necessities of passage, the world beyond had to be titrated, let in selectively, measured. Enough war news to remain vigilant and cautious, but not so much as to induce despair or madness. A psychological screen was in place, as essential and vulnerable as a curtain over a window. For as long as the inhabitants could, while the outside seeped and tore its way in, they kept waking on schedule, studying math, celebrating birthdays, marking the girls' growth on the wall. Paradoxically, adherence to conventional time and activity constituted an act of resistance. Learning French, since there is a future, brushing one's hair, since one desires to be seen. Insisting on the value of participation, of appearance even within the cloak of disappearance.

Het Achterhuis, the House Behind, the double interior of hiding and writing, is an ark of sorts, a mobile homeland.[4] Within the compressed rooms between concealing walls and within the secret spaciousness of the tiny book, the site is contained: the pictures of movie stars, the royal genealogy, the products of Brazil, French conjugations, those participatory knowledges that mark out a territory, a cultural ground. This dwelling, this *place*, can move. It is attached not to a piece of earth but to human activities, rituals, and actions that summon it, that constitute home wherever they occur, gestures from which it unfolds.

In a pocket of air

The historian who sides with the victim will be forced to remain within the event, going over it again from within. The historian who does not turn away when the walls which he or she has studied fall down can hope to overcome those limitations of history within the realm of memory.

Robert Jan van Pelt, "After the Walls Have Fallen Down",
Queen's Quarterly, number 96.

Wednesday, 29 March 1944 Bolkestein, an M. P., was speaking on the Dutch News from London, and he said that they ought to make a collection of diaries and letters after the war. Of course, they all made a rush at my diary immediately. Just imagine how interesting it would be if I were to publish a romance of the "Secret Annexe". The title alone would be enough to make people think it was a detective story. But, seriously, it would seem quite funny ten years after the war if we Jews were to tell how we lived and what we ate and talked about here. Although I tell you a lot, still, even so, you know only very little of our lives.

How scared the ladies are during the air raids. For instance, on Sunday, when 350 British planes dropped half a million kilos of bombs on Ijmuiden, how the houses trembled like a wisp of grass in the wind, and who knows how many epidemics now rage. You don't know anything about all these things, and I would need to keep on writing the whole day if I were to tell you everything in detail.

Diary writing offers psychological visibility; it involves the ongoing invention of a self who speaks and the projection of a receptive, listening other. Kitty, the invented character to whom the diary entries are addressed, is the intimate companion of a lonely adolescent who feels unseen.

But Anne is unseen on multiple levels; she lives *inside*. Her circumstances are designed expressly so that, outside of her immediate family and the four others in hiding with them, she will not be seen. She is invisible to the world. Even activities that are "public" within the life of the hiding place are "private," secret, withheld from the eyes of the world. Writing to Kitty is both the most private and most public act. Kitty is the private other, but she is also the world outside, a surrogate for all who cannot see Anne. Hiding inside for two years, Anne creates herself in the diary as she cannot in the street, as a public as well as a secret self, recording her concealed lives, the secret life of adolescence and the secret life of hiding.

This writing is public in another sense. Anne intends to publish an account of the years in hiding, and this intention pervades the diary. The seeds of that future book lie within the diary. Reading her detailed recordings of daily activities and world events, we remember that Anne considers herself a journalist in training, and her explanations of how the group manages the most mundane tasks in hiding seem clearly intended for a larger public. Her diary is ambivalent about its secrecy; it is not for her parents and the others with her to read but it is written for Kitty and for us, for those who live in the world beyond the walls of the House Behind. It is written toward the future. The absolute interiority of the diary, like the absolute interiority of the space, is compromised by the desire and the need to make contact with the world beyond, to be porous to its offerings and its dangers, to listen and to speak through the walls, the floors – through "the crack of a closed window."

The diary is open now like the opened building in the cutaway axonometric. In the revelation of their presence we feel the danger that exposure constitutes and the pain that looking in and looking back entails.

Between two pieces of paper

To be able to say: "I am in the book. The book is my world, my country, my roof, and my riddle. The book is my breath and my rest."

Edmond Jabes

Saturday, 12 February 1944   I feel as if I'm going to burst, and I know that it would get better with crying; but I can't, I'm restless, I go from one room to the other, breathe through the crack of a closed window.

In the seventeenth century, when the Dutch were most active in foreign trade and colonization, they constructed and displayed scale models of the interiors of well-appointed homes. Packed full of the necessities and pleasures of bourgeois existence, these miniatures display daily life and collect and domesticate the artifacts of the expanding world beyond the household. They are often contained within cabinets, concealed behind doors that swing open to expose the interior, revealing for our inspection the rooms, the furniture, the life inside. These replicas are monuments of self-recognition and moments of self-consciousness. They served as reminders of one's identity as participant, an identity embedded in the home. In them, the world beyond the home is gathered, internalized, returned home. In the ordering and tending of the dollhouse children practice adulthood. The Dutch models, toylike in scale, are reminders of childhood as well.

The diary is a place of appearance offering resistance to psychic disappearance. *Exile, fugitive, refugee*: these terms define a distance, a displacement from home. But the diary is a place to live, the house of the book, both internal and public, miniaturized and expansive. It is the concentric monument of the self at home in the world, if home is both the center of the world and the miniature of the world. The diary is the Dutch dollhouse.

177

Figure 10.2 Sara Rothé, Doll's house. Courtesy of the Haags Gemeentemuseum
Collection

"To live means to leave traces," wrote Walter Benjamin in 1935,[5] but even
as he wrote he was entering another world in which survival meant the
eradication of every trace. Behind all traces and all concealments lies the body
itself, the trace whose obliteration we cannot survive.

Exile, fugitive, refugee. Her handwriting is a reminder, a return.

Through the crack in a closed window

On that day in August 1944, Sergeant Silverbauer, of the Green
Police, and four subordinates were executing just another phase of
what Hitler and Goebbels had conceived as the "final solution of the
Jewish problem". Their mission was to remove the Jews from
Holland. While doing so, they were to loot and to plunder and, most
importantly, they were to leave no record or documentation of this
work.

George Stevens, preface to *Anne Frank:*
The Diary of a Young Girl

It was terrible, when I went up there. Not a soul in the place. The rooms suddenly looked so big. Everything had been turned upside down and rummaged through. On the floor lay clothes, papers, letters and school notebooks. And among the papers on the floor lay a notebook with a checked red cover. I picked it up, looked at the papers and recognized Anne's handwriting.

Miep Gies[6]

Acknowledgements

Thanks to Clare Cardinal-Pett, Steve Pett, and Thyrza Nichols Goodeve.

Notes

1 Anne Frank, 1953, *Anne Frank: The Diary of a Young Girl*, New York: Pocket Books. All subsequent extracts preceeded by dates are from this voume.
2 Charles-Edouard Jeanneret (Le Corbusier), c. 1910, "La construction des villes," unpublished manuscript; quoted in Anthony Vidler's *The Architectural Uncanny: Essays in the Modern Unhomely*, 1992, Cambridge, Mass.: MIT Press, p. 90.
3 A visitor's description of the annex, private correspondence.
4 "A mobile homeland" is George Steiner's phrase from "Our Homeland, the Text," in *Salmagundi: A Quarterly of the Humanities and Social Sciences* (Skidmore College), number 66.
5 Walter Benjamin, 1986, "Paris, Capital of the Nineteenth Century," in P. Demetz, ed., *Reflections*, New York: Schocken Books, p. 155. First brought to my attention by Beatriz Colomina in "The Split Wall: Domestic Voyeurism" in B. Colomina, ed., 1992, *Sexuality and Space*, New York, Princeton Architectural Press, p. 74.
6 Miep Gies, as quoted by Ernst Schnabel in *Anne Frank: A Portrait in Courage*, 1958, New York: Harcourt, Brace and Co., pp. 190–1.

References

Gies, Miep with Alison Leslie Gold (1987) *Anne Frank Remembered: The Story of the Woman Who Helped to Hide the Anne Frank Family*, New York: Simon and Schuster.
Jabes, Edmond (1976–1983) *The Book of Questions*, Middletown, CT: Wesleyan University Press.
Levi, Primo (1961) *Survival in Auschwitz, The Nazi Assault on Humanity*, New York: Collier.
Lifton, Robert Jay (1986) *The Nazi Doctors: Medical Killing and the Psychology of Genocide*, New York: Basic Books.
Steiner, George (1985) "Our Homeland, the Text," *Salmagundi: A Quarterly of the*

Humanities and Social Sciences, Saratoga Springs, NY: Skidmore College, 66, Winter/Spring.

Van Pelt, Robert Jan (1991) "Apocalyptic Abjection," in his *Architectural Principles in the Age of Historicism*, New Haven: Yale University Press.

3

ExcessingPlacesBodies

11

BEYOND NOMADISM

The travel narratives of a "cripple"

Michael L. Dorn

Susan Bordo (1993) notes that a current group of feminist researchers are adopting poststructuralist and deconstructivist methods, and, in the process, fleeing the grounding in the everyday advocated by an earlier generation. For Bordo and others, the proliferation of postmodernist metaphors of nomadism – hybrids, freaks, cyborgs, monsters, tricksters and *mestizas* – signals a dangerous retreat from an earlier emphasis on *embodied* located critique. In pursuing dreams of being everywhere at once, poststructuralist feminists reconceive the body: "No longer an obstacle to knowledge, . . . [but rather] the vehicle of the human making and remaking of the world, constantly shifting location, capable of revealing endlessly new points of view" (Bordo 1993: 227).

Agreeing with Susan Bordo's lament that some new feminist work refuses to remain fixed in a position for which it must accept responsibility, I intend to deepen her critique by examining ableist assumptions underling post-structuralist feminist dreams of cyborg identity and nomadic travel. I would like to show, first, how some poststructuralist feminists ignore institutional sedimentations in the built environment, assuming instead an apparently obstacle-less, frictionless plain of social interaction. Second, I intend to show how the "crippled" body exposes the ableist foundations of such postmodern renditions of nomadic forms of thought and movement, considering the life experiences of disability activist, designer and philosopher Patty Hayes.[1] Her story of coping with environmental and perceptual barriers and mastering the everyday challenges involved in learning to maneuver her wheelchair suggest the development of *geographical maturity*. By comparing Patty Hayes' story with a model of skill development advanced by Hubert and Stuart Dreyfus, I show how a "spatial dissident" exhibits a mature form of environmental sensitivity by remaining attentive and responsive to changing environmental conditions, in the process helping to chart new routes for others to follow.

Post-structuralist feminist figurations

> Unlike the hopes of Frankenstein's monster, the cyborg does not
> expect its father to save it through a restoration of the garden, that
> is, through the fabrication of a heterosexual mate, through its
> completion in a finished whole, a city and cosmos. . . . The cyborg
> would not recognize the Garden of Eden; it is not made of mud and
> cannot dream of returning to dust.
>
> (Haraway 1991: 151)

I am deeply concerned about the political implications of poststructuralist
feminists' efforts to romanticize nomad thought and hybrid embodiment.
Many poststructuralist feminist metaphors coined in the process – the
androgen, the freak, the monster and the cyborg, to name a few – help
produce a dream of multiple, fractured identities, at the expense of an
exploration of the wounds occasioned by the body and the political value of
situated identities. The literary figures of three poststructuralist feminists
– Donna Haraway, Elizabeth Grosz and Rosi Braidotti – exemplify a flight
from the messiness of disability into myth and metaphor. These writers
privilege the capacity for movement and change, arguing for a willful effacement of the boundary between nature and culture, fact and fiction. Abstract
identities, rather than hard-won standpoints, become ciphers in constant
negotiation.

In the poststructurist romance, colonialist stereotypes of the nomad life
are revalued;[2] nomads travel across trackless deserts and maintain constant
warfare against the sedentary and domestic. In order to trade between various
markets, the postmodern nomad must develop the ability to switch quickly
between different tongues. While linguistic proficiency may improve the
nomad's ease of movement and commerce, it may also invite an ambivalence
towards the sedentary and univocal. Identifying with this intellectual
nomadism, Braidotti (1994: 11, emphasis mine) notes autobiographically:

> Over the years, I have developed a relationship of great fascination
> toward monolingual people: those who were born to the symbolic
> system in the one language that was to remain theirs for the rest of
> their life. Come to think of it, I do not know many people like that,
> *but I can easily imagine them*: people comfortably established in the
> illusion of familiarity their "mother" tongue gives them.

The postmodern nomad harbors no desire for a "motherland." The only signs
of domestication that this nomad allows for are the fugitive tents pitched
against the evening cold on the desert, to be packed up in the morning: "as an
intellectual style, nomadism consists not so much in being homeless, as in
being capable of recreating your home everywhere" (Braidotti 1994: 16).

What seems so distinctive about this new-found "nomadic" flexibility is the sense of flight, movement, slipperiness, the flirting with radical non-belonging and outsideness. Yet what is the intellectual nomad trying to escape? At some points it seems that hers is a flight *from* philosophical responsibility rather than a flight toward *anywhere*. Rather than being consigned to a zone of "irrationality," the nomad seeks redefinition of "the structure and aims of human subjectivity in its relation to difference" (1994: 94). As an alternative to philosophical culpability, the nomad theorist revalorizes the colonial project of mapping frontiers. As Braidotti (1994: 16–17) writes:

> I think that many of the things I write are cartographies, that is to say a sort of intellectual landscape gardening that gives me a horizon, a frame of reference within which I can take my bearing, move about, and set up my own theoretical tent. . . . The frequency of the spatial metaphor expresses the simultaneity of the nomad status and the need to draw maps; each text is like a camping site: it traces places where I have been, in the shifting landscape of my singularity.

While using space as a textual metaphor, intellectual nomadism remains incommensurably abstracted from material conditions that would channel the travelling instinct – the lived, physical topography across which all traveling occurs (Katz 1997).

Flight from the material environment is similarly evident in post-structuralist feminists' fantasies of monstrous bodies. In describing configurations of cyborg identities, those especially tight couplings of humans and animals, humans and machines, Donna Haraway (1991) makes constant reference to the images of science fiction, arguing that they are little different from science fact. While many of Haraway's readers may have been shocked to realize the extent to which we each operate as cyborgs these days, it could be argued that this assertion would come as no surprise to disabled people, who throughout this century have found themselves wrapped tighter and tighter into the expanding bio-medical industrial complex. In fact, the extraction of surplus value from bodies *requires* their territorialization within taxonomies of difference, with differences upon differences rendered amenable to manipulation.[3] The Human Genome Project is only the most recent example of biomedical investment in technology devoted to the discernment of minute human differences for the purpose of socio-behavioral engineering.[4]

Biomedical emphasis on *preventing* disabilities is pushing the attribution and treatment for abnormality further back into the mother's womb. Braidotti (1994: 79) in her chapter "Mothers, monsters, machines" argues that new reproductive technology poses new threats to the rights of the mother.

The legal, economic, and political repercussions of the new reproductive technologies are far-reaching. The recent stand taken by the
Roman Catholic church and by innumerable "bioethics committees"
all across Western Europe against experimentation and genetic
manipulations may appear fair enough. They all invariably shift
the debate, however, far from the power of science over the women's
body in favor of placing increasing emphasis on the rights of the
fetus or of embryos. This emphasis is played against the rights of the
mother – and therefore of the woman – and we have been witnessing
systematic slippages between the discourse against genetic
manipulations and the rhetoric of the antiabortion campaigners.

Braidotti argues that groups challenging the ethics of new medical
technologies are increasingly allied with the Christian Right, fetuses, and
embryos to take abortion rights away from women. Yet it is important
to recognize that these ethical discussions take into account the very legitimate concerns of disabled people whose own congenital conditions make
them particularly interested observers and subjects of medical practices.
Many disabled people, parents and advocates feel that genetic counseling
– providing pregnant women with the results of pre-natal testing for
congenital "abnormalities" – too often takes place in a biomedical setting
with a dearth of information concerning the potential quality of life of
young people growing up with disabilities.[5] Considering that many
activists in the disability rights movement have congenital conditions
themselves, it is striking to consider what the spatial and material effects for
the 1970s political mobilization for barrier removal and access would have
been if these children had never been born or had been given "gene therapy"
in the womb until they matched an able-bodied DNA profile now being
compiled as part of the Human Genome Project.[6]

By tracing the conjunctions between mothers, monsters and machines,
Braidotti (1994: 75–94) hopes to illuminate biomedicine's ongoing investment in mechanizing the maternal function. The rationale for this investment
can be traced back to the linkage between femininity and monstrosity.
Braidotti discusses the anxiety of European patriarchs through the sixteenth
and seventeenth centuries revolving around the power of women in the realm
of reproduction. Medical doctors believed that the willful woman's imagination had the teratogenic power to obliterate "natural" resemblances between
father and child, rendering her fetus monstrous (see also Huet 1993).
Braidotti asks her readers why such belief in the power of human passions
in sexual difference did not lead, instead, to a position of recognition and
wonder? The association between the pregnant female body and the monstrous child, she claims, led to the placement of "woman" into a set of
hierarchical dichotomies of woman/man, monster/normal. "Woman" becomes
"the anomaly that confirms the positivity of the norm" (Braidotti 1994: 83).

Yet hierarchical dichotomies have not remained uniformly fixed through time and across space. At numerous junctures monstrosity has been understood as a *positive* excess (Hahn 1988). The history of disabilities provides many examples that challenge Braidotti's suppositions. According to Mikhail Bakhtin in *Rabelais and His World* (1968), the deformities of the disabled leper and the polyvocality of the mad fool exerted an erotic, subversive social force during periods of social upheaval up to the seventeenth century in Europe, evidenced, for example, in the ritual festivities of the pre-Lenten Carnival. While the carnival was eventually legislated out of the European landscape, its industrial era replacements – the dime museum, traveling freak show, and circus side-show attraction – retained popularity through the nineteenth century (Stallybrass and White 1986; Bogdan 1988).

More recently, carnival culture has been linked to the post-1968 resurgence of youth protest and continues to be vibrant in the works of science fiction and comic books (Fiedler 1978). Disabled people's dreams of liberation fermented in the "freak" culture of late 1960s and early 1970s Berkeley, California.[7] Yet poststructuralist feminists consistently neglect the role that disability typologies have played in "demonstrating" normative framings of the body politic. Thus, after mentioning the 1932 Tod Browning horror film *Freaks*, which promised its viewers that freaks were quickly disappearing from the world because of advances in medical diagnostics, Braidotti replays its apocalyptic claim: "the whole of contemporary popular culture is about freaks, just as the last of the physical freaks have disappeared" (1994: 92). I find it impossible to believe, given humans' unflagging ability to create discriminating distinctions based on appearance, that we are ever going to free ourselves from category-busting freaks and mutants. Yet Rosi Braidotti maintains that this has already happened. Almost as an afterthought she notes that "abnormally formed people have organized themselves in the handicapped political movement, thereby claiming not only a renewed sense of dignity but also wider social and political rights" (1994: 93). Braidotti makes it clear that this sense of dignity is only available to the disabled once the taint of monstrosity has been eliminated.

Elizabeth Grosz (1991, 1996) similarly finds freaks good to think with. In the article "Freaks," she explicitly rules out the possibility of considering the lives of the *banal* disabled, neither those with "more common-place bodily infirmities and deficiencies," those with "congenital abnormalities in internal or invisible organs," nor those who have experienced "accidental tragedies, in which individuals are maimed or wounded." These exclusions are rationalized by the fact that the banal disabled do not elicit the same response of simultaneous horror and fascination as found with "real" freaks. For Grosz, the *real* disabled are "unusually disadvantaged," and more likely to engender a response of *simple pity*. Grosz's idealization of freaks such as hermaphrodites and Siamese twins over the lived experience of the average disabled person serves as a striking contrast to, for example, the work of Iris Marion Young

(1990), whose "politics of difference" engages seriously with psychosocial and spatial aspects of *abjection* – the fascination overwhelmed by revulsion elicited by exposure to *any* "other" being that challenges one's own notion of bodily integrity. Grosz's freaks operate as distinct genres of hybrid identity, but are abstracted from the political economic environment in which they would be forced to live and interact. For Iris Marion Young, deformed, disabled bodies are considered in tandem with their material environments. Abjection is experienced as more than a psychic phenomenon – it also plays out at the level of the population to form distinctive socio-spatial topographies.

The patterns of social abjection traced by many disabled people's lives are being modified in the dawning cyborg society described by Donna Haraway (1991). Disabled people do not harbor dreams of escape to a mythic barrier- and deprivation-free past before the advent of ableist exploitation. Instead, they often strive to rework the biotechnological revolution to their advantage. Poststructuralist feminist dreams of deterritorialialized, dehierarchicalized bodies and the construction of hybrid relationships between humans and animals, humans and machines also draw on the everyday experience of living in post-Fordist, cyber-mediated society. Yet they are in flight from the bodies of disabled people – the bodies most embedded in these new space/power diagrams.

An alternative figure – the *spatial dissident* – can be retrieved from the writings of Julia Kristeva (1986) that rings true with my own involvement in the heterogeneous currents of the disability rights movement (Dorn 1994). Inspired by the "carnival" writings of the Russian literary critic Mikhail Bakhtin, Kristeva writes of an emergent dissident culture composed of the future unemployed, the presently unemployed and the students who will not be bought into morally bankrupt social formations in the first place. This occasion is characterized by "a new synthesis between sense, sound, gesture and colour, [when] the master discourses begin to drift and the simple rational coherence of culture and institutional codes breaks down" (Kristeva 1986: 294). The resultant crisis of representation will not be patched up from inside the academy. "From this point on, another society, another community, another body starts to emerge" (1986: 294). Kristeva's four principal proto- types for the dissident in today's society are (1) the rebel who attacks political power, (2) the psychoanalyst who "transforms the dialectic of law-and- desire into a contest between *death and discourse*" (1986: 295), (3) the writer who experiments with the limits of identity, and (4) woman with her mythic connection with the law of death and the contradictory experience of motherhood, where self-identity is experienced as doubled (1986: 296–8). I would like to extend Kristeva's definition of the dissident in order to consider the spatial experiences of the disabled person who finds new uses for commonly accepted spatial arrangements (Dorn 1994).

To articulate this point consider the difference between "dissent" and "dissidence" in their Latin roots (Field 1989). "Dissent" comes from the Latin *sentire*, to feel or to perceive. It refers to an individual's subjective experience which may or may not be expressed in popular discourse. "Dissent" does not form the basis for an oppositional standpoint, because it does not involve a major commitment of energy – "talk is cheap." "Dissidence," on the other hand, has an irreducibly spatial and physical aspect, as it comes from the Latin verb *dissidere*, to sit apart. In the ecclesiastical court of the Middle Ages, those who did not agree with court opinion would take a seat in a position removed from the majority. This carried over into the spatial organization of the English Parliament and the United States Congress, where opposing parties sit on either side of a center aisle (Field 1989). While in society dissent is generally tolerated because to eliminate it would require brainwashing, dissidence is much more threatening to the social order and therefore measures are taken to subjugate or naturalize it.

The architecture of public space enables a certain subset of the population to engage in commerce and rituals of citizenship. Other individuals find that these socio-spatial arrangements neglect their different needs and abilities. For these misfits, active dis-identification with the rituals of public power and "proper" use of public space becomes a legitimate, even sometimes unavoidable, option for resistance. The *spatial dissident* does not pass over the "trackless plains" scripted for the postmodern nomad. Nor does she follow the choreography of public ritual handed down to her. Rather, the dissident's tactic is to *doubt* these received cartographies and invent new uses for them. Her re-mappings expose new routes for others to follow.

In the next section, I will use the experiences of disability activist Patty Hayes to demonstrate what feminist analysis loses by ignoring the creative *spatial dissidence* of disability. By contrasting her unfolding self-awareness as a disabled woman with models of skill development advanced by Hubert and Stuart Dreyfus, I show how spatial dissidence requires a kind of geographical maturity. This is a form of being-in-the-world that is never complacent with the state of things, but sensitive and responsive to changing environmental conditions and willing to chart new lines of movement that others might follow.

Spatial dissidence

Discussions of expertise rarely question the equipmental context within which the development of expertise unfolds, or the various rules that enframe the activity that are called upon to accentuate the advantage of a given performer. Usually accounts of expertise take the body for granted. The potential expert's body is assumed to have a certain amount of intuitive talent that is then picked up and elaborated upon by society. But "talent" also requires an exchange between abilities and environment. Another version of talent might be the ability to adapt to inhospitable, volatile environments.

In discussions of ethical or moral maturity two opposing views prevail – those of Jürgen Habermas (1982) and Carol Gilligan (1982; 1986).[8] Their primary disagreement concerns the end point of what they deem to be moral maturity. For Habermas, moral maturity entails movement through stages of development: from lower stages of involved, situational response to ethical situations, to higher stages which assume the willingness and ability to consider moral questions from the hypothetical and disinterested perspective. Intriguingly, when the Kohlberg scale, on which Habermas's model is based, is tested on young men and women, men are found to be more mature in their moral responses than women. For Carol Gilligan, Habermas's ethical expert who speaks from a "justice" perspective is morally inferior to the young women whom she interviews. Although they may not be able to verbalize or explain their rationale for acting in a particular case, these women adopt a perspective of "care." Hubert and Stuart Dreyfus's (1990: 252) reconsideration of moral maturity reaffirms Gilligan's findings, suggesting that women develop an expertise in situational caring, one that does not require them to abstract themselves from an embodied situation. "Thus, when faced with a dilemma, the expert does not seek principles but, rather, reflects on and tries to sharpen his or her intuitions by getting more information until one decision emerges as obvious." Rather, he or she depends upon embodied responses and tactics drawn upon experiences from previous similar circumstances – a situational ethic of caring.

Hubert Dreyfus finds inspiration for his position in his reading of Heidegger (Dreyfus 1991) and in the phenomenological research that he has undertaken with his brother Stuart. By analyzing the phenomenology of such commonplace activities as learning to play chess or drive a car, Hubert and Stuart Dreyfus developed a series of "skills studies," published in their book *Mind Over Machine: The Power of Human Intuitive Expertise in the Era of the Computer* (1986), which modeled what happens to people who move beyond a level of mere competence to a level of expertise. They claim that in developing expertise in any skill a person moves through five stages (Table 11.1).

Let us now consider the situation of Patty Hayes in terms of the Dreyfus model of skill acquisition before we return to consider the relative merits of Habermas' and Gilligan's approaches to moral expertise.

Patty's story

Patty Hayes grew up in Sheboygan, Wisconsin during the 1950s in a middle-class nuclear family, a middle child between two brothers and a sister. Her role model while growing up was her strong-willed mother. Patty describes her mother as a pretty Southern Belle who aspired to bring about a second Victorian Age, "a feat almost feasible in this small Wisconsin town" (Hayes 1995). At the age of 17 Patty, a senior in high school, began to experience symptoms that appeared like MS to doctors (blurred eyesight and temporary

Table 11.1 Hubert and Stuart Dreyfus's (1990) 5-stage model of skill acquisition

Stage 1	Novice	The task is broken down into context-free rules; rigid attention to rules often leads to breakdowns.
Stage 2	Advanced beginner	Examples of meaningful additional components of environment added to the consideration of each situation; rules are broken down into maxims, applied to aspects of given situations.
Stage 3	Competence	Overwhelming numbers of features and aspects are being considered, leading to a hierarchical approach to decision-making; only relevant features and aspects of any situation are reviewed.
Stage 4	Proficiency	Going on emotions, the subject is "struck" by the plan after understanding without conscious effort what is going on.
Stage 5	Expert	Intuitive response to each situation; no need to reflect on the next appropriate action.

imbalances in her gait). Patty was misdiagnosed with MS for three years, receiving inappropriate drug therapy that exacerbated her congenital condition. Reflecting back on this period, Patty remembers feeling as though she was in a state of limbo.

After she was properly diagnosed as *ataxic* at the age of 20, Patty went on to finish college, marry and raise a family. During the first ten years after her diagnosis, Patty also attempted in vain to find a "cure" while experiencing a relentless decline of bodily functioning, due to nerve degeneration which resulted in a lack of muscle control. For the last fifteen years she has been using a wheelchair, experimenting with several different models. Now, a single parent raising two teenage children, Patty has also become an interior designer, a disability rights activist, and a graduate student in Philosophy at the University of Wisconsin, Madison. In her never-ending voyage of discovery as a "cripple," Patty Hayes describes a four-stage process of learning to interact with her environment:

1 Disableism and denial: attempting to ignore the chair;
2 Rehabilitationism: looking to experts for advice on how to make accommodations for her chair;
3 Independent functioning: designing for the *bodychair* in use;
4 Disability rights: taking her wheelchair out of the house and into the streets as a disability activist.

1 Disableism and denial

Patty's first five years in a wheelchair were ones of continual shock and surprise; she felt overcome with guilt that she had done something wrong. Contrary to common clichés, she found the wheelchair to be an eye-*closing* experience; she faced an overwhelming array of limitations: from an inability to drive a car or take public transportation, to the presence of architectural barriers preventing engagement in many arenas of public life. As limitations and barriers compounded each other, Patty found herself incapable of making plans. Unanticipated physical obstacles arose to greet her each time she arrived at a new place. She began to describe life with a motor disability as oppression in 3-D; vertical dimensions that she had previously viewed as molehills now were mountains.

Patty used to pride herself on maintaining a balance between mental and physical needs, a balance upset by her progressive disablement. She was now forced to devote attention to a third term – the chair. Something had to give:

> I used to have great respect for the separation of the mind and the body. And then along came the wheelchair. It seemed like the third wheel. Too much to deal with, so I wound up juggling. Always something was left out of the picture.
>
> (Hayes 1995)

The built social environment proved to be intrinsically *ableist*. Patty's uses the term ableism to refer to able-bodied people's fear and willful neglect of the disabled, often translating into inaccessible or uncompromising urban design. Since ableism ran strong in her own family, Patty could not retreat from a "ableist" public sphere into an "accommodating" private one. She uses the term *dis*ableism to refer to the fear and dissension *within* the ranks of the disabled brought about by denial of their own differences. For Patty, disableism kept her from recognizing others undergoing comparable difficulties and from learning through their example. Until she learned to deal with her fear, Patty was incapable of learning ways of navigating through her environment.

2 Rehabilitationism

After an initial five-year period denying her wheelchair, Patty began to accept a new image of herself and to look for ways to make physical accommodations for her wheelchair. She called upon experts to help her understand the new situation, experts who turned out to be members of the rehabilitation industry. As such, they offered pre-packaged solutions for modifying Patty's personal requirements, in a sense forcing her physically and emotionally to fit into an existing normative, ableist environment. During the next five years,

Patty adopted a rehabilitationist mindset and sought to maximize her physical functioning. Whereas in the first stage, Patty would not even deal with the chair, trying to deny it away, in the second stage she became conspicuously aware of the chair as an appendage with unwieldy spatial requirements.

Patty placed great faith in laws and institutions like the state's vocational rehabilitation system, which she saw as akin to ramps. Little concerned with the aesthetics of access, Patty accepted the built environment as intrinsically ableist, only asking that she be allowed to occupy a reasonable amount of space within it. This was when Patty bought a new house with her husband on Burleigh Street in Milwaukee, Wisconsin, and set about making it accessible. She used architectural access guidelines, for example, to design the layout of her kitchen. When it was completed, Patty could move through it with ease. Yet she had failed to consider the ergodynamics of tasks essential for cooking. Muscle strain, for example, prevented her from successfully transferring a steaming pot of spaghetti from the stove to the sink. One could say that Patty was in danger of becoming stranded in a disembodied Cartesian space, a geometrically metered container of her own design.

3 Independent functioning

Nonetheless, Patty soon experienced her "going Copernican" phase. As she became a more capable designer, Patty began to redesign her life to maximize *independent* functioning. Over the next five years, Patty finalized her divorce from her husband and began to explore her identity as a cripple. Rather than listening exclusively to experts, Patty learned to listen to her own body knowledge and intuitions. She also acquired a new wheelchair, something key to her conversion from absolute to relative understandings and negotiations of space. Whereas her first three chairs were foldable, a new chair was made according to her specifications with a rigid frame, which was more compact and responsive to her body movements. Overall, Patty felt the chair gave her more stability and confidence; she began to work with the chair rather than fight against it. This allowed her to begin to resynchronize the three terms: body, mind and chair.

> The wheelchair, if you will, has become my external skeleton, and I do rely on it. That is why it is very important to be as small and trim as possible, but at the same time to have enough protection. I realize, in other words, that I have to function *while in the chair*. It's not just accommodating the outer measures of the wheelchair. That greatly effects the way you envision design, especially in terms of kitchen and bathroom.
>
> (Hayes 1995)

193

As the chair was incorporated into Patty's self-identity, it was libidinally transformed from an unwieldy appendage into a form-fitting exo-skeleton. The body and the chair began to meld together, changing the dualistic relations in personal navigation from mind-and-body to mind-and-bodychair. Whereas before the environment was seen through an architect's Cartesian grid, a combination of regulated geometrical parameters – doorways of a certain width and hallways of a specific turning radius – now space was modified through its use.

The transition from rehabilitationism to independent functioning was reflected in the design of Patty's house on Burleigh Street. Disability access was no longer simply an afterthought or an add-on. One of the first surprises for visitors today is how open the house is; the first architectural elements to go were the interior doors, and significant changes were made in the kitchen. When first adapting the house, Patty had been content to allow gas lines and water pipes to determine the placement of the stove and the sink. But then she still had a big problem. Her *chair* would fit in her kitchen, but not *herself in the chair, making use of the kitchen*. To drain water from boiled spaghetti, Patty would have to carry the pot from the stove across to a sink at the same height, even though her motor coordination (declining due to the ataxia) left her in continual danger of being scalded. Patty's "Copernican" solution was to enlarge and restructure the kitchen, in part by converting an old pantry into a more compact cooking space. A stove top was designed with a lip, allowing her to get close to the stove by providing space for her bodychair. Moreover, she designed the space such that when she faced the stove top, the sink was positioned directly behind her. Most importantly, a drain was built into the floor, allowing her to pour out liquids with ease – of special importance when dealing with liquids of high temperature. This is just one of numerous sets of changes that Patty enacted to recast the ableist space of the traditional house and fight her own disablist complacency.

In her dissident consumption of traditional kitchen space, Patty constructed new possibilities through productive juxtapositions of use, as when she placed an industrial drain in the floor so that she could work *with* gravity rather than against it, or when she aligned cupboards and stovetop in ways that allowed her to roll up to them. Rather than a refugee in the able-bodied (AB) spaces handed down to her, Patty began to think of herself as an expert traveler and an interior/exterior designer of her own "crippled" spatiality.

4 Disability rights

After fifteen years in the wheelchair, Patty looks forward to the future and to becoming more of an activist fighting for disability rights. The third stage's goal of independent functioning doesn't hold quite the same attraction for Patty that it once did. Certain problems can never be adequately addressed if Patty relies on her own strength alone. Patty is beginning to talk about the great merits to living cooperatively, allowing one to deal communally with difficulties which individuals cannot address by themselves – like the leaking roof two stories above her. In order to stretch her zone of enablement beyond her house, Patty has become involved in community, state and national politics. Laws like the Fair Housing Act amendments (1988) and the Americans with Disabilities Act (1990) greatly benefit the disabled if they are enforced. But unfortunately, enforcement of these laws is consumer-driven. Patty sees herself as something of a pioneer venturing out of the disability ghetto of public subsidized housing to find a place for herself and her family in the private housing market. As a pioneer, she has had to fight for her access rights. She brought a suit against her previous landlord, for example, when he did not allow her to make reasonable changes that could have improved the functionality of her apartment, and did not provide access to important services at the apartment complex.[9]

Lack of access in private housing stock helps reproduce the ghettoization of disabled people. In order to fight against ghettoization, Patty is involved in a radical housing advocacy group, Concrete Change, fighting for the recognition of basic access standards in publicly and privately-built housing. Concrete Change promotes no-step entrances and 32 inch doors in *all* homes built in the United States so as to provide disabled and frail elderly people with options located outside the cramped market for subsidized public housing. Accordingly, Eleanor Smith, founder of Concrete Change, has succeeded in getting the city council of Atlanta, Georgia to pass a "no-step" ordinance, but like so many other access laws, it is not adequately enforced. Lobbying non-profit home builders, particularly the nationwide Christian builder of low-cost housing Habitat for Humanity, has been more effective (Smith 1993). These groups have come to share the realization that building homes alone does not build a strong neighborhood. It is only in their use by people who can visit each other that a neighborhood becomes a community.

Through her involvement with Concrete Change and disability rights activism, Patty is taking her bodily struggles for access and mobility out of her house and into the streets, with the goal of changing her neighborhood. Rather than being satisfied with having her house being the only one (of thirty) on her street that is accessible, she would like a disabled child moving into her neighborhood to be able to visit neighbors. Applying her design and legal experience to barriers in the broader society, Patty endeavors to open up architectural environments to difference, extending the field of barrier-free travel in Milwaukee and other cities around the country. Patty is therefore invested in actively rechanneling the flows of knowledge and power that reproduce ableist barriers. It is a fight Patty believes she has been avoiding for too long.

Geographical maturity

How does Patty's story compare to the Dreyfus model of skill acquisition? They are pretty much consonant until we reach the final Dreyfus stage of "expert."

> The proficient performer, immersed in the world of skillful activity, *sees* what needs to be done, but must *decide* how to do it. With enough experience with a variety of situations, all seen from the same perspective but requiring different tactical decisions, the proficient performer seems gradually to decompose this class of situations into subclasses, each of which have the same decision, single action, or tactic. This allows an immediate intuitive response to each situation.
>
> (Dreyfus and Dreyfus 1990: 242–43, emphasis in original)

According to their definition, Patty would probably not be considered an expert in managing the demands of wheelchair and environment: she has not developed the ability to decompose her environmental interactions into a group of subclasses, nor has she developed a large portfolio of consistent, reliable tactical responses. Even within her house, new situations arise from time to time as her body's nervous system degenerates. Operating today in accord with yesterday's intuitive and bodily responses, she finds that the high shelves of her cupboards have become inaccessible – "microshock."

If she had been able to develop nonreflective intuitive expertise, we might feel tempted to describe Patty as a postmodern nomad moving lightly across an urban wasteland. However, from Patty's testimony, she certainly does not *feel* like an expert traveler. Rather than eliminating startling situations, her body continues to propagate more sophisticated and detailed ones. Each of her travels exposes a changing set of the barriers posed by an ableist urban fabric. As a disabled traveler and spatial dissident her travel is also a *travail* or labor. Each route of movement must be secured in an environment that presents new, ever more personalized barriers. Yet for all its rigors, it is a liberating travel, since it leaves physical conditions changed in ways that make it easier for herself and fellow dissidents to maneuver in the future.

Why do difficulties arise when we try to apply Hubert and Stuart Dreyfus's model of skill development to Patty? Their model assumes an equipmental context that is pretty much taken for granted (see Dreyfus 1996). Rules and structures that serve as the equipmental context of car-driving and chess-playing, for example (the prototypes for the skill studies in Dreyfus and Dreyfus 1986) rarely change. That is, traffic laws and chess rules stay consistent from year to year; major innovations in the design of traffic arteries or chess boards don't come along that often. In Patty's case, however, she lives in an urban culture that is constantly up for grabs and contested.

Handicapped access laws are unevenly enforced, and even when they *are* enforced, laws are not guaranteed to meet the needs of all disabled people. Some disability activists talk of creating a universally accessible city where all spatial barriers to mobility are gone and disabled people are fully integrated into society (see Dorn 1994). Dreams of completely eliminating the adverse environmental implications of disability through better engineering are flawed in practice because of the volatility of human needs and abilities. In the city of "universal accessibility" we could all be postmodern nomads, indifferent to corporeal and topographical situation. But the figure of the dissident refuses such panaceas. Spatial dissidents haunt the boundaries of the normative, bearing witness to the unfinished nature of all human constructions and pointing the way to a better future.

Patty's identity as a cripple is entirely bound up in the nature of her equipmental context. Her oppression finds material, spatio-temporal expression. Large zones of her environment are unsuitable for attaining even minimal competence as a traveler. Patty's hopes to move through competent performance towards proficiency and expertise depends upon more than just her own intelligence and strength. It depends to some extent on the willingness of other disabled people and non-disabled activists to fight to change rules that prop up certain configurations of the urban terrain and its equipment.

I would like to suggest that Patty has developed a form of *geographical maturity* in dealing with the demands placed on her by her body and her environment. Geographical maturity conforms with neither of the models of moral maturity advocated by Jürgen Habermas and Carol Gilligan. Patty does not have the luxury of appealing to either universal norms of justice or dependable intuitive bodily responses. If her development into an expert of her own crippled spatiality is to continue, then she must continue to tack between an involved, deliberative and intuitive assessment of specific situations met while maneuvering her bodychair, and a kind of detached assessment of strategies required to intervene at the level of law and patterns of production in the urban space-economy. This movement between involved assessment and legal intervention is what Patty is referring to when she insists on the need to become more active in the body-politics of disability rights. Only by sharing the load with others and acting as part of an organized body-politic, advocating with the benefit of wisdom gained by living with a disability, can Patty extend her domain of travel beyond those areas, like her house, where she exercises more immediate, personal control.

The implications of Patty's story for geographers of disability

To close, I would now like to discuss the implications of Patty Hayes' travel narratives for the way we conceive of geography for the disabled (Golledge

1993). We get into trouble assuming that because geographers are supposedly "spatially aware" that they necessarily have the tools to improve the quality of life of the disabled. The dominant experience of disabled people when they place themselves in the hands of professionals is one of knowledge denial rather than knowledge enhancement (Laws 1994; Dyck 1995, 1998; Imrie 1996a) as their ways of knowing and accounting for their experience are devalued as insufficiently "dispassionate" and "objective." All too many programs and services for the disabled *presume* passivity on the part of the consumer – a blank slate upon which institutional protocols are inscribed. The denigration of crippled ways of knowing is reproduced by psychologists, philosophers and geographers who assume that geographical maturity must be demonstrated through the ability to apply abstract universal laws and Cartesian coordinates when making decisions and planning trips.[10]

Laws only live through material inscriptions in bodies and places. They can thereby be contested through spatial praxis. Patty's everyday activism is synaptic and situational, tactics deployed on the fly, with transformative effects. While to the detached viewer, Patty occupies an *ataxic* (disordered) social space far different from the norm, the engaged viewer realizes that Patty has learned to produce her own sense of purposeful travel – gesturing, moving, signifying and skillfully coping with each new obstacle. Such a dissident does not necessarily travel alone. By assembling their subjugated "crippled" ways of knowing, the disabled become part of a counter-hegemonic field. In clearing obstacles that bar her path, Patty is fully aware of the benefits derived by fellow friends and travelers. Newly accessible public spaces become sites for dialogue between the disabled and the able-bodied. Operating as a part of this field, Patty's actions resonate across wider domains, effecting broader societal changes. For the worker who becomes temporarily disabled or experiences a decline of bodily functioning with age, the access elements designed into his house because of the campaign by Concrete Change mean that he will not be forced to move due to architectural barriers. The elderly resident and the young mother carrying her child in a stroller can both use curb cuts added to the sidewalks in Patty's neighborhood. A young man with cerebral palsy is no longer forced to grow up in a nursing home due to the lack of accessible (physically and financially) housing in his city. Geographies of the disabled must respect an alternative geographical maturity, no matter how apparently *ataxic* it appears in cartographic representations.[11]

Significantly, Patty's Copernican revolution from body *and* chair to *bodychair* took place precisely when she was able to develop trust in her own instincts and inclinations, intervening and rearranging the supposedly universal, external laws, guidelines and codes of her urban and domestic environment. Yet, physically and mentally incapable of taking elements of her environment entirely for granted, Patty turns to the realm of law when she wishes to force her way into or out of a set of overly rigid spatial boundaries.

Once these boundaries have been opened, she will then tack back to the level of spatially "lawful transgressions" to go about her everyday discursive and material practices – reusing elements from the haphazard *bricolage* of current urban environments to speak a new story about herself and city in which she lives (Nast and Wilson 1994).

Patty's "challenges" are not essentially different, but rather more systematic than those faced by most of us in day-by-day activities. Levels of skillful coping become so familiar to us that they pass below the threshold of reflection. Traveling with Patty through an urban environment uproots commonsense assumptions about what it takes to make a community. In her own way, Patty is a frontier surveyor, clearing or removing hurdles from her way as she recolonizes an urban jungle. In the process of deterritorialization (getting new houses built with no-step entrances and 32" doors) and reterritorialization (redesigning the kitchen for her personal use) she lays down new crip-tracks for others to follow.

Acknowledgments

The various elements of this chapter came together during a particularly fruitful semester at the University of Kentucky in the Spring of 1995. The UK Committee on Social Theory brought a series of speakers including Elizabeth Grosz and Hubert Dreyfus to campus for public lectures and seminar discussions around issues of "incorporating" Enlightment notions of reason and rationality. I would like to thank Hubert Dreyfus for agreeing to participate in, and the graduate editorial collective of *disClosure* for sponsoring, an interview that was eventually published in their journal (Dreyfus 1996) addressing the potential for extending the Hubert and Stuart Dreyfus (1986) skill studies to an analysis of the development of ethical expertise. Heidi Nast inspired me to analyze the politics of "nomad thought" and contrasting modes of travel in the Spring 1995 women's studies/ geography graduate seminar, "Travel Writing and Imperialism" and pushed me to apply these theoretical insights to a concrete case study. Finally, I must thank Patty Hayes for welcoming me into her life in Milwaukee, Wisconsin so that I could observe her environmental interventions and interview her about the fifteen-year process of "coming out" as a disabled woman and mother. While each of these individual and collective bodies has contributed, I alone accept responsibility for the analysis in its present form.

Notes

1 Never one to sugar-coat the truth, Patty Hayes prefers to identify herself as a "cripple" rather than resort to such euphemistic, but untimately vacuous terms as "physically challenged" or "handi-capable." These terms get coined by

human service agencies to boost the "self-esteem" of their clients. In their lack of specificity and denial of the political dimension of disability, these terms are both obfuscating and condescending (for another personal discussion of the politics of "crippled" identity, see Mairs 1996).

2 These writers did not invent "nomadism." Poststructuralist feminist nomadism draws heavily on the two volume work of Gilles Deleuze and Félix Guattari, *Capitalism and Schizophrenia*: (1) *Anti-Oedipus* (1983); (2) *A Thousand Plateaus* (1987). These French political theorists fashion an "intellectual nomadism" from a wide array of empirical materials. In a detailed reading of the footnotes to *A Thousand Plateaus*, Christopher Miller (1993) demonstrates how Deleuze and Guattari build their textual authority by referring to studies of "actual" nomadic tribal peoples by colonial anthropologists. Like the feminists who appropriate their work, Deleuze and Guattari claim that their nomadism bears very little resemblance to nomadism as understood by anthropologists, while building their nomad mythology from stereotyped depictions of nomadic societies. Caren Kaplan (1996: 68) draws connections between the nomadism of Deleuze and Guattari, and that of Jean Baudrillard, in "the specificity of their modernist critical traditions along with an inability to account for the transnational power relations that construct postmodern subjectivities." According to Kaplan (1996), critics of Deleuze and Guattari's Eurocentrism and their positionality in their writings were initially marginalized in the enthusiastic American reception to the translation and publication of their work by the University of Minnesota Press.

3 The potential medical and commercial windfall for mapping and sequencing the genetic material found in the human DNA molecule is enormous, inspiring intense competition between American, European and Japanese research teams. Insurance companies will also insist that they have access to genetic information in order to *discriminate* between individuals who might represents different amounts of genetic risk (Greely 1992; Pokorski 1994). "What insurers fear most in the future is that people will learn of important, personal genetic information outside the context of insurance and then successfully use this medical knowledge to gain an undue advantage in the application process" (Pokorski 1994: 108).

4 Charles Benedict Davenport, a prominent American biologist and eugenicist who served as head of the Carnegie Institution of Washington's Department of Genetics at Cold Spring Harbor on Long Island, New York, found amongst certain family genealogies in his registry a recessive trait of "nomadism." His understanding of nomad ways was drawn on contemporary anthropological accounts of "real" nomads (Davenport 1915). Such evidence provided by eugenics researchers helped convince twenty-nine American states and one territory to pass laws legalizing forced sterilization of the "feeble-minded" before the Second World War (Pepenoe 1938: 202).

5 The literature featuring this perspective is now extensive. Essential works include Asch and Fine (1984), Saxton (1984), Asch (1989), Finger (1990), Bérubé (1996). Disability magazines and journals have devoted special issues to the topic: *Disability Studies Quarterly* (Asch 1993); *The Disability Rag* (Asch 1994; Blumberg 1994; Shaw 1994; Woodward 1994).

6 A healthy debate over the practice of prenatal diagnosis and selective abortion

is taking place between activists in the disability rights movement and bioethicists (Wertz and Fletcher 1989, 1993; Steinbock and McClamrock 1994; Botkin 1995; Asch and Geller 1996). The 1997 meeting of the Society for Disability Studies featured an historic congress of activists/academics in disability studies and bioethicists working with the Hastings Center Project on Prenatal Testing for Genetic Disabilities (Pfeiffer 1997). Their discussions were oriented toward characterizing the effects of this practice on the goals of people with disabilities to achieve full social equality. "If it is harmful, in what way is it harmful? Is there a way to respect the dignity of a life with disability and to respect the desires of prospective parents to make reproductive choices?" (Asch 1997). These dilemmas of biomedicalized society will not be easily brushed aside, as disabled people around the world increasingly bring them to the theater of public and medical debate.

7 As Lesley Fiedler (1978) notes, the "new mutants" of the 1960s expunged the boundaries between political radicalism and personal deviancy. Freak out music and freak comic books celebrated the new identity and developed the new language; on 18 April, 1969, the freaks decided to open up the window and let some fresh air into the heart of Berkeley and the University of California: lot 1875, the People's Park. "We want the park to be a cultural, political, freak-out rap center of the Western World" (Mitchell 1992). These lessons of spatial re-territorialization were not lost on a group of physically disabled students at the University of California known as the Rolling Quads. In 1972, after success in developing the Physically Disabled Students' Program on campus, group members Ed Roberts and Phil Draper moved off-campus to found the world's first Center for Independent Living (CIL) in Berkeley (Levey 1988; Shapiro 1993).

8 These opposing models are described in Dreyfus and Dreyfus (1990).

9 "Access" for disabled people may be constrained either *physically* or *financially*. The specific physical access needs vary between disability groups, but a few are crucial for the motor impaired. Barriers to be removed may be as simple as a single step at the building's entrance or overly narrow doors to the restrooms. Physical access may include a well-funded public transit and/or paratransit system in the community. Financial barriers may compound the situation. Often the least expensive houses are also the least accessible ones.

10 Based on research in cognitive psychology (Piaget 1960; Golledge et al. 1979), behavioral geographers have recently begun to develop wayfinding systems for the visually impaired (Golledge 1993; Loomis et al. 1993). Cognitivist beliefs that behavior follows a set of rules that can be mapped, and that by building upon these insights human reasoning capacities may eventually be exhibited by artificial intelligence (AI) systems have been critiqued by Hubert Dreyfus (1990, 1996). The assumptions underlying the behavioral geography approach advocated by Golledge (1993) – research for the disabled – has been severely criticized by a younger generation of geographers advocating research *with* and *by* the disabled (Butler 1994; Gleeson 1996; Imrie 1996b; Kitchin 1997; Parr 1997; for replies, see Golledge 1994, 1996).

11 Several examples can be cited of ataxic mapping by geographers. Some, like Golledge (1990), apply cognitivist, Cartesian norms in studying the travels of persons with mild mental retardation and vision impairment. Others, however,

explore more dialogically the cartographic needs and environmental values of wheelchair users (Vujakovic and Matthews 1994; Matthews and Vujakovic 1995), and the artistic visions of the institutionalized mentally disabled (Park *et al.* 1994).

References

Asch, Adrienne (1989) "Reproductive technology and disability" in Sherrill Cohen and Nadine Taub (eds) *Reproductive Laws for the 1990s* (Humana Press: Clifton, New Jersey), pp. 69–124.

Asch, Adrienne (1993) "The human genome project and disability rights: thoughts for researchers and advocates" *Disability Studies Quarterly* 13(3): 3–5.

Asch, Adrienne (1994) "The human genome and disability rights" *The Disability Rag* Jan/Feb: 12–15.

Asch, Adrienne (1997) President's letter to members of the Society for Disability Studies.

Asch, Adrienne and Michelle Fine (1984) "Shared dreams: a left perspective on disability rights and reproductive rights" *Radical America* 18(4): 51–8. Reprinted in Michelle Fine and Adrienne Asch (eds) *Women With Disabilities: Essays in Psychology, Culture and Politics* (Temple University Press: Philadelphia, 1988), pp. 297–305. Reprinted in M. G. Fried (ed.) *From Abortion to Reproductive Freedom: Transforming a Movement* (South End Press: Boston, 1990), pp. 233–40.

Asch, Adrienne and Gail Geller (1996) "Feminism, bioethics and genetics" in Susan M. Wolf (ed.) *Feminism and Bioethics: Beyond Reproduction* (Oxford University Press: New York), pp. 318–50.

Bakhtin, Mikhail (1968) *Rabelais and His World*, Helen Islowsky, trans. (MIT Press: Cambridge, MA).

Bérubé, Michael (1996) *Life as We Know It: A Father, a Family, and an Exceptional Child* (Pantheon: New York).

Blumberg, Lisa (1994) "Eugenics vs. reproductive choice" *The Disability Rag* Jan/Feb: 3–11.

Bogdan, Robert (1988) *Freakshow: Presenting Human Oddities for Amusement and Profit* (University of Chicago Press: Chicago, IL).

Bordo, Susan (1993) *Unbearable Weight: Feminism, Western Culture and the Body* (University of California Press: Berkeley, CA).

Botkin, J. R. (1995) "Fetal privacy and confidentiality" *Hastings Center Report* 25(5): 32–9.

Butler, Ruth (1994) "Geography and vision-impaired and blind populations" *Transactions of the Institute of British Geographers* 19: 366–8.

Braidotti, Rosi (1994) *Nomadic Subjects: Embodiment and Sexual Difference in Contemporary Feminism* (Columbia University Press: New York).

Davenport, Charles Benedict (1915) *The Feebly Inherited; Nomadism, or the Wandering Impulse. . . .* (Carnegie Institution of Washington: Washington, DC).

Deleuze, Gilles and Félix Guattari (1983) *Anti-Oedipus: Capitalism and Schizophrenia*, Michel Foucault, preface, Robert Hurley, Mark Seem, and Helen R. Lane, trans. (University of Minnesota Press: Minneapolis, MN).

Deleuze, Gilles and Félix Guattari (1987) *A Thousand Plateaus: Capitalism and*

Schizophrenia, Brian Massumi, trans. (University of Minnesota Press: Minneapolis, MN).

Dorn, Michael L. (1994) "Disability as spatial dissidence: a cultural geography of the stigmatized body" unpublished masters thesis from the Department of Geography, The Pennsylvania State University, University Park, PA.

Dreyfus, Hubert L. (1991) *Being-in-the-World: A Commentary on Heidegger's* Being and Time, *Division 1* (MIT Press: Cambridge, MA).

Dreyfus, Hubert L. (1996) "Becoming skilled in doing what's appropriate: the non-reflective rationality of ethical expertise" an interview conducted by Robert Shields, Michael L. Dorn, Dan Schuman and Dorothee Seifen *disClosure: a journal of social theory* 5: 3–26.

Dreyfus, Hubert L. and Stuart Dreyfus (1986) *Mind over Machine: The Power of Human Intuitive Expertise in the Era of the Computer* (Free Press: New York).

Dreyfus, Hubert L. and Stuart E. Dreyfus (1990) "What is morality? A phenomenological account of the development of ethical expertise" in David Rasmussen (ed.) *Universalism vs. Communitarianism: Contemporary Debates in Ethics* (MIT Press: Cambridge, MA), pp. 237–64.

Dyck, Isabel (1995) "Hidden geographies: the changing lifeworlds of women with multiple sclerosis" *Social Science and Medicine* 40(3): 307–20.

Dyck, Isabel (forthcoming) "Women with disabilities and everyday geographies: home space and the contested body" in R. A. Kearns and W. M. Gesler (eds) *Putting Health into Place: Landscape, Identity and Wellbeing* (Syracuse University Press: Syracuse, NY).

Fiedler, Leslie (1978) *Freaks: Myths and Images of the Secret Self* (Simon and Schuster: New York).

Field, M. G. (1989) "Dissidence as disability: the medicalization of dissidence in Soviet Russia" in W. O. McCagg and L. Siegelbaum (eds) *The Disabled in the Soviet Union: Past and Present, Theory and Practice* (University of Pittsburgh Press: Pittsburgh, Pennsylvania), pp. 253–76.

Finger, Anne (1984) "Claiming all of our bodies" in R. Arditti, R. Klein and S. Minden (eds) *Test-Tube Women: What Future for Motherhood?* (Boston: Pandora Press).

Finger, Anne (1990) *Past Due: A Story of Disability, Pregnancy, and Birth* (Seal Press: Seattle, Washington).

Gilligan, Carol (1982) *In a Different Voice: Psychological Theory and Women's Development* (Harvard University Press: Cambridge, MA).

Gilligan, Carol (1986) "On *In a Different Voice*: an interdisciplinary forum" *Signs: Journal of Women in Culture and Society* 11(2): 327.

Gleeson, Brendan J. (1996) "A geography for disabled people?" *Transactions of the Institute of British Geographers* 21: 387–96.

Golledge, Reginald G. (1990) "Special populations in contemporary urban regions" in John Fraser Hart (ed.) *Our Changing Cities* (John Hopkins University Press: Baltimore), pp. 146–69.

Golledge, Reginald G. (1993) "Geography and the disabled: a survey with special reference to vision impaired and blind populations" *Transactions of the Institute of British Geographers* 18: 63–85.

Golledge, Reginald G. (1994) "A response to Butler" *Transactions of the Institute of British Geographers* 19: 369–72.

Golledge, Reginald G. (1996) "A response to Gleeson and Imrie" *Transactions of the Institute of British Geographers* 21: 404–11.

Golledge, Reginald G., J. J. Parnicky and J. N. Rayner (1979) "An experimental design for assessing the spatial competence of mildly retarded populations" *Social Science and Medicine* 13D: 292–5.

Greely, Henry T. (1992) "Health insurance, employment discrimination, and the genetics revolution" in Daniel J. Kevles and Leroy Hood (eds) *The Code of Codes: Scientific and Social Issues in the Human Genome Project* (Harvard University Press: Cambridge, MA), pp. 265–80.

Gregory, Derek (1994) *Geographical Imaginations* (Blackwell: London).

Grosz, Elizabeth (1991) "Freaks" *Social Semiotics* 1(2): 22–38.

Grosz, Elizabeth (1994) *Volatile Bodies: Towards a Corporeal Feminism* (Indiana University Press: Bloomington, IN).

Grosz, Elizabeth (1996) "Intolerable ambiguity: freaks as/at the limit" in Rosemarie Garland Thomson (ed.) *Freakery: Cultural Spectacles of the Extraordinary Body* (New York University Press: New York), pp. 55–66.

Habermas, Jürgen (1982) "A reply to my critics" *Habermas Critical Debates* (MIT Press: Cambridge, MA), pp. 218–83.

Hahn, Harlan (1988). "Can disability be beautiful?" *Social Policy* 18 (3): 26–32.

Haraway, Donna (1991) "A cyborg manifesto: science, technology and socialist feminism in the late 20th century" in *Simians, Cyborgs, and Women: The Reinvention of Nature* (Routledge: New York), pp. 149–81.

Hayes, Patty (1995) Personal interview with Michael L. Dorn, Milwaukee, Wisconsin, March 12.

Huet, Marie-Hélène (1993) *Monstrous Imagination* (Harvard University Press: Cambridge, MA).

Imrie (1996a) *Disability and the City: International Perspectives* (St. Martin's Press: New York).

Imrie, Rob (1996b) "Ableist geographers, disablist spaces: towards a reconstruction of Golledge's 'Geography and the disabled'" *Transactions of the Institute of British Geographers* 21(2): 387–96.

Kaplan, Caren (1996) *Questions of Travel: Postmodern Discourses of Displacement* (Duke University Press: Durham, NC).

Katz, Cindi (1997) "Creating safe spaces and the materiality of the margins" an interview conducted by Vincent DelCasino, Mike L. Dorn and Carole Gallaher, *disClosure* 6: 37–55.

Kitchin, Robert M. (1997) "A geography *of, for, with* or *by* disabled people: reconceptualizing the position of geographer as expert" SARU Working Paper 1, School of Geosciences, Queens University of Belfast, Northern Ireland.

Kristeva, Julia (1986) "A new type of intellectual: the dissident," S. Hand, trans. in T. Moi (ed.) *The Kristeva Reader* (Columbia University Press: New York), pp. 292–300.

Laws, Glenda (1994) "Oppression, knowledge and the built environment" *Political Geography* 13(1): 7–32.

Levey, C. W. (1988) *The People's History of the Independent Living Movement* (Research and Training Center on Independent Living: Lawrence, Kansas).

Loomis, Jack M., Roberta L. Klatsky, Reginald G. Golledge, Joseph G. Cicinelli, James W. Pellegrino and Phyllis A. Fry (1993) "Nonvisual navigation by blind and sighted: assessment of path integration ability" *Journal of Experimental Psychology* 122(1): 73–91.

Mairs, Nancy (1996) *Waist-high in the World: A Life Amongst the Nondisabled* (Beacon Press: Boston).

Matthews, M. P. and Vujakovic, P. (1995) "Private worlds and public places: mapping the environmental values of wheelchair users" *Environment and Planning A* 27: 1069–83.

Miller, Christopher L. (1993) "The postidentitarian predicament in the footnotes of *A Thousand Plateaus*: nomadology, anthropology, and authority" *Diacritics* 23 (3): 6–35.

Mitchell, Don (1992) "Iconography and locational conflict from the underside: free speech, People's Park and the politics of homelessness in Berkeley, California" *Political Geography* 11(2): 152–69.

Nast, Heidi and Mabel Wilson (1994) "Lawful transgressions: this is the house that Jackie built" *Assemblage: A Critical Journal of Architecture and Design Culture* 24: 48–55.

Park, D. C., P. Simpson-Housley and A. deMann (1994) "To the infinite spaces of creation: the interior landscape of a schizophrenic artist" *Annals of the Association of American Geographers* 84(2): 192–209.

Pepenoe, P. (1938) "Sterilization in practice" *Survey Midmonthly: Journal of Social Work* 74(6): 202–4.

Pfeiffer, David (1997) Program for 10th Annual Meeting of the Society for Disability Studies, Minneapolis, Minnesota, May 22–25, 1997.

Piaget, Jean (1960) *The Moral Judgement of the Child*, Marjorie Bagain, trans. (Free Press: Glencoe, Illinois).

Pokorski, Robert J. (1994) "Use of genetic information by private insurers" in Timothy F. Murphy and Marc A. Lappé (eds) *Justice and the Human Genome Project* (University of California Press: Berkeley), pp. 91–109.

Saxton, Marsha (1984) "Born and unborn: the implications of reproductive technologies for people with disabilities" in Rita Arditti, Renate Duelli Klein and Shelley Minden (eds) *Test-Tube Women: What Future for Motherhood?* (Pandora Press: Boston), pp. 298–313.

Shapiro, Joseph P. (1993) *No Pity: People with Disabilities Forging a New Civil Rights Movement* (Times Books: New York, NY).

Shaw, Barrett (1994) *The Disability Rag* Jan/Feb: 23–5.

Smith, Eleanor (1993) Presentation at the first national organizing meeting of Concrete Change, Nashville, TN, September 25.

Stallybrass, Peter and Allon White (1986) *The Politics and Poetics of Transgression* (Cornel University Press: Ithaca, NY).

Steinbock, B. and R. McClamrock (1994) "When is birth unfair to the child?" *Hastings Center Report* 24 (6): 15–21.

Vujakovic, P. and M. H. Matthews (1994) "Comforted, folded, torn: environmental values, cartographic representation, and the politics of disability" *Disability and Society* 9: 359–75.

Wertz, D. and J. Fletcher (1989) "Fatal knowledge? Prenatal diagnosis and sex selection" *Hastings Center Report* 19(3): 21–7.

Wertz, D. and J. Fletcher (1993) "A critique of some feminist challenges to prenatal diagnosis" *Journal of Women's Health* 2(2): 183–8.

Woodward, John R. (1994) "It can happen here" *The Disability Rag* Jan/Feb: 16–18.

Young, Iris Marion (1990) *Justice and the Politics of Difference* (Princeton University Press: Princeton, NJ).

12

ENCOUNTERING MARY

Ritualization and place contagion in postmodernity

Angela K. Martin
and
Sandra Kryst

Introduction: "postmodern Marian pilgrimage"

There have been over two hundred reported apparitions of the Virgin Mary worldwide in the past sixty years (Miller and Samples 1992: 82). As her appearances have become more frequent in the late twentieth century, her pilgrimage sites have also begun appearing and spreading into lands and territories not typically associated with devotion to Mary, or even with the Roman Catholic faith in general. It seems that Mary's activities are changing the mythical landscape of the Christian world. In their study of historic and (then) contemporary pilgrimage to Marian and other Chrisitan devotional sites, anthropologists Edith and Victor Turner termed the phenomenon "postindustrial Marian pilgrimage" (Turner and Turner 1978; 1982), although they made no attempt to account for the mechanism or meaning of the spread of present-day apparition sites.

In 1988, the Virgin Mary began appearing in the southern United States in Conyers, Georgia. Her public apparitions continued there every month until 1994. In 1992 she also began appearing at two sites in northern Kentucky, in the heart of the Baptist bible belt. As anthropologists we were intrigued by Mary's odd choice of place, considering the local cultural context, which was predominantly Protestant. We were also concerned with what people experienced at these sites and hoped that our attempts to interpret their experiences of Marian visitation might provide us with some insight into how apparition sites emerge and spread in what would today be called the "postmodern" world. In this chapter, we critically examine the phenomenon of postmodern Marian pilgrimage, and attempt to characterize the production of pilgrim bodies and sacred places in the context of Marian visitation and the apparition performance.[1]

The three apparition sites we studied (Cold Spring and Falmouth, KY and Conyers, GA) each initially came into being via a kind of "place contagion." The "carriers" of this "Marian contagion" are all individuals who have visited the most frequented and most famous active apparition site in the world today – Medjugorje in the former Yugoslavia. It is not uncommon for pilgrims who visit this site to undergo a spiritual conversion experience as a result of the apparition performance in which they participate. Their bodies are ritualized[2] as a result of the process and they not only carry away with them a new feeling about their spiritual relationship with Christ, God, or Mary, they also embody a new habitus[3] or way of being in the world that restructures existence in their more commonly frequented locales. As a result, it is sometimes the case that once people who have visited Medjugorje return home, they, or individuals close to them, begin to also experience apparitions of Mary.[4]

Ritualization and the mutual constitution of bodies and places

Our data suggest that mimesis and both sympathetic and contagious principles of magical practice constitute the transformative or embodying aspects/mechanisms of apparition performances.[5] Marian devotion fundamentally involves notions of immitation, mimicry and contagion, particularly with respect to the relationship between Mary and her female devotees. Most studies of Marian devotion in particular cultural contexts have referred to the various ways in which women are encouraged, and attempt, to emulate Mary (see, for example, Rodriguez 1994; Condren 1989; Collier 1986; Campbell 1982; Warner 1976). Although it has not often been problematized as such (see Martin 1995; forthcoming), emulation of Mary is really an example of a mimetic process.

We can see mimesis at work when women mimic Mary in their performance of self-sacrificing labor in the home for husband and children, in their acceptance of physical and emotional suffering without complaint, and in the constitution of feminine selves in such a way that gendered notions of motherhood, suffering, sacrifice and sexual purity are central components of their identities. This particular way of negotiating the world is also fundamental to Marian pilgrimage and to the experience of the apparition place as meaningful.

In this chapter we show that the process of ritualization (see Bell 1992), as it takes place in the context of apparition performances, involves the production of ritualized bodies and places. Pilgrim bodies are produced in the moment of practice as gendered and sexualized entities, each embodying a knowledge of sacred places which they carry with them into more mundane spaces. This embodiment accounts for the phenomenon of "place contagion" described here as the means by which apparitions of Mary spread and manifest themselves in various locations around the world. The process of ritualization is shown to gain efficacy through a play of principles of sympathy, contagion, and mimesis at apparition sites. We begin by setting the scene.[6]

Encountering Mary: performances of sympathy and contagion

Early in the morning on the eighth day of each month, thousands of pilgrims begin to arrive at Our Lady's Farm in Falmouth, Kentucky (Figure 12.1) carrying their burdens of lawnchairs and coolers, blankets, cameras, babies and rosaries as close as possible to the grotto of Our Lady of Light at the center of the farm. Surrounded by rolling hills, reached by a winding and narrow highway, the farm seems like it's in the middle of nowhere, even for those who live only an hour away. From the outside, it's hard to distinguish this particular farm from those that surround it; but once one enters, whether by car, tour bus, on foot or in a wheelchair, differences begin to materialize.

Very few farms in Kentucky are arranged and organized as places of worship. At Our Lady's Farm, barns stand empty, ponds abandoned by livestock; there are no fields under plow or crops to be harvested. Rather, this

Figure 12.1 The entrance to Our Lady's farm in Falmouth, Kentucky on the day of Mary's monthly visitation to a visionary at this site. Many people journey to the site on the eighth day of every month, some in specially chartered tour buses like the one shown, to witness her visitation to Sandy, the Falmouth visionary.
(Photo credit: Sandra Kryst and Angela Martin)

farm is organized for the mass harvesting of souls to the cause of the Virgin Mary. Under these circumstances, the contours of the land (ridges, hills, valleys, slopes and vistas) become part of the organization and mediation of religious experience. A sympathetic relationship exists between the material manipulation of the site by those involved in the apparition performance and the symbolic content of the performance itself.

Sympathetic magic works on the principle of similarity. Material objects are thought to gain a kind of magical potency through their similarity to a person, animal or other important object or locale. A so-called "voodoo doll" is the perfect example: According to sympathetic beliefs, producing an effigy of an individual will allow one to harm that individual through abuse of her or his representation. In the Catholic tradition, sympathetic notions are found in the belief that icons or representations of saints or Mary, crucifixes, and statues are imbued with particular holiness or somehow bring one closer to the holy figure or place represented.

At Falmouth, a kind of sympathetic belief is also found in manipulations of the "natural" lay of the land. The apparition performance is situated so, and land has been cleared in such a way, that "natural" material aspects of the terrain lend structure and potency to the performance. Manipulations of high and low (Mary appears in the sky above the main (and highest) hilltop where the grotto has been situated), near and far (the crucifixion scene at Calvary is reproduced and situated on a distant ridge visible only from the grotto top), and sloping versus flat (the crowd is seated mostly on cleared, sloping terrain) create an outdoor theater effect with supernatural overtones.

The effectiveness of manipulation of the principle of sympathy can be seen here in the statements people make when asked why Mary is coming to this particular location. One pilgrim said, "Mary said this [Falmouth] is new holy ground because of the faith in the area – it's peaceful and idyllic." Another said, "She's coming here because it's a beautiful place." In these statements one can see that the supernatural aspects or "holiness" of the place are not separated from the material or "natural" aspects. Pilgrims also often talk about and discuss among themselves the special "feel" of apparition sites. They indicate that these places are experienced affectively or through embodied feelings of holiness or peace. Being in this special place, waiting for Mary to appear, one begins to get the "feel" of the site. Yet in their descriptions of this feeling pilgrims do not separate the materiality of the site (its vistas, location and topography) from its feeling of holiness or closeness to Mary. As one woman said, "I came to Our Lady's Farm to get as close as possible to Mary – you can't see her, but it's as close to heaven as you can get."

Upon entering the farm, one is struck by the various forms of labor employed in the organization of such large numbers of people – unseen workers have laid gravel, roped off pathways, and erected special tents for the handicapped and ill in central and easily accessible locations; cars and buses

210

numbering in the hundreds are directed by men wearing rosaries into complex, but smoothly flowing arteries; people stand docilely at the port-o-johns and medical stations located at strategic points around the farm; volunteers hand out free rosaries and medals and keep a steady stream of pilgrims moving through the grotto area where each takes his or her turn gazing at and touching the near lifesize statue of Our Lady. It's here, on the top of a round and gently sloping hill near the center of the farm, that Our Lady has chosen to appear to Sandy on the eighth day of every month for the last few years. The top of this hill is also what is known as the grotto area – the grotto of Our Lady of Light (Figure 12.2).

The sympathetic construction of the farm at Falmouth is also evidenced by the fact that it is mimetic of other Marian visitation places. For example, two apparition sites in northern Kentucky, as well as Our Lady's Farm, have Italian-made statues of Mary that are replicas of her statue found at Medjugorje, the world's premier apparition site. In addition, efforts have been made to reproduce at Our Lady's Farm the qualities of place felt at other apparition sites by moving materials between them. Requests were made over the loudspeaker at Falmouth for rocks transported from other apparition sites, such as Conyers, GA, Medjugorje, and Lourdes. The grotto at Our Lady's Farm and the platform on which Mary's statue rests have been constructed out of these rocks. The assumption underlying this practice is one of contagion – these materials somehow carry or embody quality of place.

The principle of contagious magic involves the idea that one can gain access to or experience the supernatural qualities contained in special objects by direct contact with them, or that special qualities can be imparted to objects via direct physical contact with other venerated objects, places, or people. For example, one woman on the e-mail network describes the process whereby her holy medals took on a special quality at Conyers:

> I had brought some really inexpensive medals with me, and I asked her [Nancy, the visionary] to hold them in her hands for me. She told me that she had no special powers, and I replied that I knew that, but it would make people happy to have something she had touched. She took them, said prayers, and even reached into my bag to touch my other things. Once she got started, she seemed as if [she] liked the idea of doing this. . . . Soon I went on to the Holy Hill [at Conyers] to put my medals on the altar and prayed. I had already asked the Bishop [of Knoxville, Tn.] to bless them before I left. . . . I think it's neat to have a medal blessed by a holy bishop, held by Nancy Fowler, and placed on the altar at the Holy Hill.
>
> (Correspondence dated 28 March 1994)

The same principle of contagious magic lies behind the tendency of pilgrims to file through the grotto area to touch the statue of Mary at Falmouth,

Figure 12.2 The statue representing Our Lady of Light situated in the "Grotto Area" where Mary's visitation occurs.
(Sandra Kryst and Angela Martin)

especially after the apparition has occurred (Figure 12.3). Like the stones carried from Medjugorje, Lourdes, or Conyers, this statue has special qualities, not only because of its sympathetic mimicry of Mary and other apparition site statues, but because of its direct association with Mary's appearance at this site and the special stone altar on which it stands. There is a constant play in this example between principles of contagion and sympathy.

Although Mary always appears between 1:30 and 2:00 in the afternoon, the pilgrims arrive early with the aim of claiming a space for their chairs and blankets as close as possible to the grotto. This is where Mary appears to Sandy, in the sky above her own likeness. It is Sandy rather than Mary that everyone strains to see, especially at the moment of the visitation. Sandy herself constitutes the visual evidence and indication of Mary's presence.

Yet most of the long morning and early afternoon are spent in preparation for Mary's arrival. The loud speaker system at the farm is constantly playing religious music, announcements, or, more frequently, the recurring recitation of the rosary interspersed with a short chorus of "Ave Maria." The pilgrims spend their time praying or wandering about the farm, filling out petitions to Mary or performing the Stations of the Cross. Most join the constant movement of bodies through the grotto area where they pause to study, touch, and give prayer at the statue of Our Lady.

Around 1:30 in the afternoon, the grotto area at Our Lady's Farm is closed to pilgrim traffic. The time for Mary to appear to Sandy is drawing near. The music blaring over the loudspeaker is silenced and a voice announces that "Sandy is approaching the grotto area." Excitedly everyone turns to get a glimpse of her and a number of pilgrims try to ride the momentum of her entourage into the grotto area, where they are caught and turned away by volunteers. Shortly thereafter, Sandy and her spiritual advisor, Fr. Smith, kneel at the foot of the statue of Our Lady and lead the pilgrims in recitation of the rosary.

Mary and earthly women are conceptually linked to one another via the apparition performance. The constant praying of the Rosary in unison, and the substantial number of Hail Marys embedded within it, produce a highly ritualized, verbal commentary on Mary's link to women that is powerfully sanctioned by thousands of reverberating voices:

Hail Mary, Full of Grace. The Lord is with Thee. Blessed Art Thou Amongst Women and Blessed is the Fruit of Thy Womb, Jesus.

Moreover, prohibitions against abortion and the idealization of Mary's role as mother strongly encourage women to emulate her. As one man noted when asked why Mary was appearing to people:

I would have to say abortion is the biggest reason Mary's coming – it's a wonder she's not chasing us around with a broom! Oh, gosh,

excuse me for saying that because it has dark connotations . . . .
What I'm saying is, it's a wonder she's not chasing us around like an
angry mother trying to get us to wake up before the tough times
come.

(6/8/1994)

Pilgrims constantly make associations between mothers, Mary and women at
this site and elsewhere. Especially important in solidifying the associative or
metonymic relationship between Mary and earthly women, however, is
Sandy's interaction with Mary, Fr. Smith, and the pilgrims during the
apparition performance.

Significant in this respect is the gendered division of labor that clearly
exists between Sandy as visionary and Fr. Smith as her spiritual advisor. Sandy,
like all of the visionaries of Mary we have studied in the southeast, is the one
who interacts directly with Mary, the one who receives her messages and
relays them to the public. Thus, Sandy gains her spiritual significance directly
from Mary. Her power is the product of an oscillation between contagion
and sympathy – she is special, first, because Mary appeared to her, chose her,
touched her, and, second, because, as we will see, Sandy's apparition performance places her and other visionaries in a sympathetic position relative
to Mary. Sandy, however, like other visionaries, has a male spiritual director,
who is also a priest, and whose privilege it is to interpret Mary's messages
while he keeps an eye on Sandy's spiritual health, serving as her confessor and
advisor.

Mary often appears during the recitation of the rosary, but only Sandy
knows that she has arrived. Sandy indicates that the apparition has appeared
and Fr. Smith stops the rosary and notifies the pilgrims over the PA system.
Everyone strains to see Sandy, shifting their vision back and forth between the
visionary who gazes up at the sky, lips sometimes moving and head nodding,
and the place above the statue on which Sandy's own eyes are focused.

Once Mary's visit is completed, Sandy writes down the message. Once
finished, Sandy hands Mary's words to Fr. Smith. He reads them and approves
the contents before Sandy reads the message aloud over the loud speaker to the
large and expectant crowd of pilgrims (Figure 12.4). The visionary is
constructed via this performance as the passive instrument, or conduit
through which Mary speaks. Fr. Smith, as spiritual advisor, is the evaluator,
the censor and director of action. Indeed, visionaries who do not defer to
priestly spiritual advisors are seen as worthy of suspicion and usually lack
legitimacy.[7]

Through the apparition performance, the visionary is placed in much the
same position in relation to her spiritual director as Mary is placed in relation
to Jesus in institutionalized Church dogma. Sandy is mimetically constructed
as a passive conduit or interface between Mary and the priest who then
chooses what can and cannot be revealed to the pilgrims. Her position

Figure 12.3 Many pilgrims enter the Grotto Area before and after Mary's visitation to pray and touch the statue.
(Sandra Kryst and Angela Martin)

Figure 12.4 Sandy, the Falmouth visionary stands with a microphone and reads Mary's message to the pilgrims shortly after her visitation. (Sandra Kryst and Angela Martin)

corresponds structurally to Mary's position in popular Church dogma. Mary is the passive recipient of God's will, acting as the vessel through which the Son of God was and may again be born into the world. Today, at apparition sites, she comes as intercessor or mediator between worlds, and only with her Son's permission, bringing His will to the pilgrims. Positioned similarly as a result of the apparition performance, Sandy is mimetic of Mary and, as visionary, becomes the center of the Marian experience.

Despite the fact that everyone present participates in and creates the apparition performance, both women and men alike, the female subject positions are infused with more symbolic capital. As the above description and analysis suggests, a gendered divison of labor does exist between Sandy and her spiritual advisor, Fr. Smith, throughout the apparition performance. Michael Taussig would more specifically characterize it as a "gendered division of mimetic labor." As he notes, in "spiritual politics . . . image-power is an exceedingly valuable resource" (Taussig 1993: 177). In the apparition performance female sexuality, female bodily boundaries, and woman as Mother are foregrounded as a result of the discursive practices that constitute devotion to the Virgin Mother of God. Women, especially the visionary, carry much more of the symbolic labor or load of the performance. Women are more intensely disciplined than men, as constructions of femininity are more hotly contested than constructions of masculinity, which are defined by extension or juxtaposition.

The mimetic or sympathetic relationship between Mary and her visionaries is also evidenced by the corporeal and symbolic importance of suffering in the lives of the visionaries. Those participating in the apparition e-mail network have reflected a number of times on the tendency of visionaries around the world to be sufferers of chronic physical pain. Their physical suffering is such that their followers sometimes call them "victim souls." For Nancy Fowler, the visionary at Conyers who receives regular messages from Mary and Jesus, her own physical suffering is always linked directly to Jesus, who asks her to endure physical suffering for specific reasons. On the occasion of the 14 January 1994 public appearance of Mary at Conyers, Nancy explained to the crowd of pilgrims:

> There is merit in suffering . . . join your suffering to Christ. At this particular time I'm being asked to suffer for troubled families. He said for all . . . so it's for all families. I'm sure for mine too.

Earlier in the same question and answer session, Nancy also explained to the pilgrims that Mary is appearing in the US at this time because, "A mother does not want her children to suffer. A mother is concerned and loving and caring toward her children." For Nancy, gendered prescriptions about motherhood and the family go hand-in-hand with commentary on suffering, but they are also reinforced in her direct bodily experience of pain and emotion. It is significant that Nancy is free of pain only when she is on her knees before the vision of Mary – the one who suffers as a mother and thereby takes away the suffering of her children. Is Nancy a mother when she suffers for her own family and all families? Or is she a child when Mother Mary momentarily relieves Nancy of her physical, emotional and spiritual pain? Certainly she is both.

Suffering seems to be a near prerequisite for experiencing visions of Mary. Both Sandy and another anonymous visionary in her area (the Batavia visionary connected with the apparition site at Cold Spring) carry their own physical burdens and have a personal legacy that directly connects them to sickness and suffering in others. All three of the visionaries either were or are practicing nurses. Book II of the Our Lady of Light Center which describes the apparitions at Cold Spring and Falmouth characterizes the anonymous Batavia visionary in this way:

> As further background, this messenger, like most others with special gifts, encounters the kind of suffering that seems to go with special spiritual faculties.
>
> (Ross 1993: 6)

These women were chronic sufferers of pain before encountering Mary. In the context of the apparition performance, pain and suffering take on new meaning – Nancy, for example, is no longer simply a middle-aged woman with chronic physical problems. Rather, she is spiritually transformed and brought closer to Mary, discursively linked with the suffering and self-sacrifice of Mary via her own physical experience. Physical manifestations of pain in the context of the apparition performance may be the ultimate expression of the sympathetic structural links between Mary and her visionaries.

We would characterize apparition performances at Falmouth and elsewhere as "ideal mimetic moments." The sympathetic relationship between the visionaries, women and Mary produced at active apparition sites actually constitutes a material relationship between Mary and women. The stuctural symbolic commentary that we have "read" at these sites amounts to a moral habitus – a commentary on femininity and the centrality of motherhood to the definition of Woman as defined by the ideal family context in the near southern United States. This moral habitus is not merely symbolic, it does not exist "out there" as some disembodied abstract structure. Rather, as the example of physical suffering and its transformation through the workings of the apparition performance indicates, this habitus is embodied in the process of its production. Through the workings of the apparition performance, the pilgrims, the visionary, and Mary all produce each other.

Symbolism and the habitus

The fundamental importance of mimetic processes and the play of sympathy and contagion in the constitution of apparition sites are both symptomatic of the major cultural themes that help make up the moral habitus produced at these sites. Mimesis and the principles of sympathy and contagion all involve a play with, or blurring of, different kinds of bodily boundaries: boundaries between self and other; spirit and materiality; bodies and places. The major moral issues contested at the sites are also what we would characterize as boundary issues: abortion; the breakdown of the family; school prayer; eroding morals in the modern world. A blurring of boundaries can also be found in the fundamental importance of healing in pilgrimage to, and participation in, Marian visitations, as well as in Mary's appearance itself.

The blurring of different kinds of bodily boundaries relates to the fact that these boundaries are actually produced and contested as a product of the enactment of a Marian visitation. Mimesis, sympathy and contagion all "work" because where the self begins and ends is contested, becomes unclear, and then is reaffirmed. Examples of the fuzziness of bodily boundaries are numerous at apparition sites. They consist of everything from the experience of the "feel" of the site, the infusion of the special power that results from Mary's visit through the contagious touching of her statue, to the particularly contested nature of the visionary's body. The visionary's subject position is constantly shifting: she is the focus of the apparition performance and is the material avenue via which most of those present experience Mary's visit. She is simultaneously mimetic of Mary, Everywoman, Mother and the martyrdom of Christ. This is especially evident in the multiple implications of the visionary's suffering and Mary's ability to temporarily ease that suffering. When Mary appeared to Nancy, momentarily easing the suffering Nancy was enduring for all families, Nancy was multiply placed in differing subject positions: she was mother, child, Mary and Christ. Her weak physical integrity contributed to her malleability, as her own bodily boundaries were redefined via her interaction with Mary – she was made whole and complete. Mary and her visionary simultaneously produced each other.

It is perhaps more obvious that the major political issues focused on at apparition sites are all boundary issues in one respect or another. The issue of abortion is about control over women's bodies and maintaining the integrity of these bodies as whole reproductive vessels. For the pilgrims that visit these sites, abortion itself is about the improper, amoral penetration of women's bodies as Mother. This penetration amounts to the absolute destruction of those bodies. Of course, women's bodies as reproductive vessels, as Mother, are strongly discursively linked to Mary as pure, virginal, unpenetrated Mother – this link is what makes the denouncing of abortion at apparition sites such a powerful message. It is indeed a message which helps to constitute women as appropriately feminine, as Mother-writ-large, via the apparition performance (Figure 12.5).

Figure 12.5 Many pilgrims take photographs at apparition sites, including photos of the sun, in the hope of catching an image of Mary on film which they cannot detect with the naked eye. This polaroid was taken at Conyers, GA, and circulated over an apparition listserve. It includes a superimposed image of Mary extending her mantle across a woman who is kneeling and praying at the foot of a crucifix erected at the site. Written material circulated with the photo asserts that the woman was pregnant at the time the picture was taken.
(Photo credit: Anonymous)

The breakdown of the family and "family values" go along with school prayer in that they both refer to eroding moral and political boundaries and increasing contestation in modern and postmodern societies over who has the right to define such boundaries. In the apparition perfomance, the proper location of the power of definition is clearly enacted as Father Smith disciplines Sandy's interaction with Mother Mary, and indeed, Mary's message to her followers. The following metonymic associations are established via this enactment:

Mary – Mother – Woman
Christ – Father – Man

The performance is thus simultaneously a commentary on both the amoral breakdown of the modern family and a prescriptive rite delineating appropriately masculine and feminine roles – roles which are highly heterosexualized and which reinforce the authority of the Father and Heteropatriarchal Law (see Nast and Wilson 1994).[8]

That reinforcing the authority of the Father in the heteropatriarchal family is a primary product of the apparition performance is evidenced by a number of healing narratives told by male pilgrims in particular at Falmouth. One man said:

> Mary is appearing now because of the state of our society. We're in real spiritual warfare right now, you see what's happening to families, to society? You see so much evil. . . . I've got a stepdaughter – I'm getting a divorce – this is a letter she wrote me. . . . See, I've caught her with a satanic book and I know my statues and stuff was desecrated by her. This is a letter[9] I got from her a week or two after I was here [at Falmouth where he was healed]. It made me sad, caused a lot of tears for her to say I should ask for God's forgiveness. This kid's been nothing but a nightmare! That's why I'm getting a divorce, because she's, I've caught her with dope and she's pulled a lot of shit on me before – so I think it's a little too late for me and her mother. I feel like I've been healed in some ways.
>
> (6/8/94)

In this scenario, the troublesome stepdaughter who challenges the authority of the father is demonized. The man's healing story is placed by him within the discursive context of "family breakdown" and "spiritual warfare." His own healing involves a rejection of a family situation in which he finds he cannot effectively discipline his daughter, and obviously his wife, as his problems with the daughter have resulted in divorce.

Another man at Falmouth related this healing narrative to us:

> Four years ago I went to Medjugorje. I really went over there for my
> wife's conversion and came home to a letter from a locutionist[10] that
> basically it was four pages of "Roger, clean up your act and your wife
> will convert." Four pages of this, and the last line was "and your wife
> will convert." So she became Catholic about three weeks ago and
> she only decided about six weeks ago. I put together a Medjugorje
> conference about six weeks ago in West Virginia and that kindof
> brought it all together. She reluctantly came and reluctantly met
> people and reluctantly had the story pounded into her head from
> other people.
>
> (6/8/94)

This man's healing story was about bringing his wife into the faith rather
than about his own physical and/or emotional healing, for example. Moreover,
his wife's conversion as a process was constructed for him as a result of his visit
to another apparition site as something that was within his own control
– "cleaning up his act" would eventually result in his wife's own personal
conversion.

These male healing narratives also converge with not-so-subtle commentary at apparition sites and across the e-mail network on the relationship
between Jesus and his mother, Mary. In all instances it is emphasized that
Mary appears only with her Son's permission and then with the mission
of doing His will, to prepare the people for His second coming. At the end of
every apparition event, for example, Nancy, the Conyers visionary, notes
that Mary always ends her messages with the words "Thank my Son for
allowing me to come." In a similar vein, e-mail subscribers often end their
transmissions with the epithet "In Jesus through Mary."

Finally, a blurring or breakdown of boundaries can be found in the
fundamental principles underlying Mary's ability to appear to earthly, mortal
individuals (see Martin 1993). As Leach notes:

> To be useful, gods must be near at hand, so religion sets about
> reconstructing a continuum between this world and the other world.
> But note how it is done. The gap between the two logically distinct
> categories, this world/other world, is filled with a tabooed ambiguity.
> The gap is bridged by supernatural beings of a highly ambiguous
> kind – incarnate dieties, virgin mothers, supernatural monsters
> which are half man/half beast. These marginal, ambiguous creatures
> are specifically credited with the power of mediating between gods
> and men. They are the object of the most intense taboos, more sacred
> than the gods themselves. In an objective sense, as distinct from
> theoretical theology, it is the Virgin Mary, human mother of God,
> who is the principal object of devotion in the Catholic church.
>
> (Leach 1964: 39)

Conceptually, the symbol of the Virgin Mary exists at the boundaries, on the edges of the world that human beings are able to know (Leach 1976: 35). Mary belongs in Heaven, but somehow she keeps slipping physically back into the lives of individuals on earth (see Turner and Turner 1982). Mary's ability to appear involves a blurring of boundaries between this world and the other world, between sacred and profane places. Moreover, her appearance to certain individuals in certain places results in a tranformation of place. Apparition sites are unlike other earthly places. How they differ has to do with our original questions surrounding place contagion: How does this contagion work? and What is the relationship between the production of ritualized bodies and ritualized places?

Conclusions: absolute space, ritualization and place contagion

How does the ritual process of public Marian visitation, of the apparition performance, result in the production of ritualized bodies and places? Bourdieu's notion of bodily hexis[11] can be used to interpret the apparition experience detailed above. Via the apparition performance appropriately masculine and feminine bodies are produced – "a political mythology is realized" materially (Bourdieu 1990: 69–70). This production is especially evident in the gendered division of mimetic labor inherent to the apparition performance and in the heterosexualizing of pilgrim bodies as ideal Mother and Father figures. This transformation is most clear in the strategic suffering of visionaries and in the male pilgrims' healing narratives.

The effectiveness of the production of these bodies in the ritual context of Marian visitation can be attributed to the play of sympathy and contagion and the fundamental importance of mimetic processes at the sites. The "feel" of the place that nearly all pilgrims mention is inseparable from the material qualities of the site. The spatiality of pilgrimage experience is characterized by blurred boundaries between the forces believed to be contained in materials such as rocks, and the bodily experience of a location as a Marian pilgrimage place – a place with a special feel. Principles of sympathy and contagion combine to produce this effect. Critiqueing Callois (1984), Grosz notes how the process of mimesis

> is particularly significant in outlining the ways in which the relations between an organism and its environment are blurred and confused – the way in which its environment is not distinct from the organism but is an active internal component of its "identity."
>
> (Grosz 1994: 46)

Mimetic practice with respect to place in Marian pilgrimage is not the kind of representational practice normally critiqued by cultural studies theorists, for example. The performance is not representational of meaning or a "political mythology" as Bourdieu would say, rather the performance is *enacted*, it is *lived* through the pilgrim's experience of her or his body *in place*. As Bourdieu notes:

> The body believes in what it plays at: it weeps if it mimes grief. It does not represent what it performs, it does not memorize the past, it *enacts* the past, bringing it back to life. What is "learned by the body" is not something that one has, like knowledge that can be brandished, but something that one is.
>
> (Bourdieu 1990: 73, emphasis in original)

The gendered division of mimetic labor that is a part of the structure, the habitus of the apparition performance, does not *represent* heteropatriarchy to all those present, rather it insists that they live it and that they *produce* it as they live it.

Practice at Marian apparition sites is not representational in the sense that it does not create a space between subject and object, between meaning and material. Meanings are not merely objectified in symbols or places, rather they are given body or em-bodied. Mimesis confuses boundaries between self and environment, allowing one to take on, experience, feel, become, embody certain qualities of place. Ritualization (Bell 1992) is this process – the simultaneous production of certain kinds of bodies and places. This characterization of the mimetic relationship between pilgrim bodies and apparition places is remarkably similar to Lefebvre's description of "absolute space" in his book *The Production of Space* (1991). Lefebvre uses the concept of absolute space to describe places where symbols have not been separated from their material existence, there is no separation between meaning and material, between signifier and signified. These are also the principles which underlie notions of sympathy and contagion (both of which commonly flourish in so-called sacred spaces).

Yet the pilgrim self does not dissolve in its experience of the sacred or absolute, rather, as the sacred is enacted, the pilgrim self is also enacted, opened to a process of creation, negotiation and solidification of boundaries between itself and others. This redefinition of boundaries takes place as a result of the apparition performance. We certainly are not the first to note the importance of mimesis in the definition of the self or identity (see also Grosz 1994; Taussig 1993; Boddy 1993 and Bourdieu 1990 to list only a few). Indeed, although mimicry may temporarily call into question differences or boundaries between oneself and others, it is also fundamental to the negotiation of identity, particularly in home spaces. As we have seen, in the apparition performance, Sandy is mimetic of Mary and in this "ideal mimetic moment," Mary and her visionary serve to produce each other within certain discursive limits. Sandy's performance of Mary's visitation, enacted with

Fr. Smith and all those present at the site, involves a redefinition, a production of gendered and sexualized pilgrim selves. The ultimate result of this performance is not the dissolution of the self, but the simultaneous production of ritualized bodies and places.

Although Bourdieu's notion of bodily hexis was designed primarily to analyze embodiment via mimetic practice in home spaces, it also applies well to ritualized contexts. Bell (1992: 90) has suggested that ritual practices are particularly efficacious in the embodiment of political mythologies because they serve to contrast meanings and ways of acting. Apparition performances are an effective means of the production of ritualized bodies, in this case highly gendered and sexualized bodies. But what makes mimesis in ritual spaces, like apparition sites, distinct from mimesis and bodily hexis in home spaces? The fact of spatial removal of these sites from everyday locales, the need of pilgrimage or travel to these sites, adds to their potency.[12] Although we have characterized apparition places as absolute spaces, here we do find a symbolism-writ-large or intensified political mythology. Its intensification is effected via play with principles of sympathy and contagion. Material aspects of the performance and the place are hierarchically ordered, certain oppositions are drawn out and enacted. The taken for granted, or practical knowledge of gendered and sexualized oppositions in the home, is played out in a "foreign," unnaturalized territory where the re-naturalization of the obvious, and the inscriptive production of the site itself, lends renewed, visceral credibility to the original symbolic scheme or habitus.

The question remains of how the "place contagion" that seems to characterize the spread of contemporary Marian apparition sites actually works. Our analysis is suggestive of an answer: as bodies and places are ritualized, bodies come to em-body qualities of place. As a result of the play of sympathy and contagion in the apparition performance, the apparition site is constructed as an absolute space, a place where there is no objectification of meanings from materiality. Ritualized bodies live the absolute and are inculcated with this way of being in the world via the enactment of the apparition performance. Bell notes that ritualized bodies carry the political mythology they inhabit as a result of ritualization with them as they move out of sacred and into more mundane contexts. Thus, pilgrims to Medjugorje embody a new habitus that actually restructures existence in their more commonly frequented locales. These bodies are also more likely to experience or re-enact visitations from Mary than any others.

As noted in the introduction, we think it more apt to label the type of Marian pilgrimage described in this paper as one that is particularly postmodern. As absolute spaces, apparition sites certainly represent ruptures in the abstract spaces of late capitalism. Place contagion itself is clearly a "transnational" phenomenon and the experience of Marian devotion in cyberspace involves a compression of space and time never before associated with religious or ritual practice. Additionally, one has to question if the preoccupation with boundary issues described here is itself a manifestation of the angst relating to postmodern processes of deterritorialization and the breakdown of signifying chains of meaning in late capitalism described by certain theorists (see Harvey 1989; Jameson 1992). Perhaps a comparative study of premodern, modern, and postmodern apparition events would shed some light on the uniqueness of contemporary Marian apparitions and the phenomenon of place contagion.

Notes

1 In our analysis we draw on data gathered using a number of different strategies. We collected ethnographic data at four different apparition sites in northern Kentucky, southern Ohio and Georgia. During our fieldwork at these locations we conducted 31 interviews, speaking with 55 different individuals. We attended and participated in a healing service and five visitations of Mary at two different apparition sites. In addition to taking many photographs at these sites, we videotaped the apparition events. We also collected massive amounts of textual commentary on appearances of Mary around the world from an apparition e-mail network monitored by Angela Martin through all of 1994.

2 Catherine Bell (1992) has applied Bourdieu's theory of practice (see Bourdieu 1990, 1977[1972]) to ritual contexts. She uses the term "ritualization" to refer to the process by which bodies are constituted in ritual practice. Intrinsic to ritual practice is the necessity to distinguish this type of practice from that which takes place in other, non-sacred kinds of places, such as the home. As Bell notes,

> Viewed as practice, ritualization involves the very drawing, in and through the activity itself, of a privileged distinction between ways of acting, specifically between those acts being performed and those being contrasted, mimed, or implicated somehow.
>
> (1992: 90)

The end product of these enacted distinctions or contrasts is ritualized bodies, or ones invested with a "sense of ritual" (ibid: 98).

3 Bourdieu uses the term "habitus" to refer to "a matrix of perceptions, apperceptions, and actions" (1977[1972]: 82–3, emphasis removed). More recently he has defined it as:

> systems of durable, transposable dispositions, structured structures predisposed to act as structuring structures, that is, as principles which generate and organize practices and representations that can be objectively adapted to their outcomes without presupposing a conscious aiming at ends or express mastery of the operations necessary in order to attain them.
>
> (1990: 53)

4 The apparitions at Falmouth and Cold Spring, Kentucky are both associated in some way with Fr. Leroy Smith. Following his trip to Medjugorje in October of 1988, Fr. Smith formed a prayer group in his parish that eventually attracted people from the surrounding tri-state area. One of these was Sandy, the public

visionary at Falmouth, Kentucky. Likewise, Nancy Fowler, the visionary at Conyers, Georgia, also traveled to Medjugorje in 1988. Shortly thereafter she too began experiencing visitations from Mary.

5 It is important to note that we did not begin our study with the aim of resurrecting the old anthropological distinction between sympathetic and contagious forms of magical practice. Rather, our approach was inductive in the sense that it was only after repeated exposure to similar principles at apparition sites that we decided to adopt their use for whatever insight might be gained. These analytical concepts have also been newly elaborated by Michael Taussig in his book *Mimesis and Alterity* (1993).

6 Two of the apparition sites we are addressing in this paper are located in northern Kentucky near I–75. These apparition sites are Our Lady's Farm outside of Falmouth, Kentucky and St. Joseph's Catholic Church in Cold Spring, Kentucky. Also in this area and associated with these two sites is the Our Lady of the Holy Spirit Center in Norwood, Ohio, just north of Cincinnati. The third apparition site referred to is located in Conyers, Georgia, about thirty minutes east of Atlanta by car.

Our Lady's Farm near Falmouth is situated on a high ridgetop in rolling country. A gravel road leads to the center of the site which is called the "Grotto" and consists of a roped-off area that contains a lifesized statue of Mary, along with two smaller shrines, a thorn tree, and rows of folding chairs. Our Lady appears in the sky above this Grotto to a visionary named Sandy [last name unknown] on the eighth day of every month. Sandy is generally accompanied by Fr. Smith, formerly a priest at St. Joseph's church and now director of the Our Lady of the Holy Spirit Center in Norwood. Falmouth is the most active apparition site in the area and attracts the most visitors every month, sometimes as many as 50,000 people.

The apparition site at Conyers, Georgia, is the most famous American site considered in this paper. The Virgin Mary began to appear there in 1988 to Nancy Fowler, an unemployed nurse and housewife. After that she appeared to Nancy on the 13th of every month in a farmhouse surrounded by thousands of pilgrims from all over the world. Public, monthly apparitions of Mary ended here in October of 1994.

7 Likewise, according to many of Nancy Fowler's comments to the pilgrims at Conyers, she has always been cautioned against interpreting any of the messages she recieves from Mary and Jesus, and also against evaluating any of the miracles said to have occurred at Conyers. She has mentioned to pilgrims that whenever she is tempted to place her own interpretation on anything from Christ's words to the meaning of one of her own dreams she "gets in big trouble" with her spiritual advisor.

8 "Heteropatriarchal Law" refers to a configuration of practices (sometimes codified in law) which help to manifest compulsory heterosexuality. As part of a larger habitus, they help to structure practice and are in turn reinforced via these same practices. Constructions of heterosexuality are culturally and historically specific. The construction referred to here is one which coheres around proto-typical Father and Mother figures.

9 At this point Sandra Kryst read the letter. It seemed perfectly normal and pleasant. His daughter said that she was sorry they had had a misunderstanding,

but she was offended that he thought she was involved with satanism.

10 A locutionist is someone who receives messages from Mary or God or some other figure as a disembodied voice.

11 Bourdieu defines "bodily hexis" as the process by which a "political mythology is realized, *em-bodied*, turned into a permanent disposition, a durable way of standing, speaking, walking, and thereby of feeling and thinking" (1990: 69–70, emphasis in original). At the same time as bodies are constituted via practice in culturally defined spaces, such as the home, these spaces are simultaneously constituted via bodily practices which are inscriptive of meaning.

12 Apparition sites may also be seen as "border zones," or areas which represent and are experienced as neither truly mundane and earthly nor mythical, other-worldly, or heavenly. Instead, they exist at the borders between this world and the next (see Leach 1964: 39).

References

Bell, Catherine (1992) *Ritual Theory, Ritual Practice*, New York: Oxford University Press.

Boddy, Janice (1993) "Aesthetics, Politics, and Women's Health in Northern Sudan and Beyond," a paper presented at the Annual Meeting of the American Anthropological Association.

Bourdieu, Pierre (1990) *The Logic of Practice*, Bloomington: Stanford University Press.

Bourdieu, Pierre (1977[1972]) *Outline of a Theory of Practice*, Cambridge: Cambridge University Press.

Callois, Roger (1984) "Mimicry and Legendary Psychasthenia", *October* 31 (Winter): 17–32.

Campbell, Ena (1982) "The Virgin of Guadalupe and Female Self-Image: A Mexican Case History", in J. L. Preston (ed.) *Mother Worship*, Chapel Hill: University of North Carolina Press.

Collier, Jane F. (1986) "From Mary to Modern Woman: The Material Basis of Marianismo and its Transformation in a Spanish Village," *American Ethnologist* 13: 100–7.

Condren, Mary (1989) *The Serpent and the Goddess: Women, Religion, and Power in Celtic Ireland*, San Francisco: Harper & Row.

Grosz, Elizabeth (1994) *Volatile Bodies: Toward a Corporeal Feminism*, Bloomington: Indiana University Press.

Harvey, David (1989) *The Condition of Postmodernity*, Oxford: Basil Blackwell.

Jameson, Frederick (1992) *Postmodernism, or the Cultural Logic of Late Capitalism*, London: Verso.

Leach, Edmund (1976) *Culture and Communication: The Logic by which Symbols are Connected*, Cambridge: Cambridge University Press.

Leach, Edmund (1964) "Anthropological Aspects of Language: Animal Categories and Verbal Abuse", in E.H. Lenneberg (ed.) *New Directions in the Study of Language*, Cambridge, MA: MIT Press.

Lefebvre, Henri (1991[1974]) *The Production of Space*, Oxford: Basil Blackwell.

Martin, Angela K. (1993) *The Virgin Mary: Gender, Religion and Politics in*

Contemporary Ireland, unpublished Master's thesis, Lexington, Kentucky: University of Kentucky.

Martin, Angela K. (1995) "The Virgin Mary and the Gendered Division of Mimetic Labor in Contemporary Ireland," a paper presented at the 1995 American Association of Geographers Annual Meeting.

Martin, Angela K (forthcoming) "Death of a Nation: Transnationalism, Bodies and Abortion in Late Twentieth Century Ireland," in T. Mayer (ed.) *Exposing Tensions: Nationalism, Gender, and Sexuality in International Perspective*, New York: Routledge.

Martin, Angela K. and Sandra Kryst (1994) "Disciplining Performances: Marian Pilgrimage, Healing and the Gendered Body", a paper presented at the Texts and Regions Conference, University of Kentucky, Lexington, KY.

Miller, Elliot and Kenneth R. Samples. (1992) *The Cult of the Virgin Mary: Catholic Mariology and the Apparitions of the Virgin Mary*, Grand Rapids, MI: Baker Book House.

Nast, Heidi J. and Mabel Wilson (1994) "Lawful Transgressions: This is the House that Jackie Built," *Assemblage* 24: 48–55.

Rodriguez, Jeanette (1994) *Our Lady of Guadalupe, Faith and Empowerment Among Mexican American Women*, Austin: University of Texas Press.

Ross, Gerald (1993) *More Personal Revelations of Our Lady of Light, Volume Two*, Ft. Mitchell, KY: Our Lady of Light Publications.

Taussig, Michael (1993) *Mimesis and Alterity*, London: Routledge.

Turner, Victor and Edith Turner (1982) "Postindustrial Marian Pilgrimage," in J. L. Preston (ed.) *Mother Worship*, Chapel Hill: University of North Carolina Press.

Turner, Victor and Edith Turner (1978) *Image and Pilgrimage in Christian Culture*, New York: Columbia University Press.

Warner, Marina (1976) *Alone of All Her Sex: the Myth and Cult of the Virgin Mary*, New York: Knopf.

13

PERVERSE DESIRE

The lure of the mannish lesbian

Teresa de Lauretis

Lesbian scholarship has not had much use for psychoanalysis. Developing in the political and intellectual context of feminism over the past two decades, in the Eurowestern "First World," lesbian critical writing has typically rejected Freud as the enemy of women and consequently avoided consideration of Freudian and neo-Freudian theories of sexuality. Certainly, the feminist mistrust of psychoanalysis as both a male-controlled clinical practice and a popularized social discourse on the "inferiority" of women has excellent, and historically proven, practical reasons. Nevertheless, some feminists have persistently argued that there are also very good theoretical reasons for reading and rereading Freud himself. All the more so for lesbians, I suggest, whose self-definition, self-representation, and political as well as personal identity are not only grounded in the sphere of the sexual, but actually constituted in relation to our sexual difference from socially dominant, institutionalized, heterosexual forms.[1]

One direction of my current work, of which this chapter presents a small but pivotal fragment, is to reread Freud's writings against the grain of he dominant interpretations that construct a positive, "normal," heterosexual and reproductive sexuality, and to look instead for what I would call Freud's negative theory of perversion. For it seems to me that, in his work from the *Three Essays on the Theory of Sexuality* (1905) on, the very notions of a normal sexuality, a normal psychosexual development, a normal sexual act are inseparable – and indeed derive – from the detailed consideration of their aberrant, deviant or perverse manifestations and components. And we may recall, furthermore, that the whole of Freud's theory of the human psyche, the sexual instincts and their vicissitudes, owes its foundations and development to psychoanalysis, his clinical study of the psychoneuroses; that is to say, those cases in which the mental apparatus and instinctual drives reveal themselves in their processes and mechanisms, which are "normally" hidden or unremarkable otherwise. The normal, in this respect, is only conceivable by approximation, more in the order of a projection than an actual state of being.

What is the advantage of such a project to a lesbian theorist? For one thing, in the perspective of a theory of perversion, lesbian sexuality would no longer have to be explained by Freud's own concept of the masculinity complex, which not only recasts homosexuality in the mould of normative heterosexuality, thus precluding all conceptualization of a female sexuality autonomous from men; but it also fails to account for the non-masculine lesbian, that particular figure that since the nineteenth century has consistently baffled both sexologists and psychoanalysts, and that Havelock Ellis named "the womanly woman," the feminine invert.[2] Second, if perversion is understood with Freud as a deviation of the sexual drive (*Trieb*) from the path leading to the reproductive object, that is to say, if perversion is merely another path taken by the drive in its cathexis or choice of object, rather than a pathology (although, like every other aspect of sexuality it may involve pathogenic elements), then a theory of perversion would serve to articulate a model of perverse desire, where perverse means not pathological but rather non-heterosexual or non-normatively heterosexual.[3]

In one of the rare attempts to look at lesbianism in a feminist and psychoanalytic perspective, a recent article by Diane Hamer suggests that lesbianism, for some women, may be "a psychic repudiation of the category 'woman,'" and sees a direct correspondence between feminism as "a political movement based on a refusal to accept the social 'truth' of men's superiority over women" and lesbianism as "psychic refusal of the 'truth' of women's castration." In this context, she remarks "it is interesting to note that Freud referred to both his homosexual women patients as 'feminists.'"[4] Even more interesting, to me, is to see a lesbian theorist decisively and explicitly re-appropriate, in feminist perspective, this most contended of Freud's notions, the masculinity complex in women. For, in taking this step – a very important one, in my opinion, without which our theorizing may just keep on playing in the pre-Oedipal sandbox – Hamer has left behind years of debates on Freud's sexism and feminist outrage, and volumes on Freud's historical limitations and feminist exculpation (debates and volumes, I may add, to which I have myself contributed in some measure). But when she then attempts to define lesbian desire, in Lacanian terms, she runs aground of the corollary to the masculinity complex, namely, the castration complex. This latter, she states, we must refuse:

> Classically, lesbians are thought to pretend possession of the phallus ...and are thus aligned, albeit fraudulently, on the side of masculinity. In this rather simplistic account lesbian desire becomes near impossible; desire cannot exist *between lesbians*, since they are both on the same side of desire, or, if a lesbian does experience desire, it is bound to be towards a feminine subject who could only desire her back as though she were a man. However, as I have suggested, lesbianism is less a claim to phallic possession (although it may be

this too) than it is a refusal of the meanings attached to castration. As such it is a refusal of any easy or straightforward allocation of masculine and feminine positions around the phallus. Instead it suggests a much more fluid and flexible relationship to the positions around which desire is organized.

(p. 147)

The problem with this solution – the "refusal of the meanings attached to castration" – is that it begs the question: in the Lacanian framework, symbolic castration is the condition of desire and what constitutes the paternal phallus as the "allocator" of positions in desire. In other words, castration and the phallus as signifier of desire go hand in hand, one cannot stir without the other. Thus, to reject the notion of castration (to refuse to rethink its terms) is to find ourselves without symbolic means to signify desire.

In this paper, I will up Hamer's defiant gesture and, just as she re-appropriates the masculinity complex, I want to reappropriate castration and the phallus for lesbian subjectivity, but in the perspective of Freud's negative theory of perversion. I will propose *a model of perverse desire* based on the one perversion that Freud insisted was not open to women – fetishism.

I take as my starting point a classic text of lesbianism, the classic novel of female sexual inversion, Radclyffe Hall's *The Well of Loneliness*, which, from its obscenity trial in London in 1928 to well into the 1970s, has been the most popular representation of lesbianism in fiction.[5] Thus, it needs no other introduction, except a word of warning: my reading of a crucial passage in the text – crucial because it inscribes a certain fantasy of the female body that works against the grain of the novel's explicit message – is likely to appear far fetched. This is so, I suggest, because my reading also works against the heterosexual coding of sexual difference (masculinity and femininity) which the novel itself employs and in which it demands to be read.

The scene at the mirror

The passage I selected occurs during Stephen's love affair with Angela Crosby, at the height of her unappeased passion and jealousy for the woman who, Stephen correctly suspects, is having an affair with Roger, her most loathed rival. The only things in which Stephen is superior to Roger are social status and, even more relevant to Angela, wealth: Stephen is an independently rich woman at the age of 21 and some day will be even richer. Though bothered by this "unworthy" thought Stephen nevertheless seeks to use her money and status to advantage; to impress Angela, she buys her expensive presents and orders herself "a rakish red car" as well as several tailor-made suits, gloves, scarves, heavy silk stockings, toilet water and carnation-scented soap. "Nor could she resist," remarks the narrator, "the lure of pyjamas made of white crêpe de Chine [which] led to a man's dressing-gown of brocade – an amazingly ornate garment" (p. 186). And yet, "on her way back in the train to Malvern, she gazed out of the window with renewed desolation. Money could not buy the one thing that she needed in life; it could not buy Angela's love." Then comes the following short section (book II, chapter 24, section 6):

> That night she stared at herself in the glass; and even as she did so she hated her body with its muscular shoulders, its small compact breasts, and its slender flanks of an athlete. All her life she must drag this body of hers like a monstrous fetter imposed on her spirit. This strangely ardent yet sterile body that must worship yet never be worshipped in return by the creature of its adoration. She longed to maim it, for it made her feel cruel; it was so white, so strong and so self-sufficient; yet withal so poor and unhappy a thing that her eyes filled with tears and her hate turned to pity. She began to grieve over it, touching her breasts with pitiful fingers, stroking her shoulders, letting her hands slip along her straight thighs – Oh, poor and most desolate body!
>
> Then, she, for whom Puddle was actually praying at that moment, must now pray also, but blindly; finding few words that seemed lworthy of prayer, few words that seemed to encompass her meaning – for she did not know the meaning of herself. But she loved, and loving groped for the God who had fashioned her, even unto this bitter loving.
>
> (pp. 186–7)

The typographical division that separates the last sentence of the first paragraph, describing the movement of Stephen's hands and fingers on her own body, from the second and last sentence of the second paragraph cannot disguise the intensely erotic significance of the scene. At face value, the paragraph division corresponds to the ideological division between body and mind, or "spirit", announced in the first paragraph ("all her life she must drag this body of hers like a monstrous fetter imposed on her spirit"), so that the physical, sexual character of Stephen's unappeased love and thwarted narcissistic desire is displaced onto an order of language which excludes her – the prayer to a distant, disembodied God by one who can pray to him because she also has no body, that is, Puddle, Stephen's tutor and companion, and her desexualized double. While in the first paragraph Stephen "stares" at her own body in the mirror, in the second she is blind, groping – a sudden reversal of the terms of vision which recalls the "nothing to see" of the female sex in psychoanalysis and, in a rhetorical sleight-of-hand, forecloses its view, its sensual perception, denying its very existence.

But a few words belie the (overt) sublimation and the (covert) negation of the sexual that the second paragraph would accomplish: "Then", the first word in it, temporally links the movement of the hands in the preceding paragraph to the final words of the second, "even unto this bitter loving", where the shifter "this" relocates the act of loving in a present moment that can only refer to the culmination or conclusion of the scene interrupted by the paragraph break, the scene of Stephen in front of the mirror touching her breasts with pitiful fingers, stroking her shoulders, letting her hands slip along her straight thighs [and, if we might fantasize along with the text, watching in the mirror her own hands move downward on her body] . . . even unto this bitter loving". No wonder the next paragraph must rush in to deny both her and us the vision of such an intolerable act.

The message of the novel is clear: Stephen's groping blind and wordless toward an other who should provide the meaning, but does not, only leads her back to the reality her body, to a "bitter" need which cannot accede to symbolization and so must remain, in Lady Gordon's words, "this *unspeakable* outrage that you call love" (p. 200, emphasis added). As the passage antici- pates, the narrative resolution can only be cast in terms of renunciation and salvation in an order of language that occludes the body in favor of spirit and, with regard to women specifically, forecloses the possibility of any autonomous and non-reproductive female sexuality. Stephen's "sacrifice" of her love for Mary – and, more gruesome still, of Mary's love for her – which concludes Radclyffe Hall's "parable of damnation" (in Catharine Stimpson's words) will ironically reaffirm not just the repression, but indeed the fore- closure or repudiation of lesbianism as such; that is to say, the novel cannot conceive of an autonomous female homosexuality and thus can only confirm Stephen's view of herself as a "freak", a "mistake" of nature, a masculine woman.[6]

The passage, however, contains another, ambiguous message. The scene represents a fantasy of bodily dispossession, the fantasy of an unlovely/ unlovable body – a body not feminine or maternal, not narcissistically cherished, fruitful or productive, nor, on the other hand, barren (as the term goes) or abject, but simply imperfect, faulty and faulted, dispossessed, inadequate to bear and signify desire. Because it is not feminine, this body is inadequate as the object of desire, to be desired by the other, and thus inadequate to signify the female subject's desire in its feminine mode; how- ever, because it is masculine but not male, it is also inadequate to signify or bear the subject's desire in the masculine mode. Stephen's body is not feminine, on the stereotypical Victorian model of femininity that is her mother Anna. It is "ardent and sterile," and its taut muscular strength, whiteness and phallic self-sufficiency make Stephen wish to "maim" it, to mark it with a physical, indexical sign of her symbolic castration, her captivity in gender and her semiotic dispossession ("she did not know the meaning of herself") by the Other, the God who made her "a freak of a creature." For she can "worship" the female body in another but "never be worshipped in return." If she hates her naked body, it is because that body is masculine, "so strong and so self-sufficient," so phallic. The body she desires, not only in Angela but also autoerotically for herself, the body she can make love to, is a feminine, female body. Paradoxical as it may seem, the "mythic mannish lesbian" (in Esther Newton's wonderful phrase) wishes to have a feminine body, the kind of female body she desires in Angela and later in Mary – a femme's body. How to explain such a paradox?

The fantasy of castration

I want to argue that this fantasy of bodily dispossession is subtended by an original fantasy of castration, in the sense elaborated by Laplanche and Pontalis, with the paternal phallus symbolically present and visible in the muscular, athletic body of Stephen who "dares" to look so like her father.[7] It is that paternal phallus, inscribed in her very body, which imposes the taboo that renders the female body (the mother's, other women's, and her own) for ever inaccessible to Stephen, and thus signifies her castration. But before I discuss in what ways, and in what sense, the notion of castration may be reformulated in relation to lesbian subjectivity, I want to point out how the paradox in the passage cited above contradicts, or at least complicates, the more immediate reading of Stephen's masculinity complex. For on the one hand, Stephen's sense of herself depends on a strong masculine identification; yet, on the other hand, it is precisely her masculine, phallic body which bears the mark of castration and frustrates her narcissistic desire in the scene at the mirror. So, in this case, it is not possible simply to equate the phallic with the masculine and castration with the feminine body, as psychoanalysis would have it. And hence the question: What does castration mean in relation to lesbian subjectivity and desire?

The difficulty of the psychoanalytic notion of castration for feminist theory is too well known to be rehearsed once again. To sum it up in one sentence, the problem lies in the definition of female sexuality as *complementary* to the physiological, psychic, and social needs of the male, and yet as a *deficiency vis-à-vis* his sexual organ and its symbolic representative, the phallus – a definition which results in the exclusion of women not from sexuality (for, on the contrary, women are the very locus of the sexual), but rather from the field of desire. There is another paradox in this theory, for the very effectiveness of symbolic castration consists precisely in allowing access to desire, the phallus representing at once the mark of difference and lack, the threat of castration, and the signifier of desire. But access to desire through symbolic castration, the theory states, is only for the male. The female's relation to symbolic castration does not allow her entry into the field of desire as subject, but only as object.

This is so, Freudians and Lacanians join forces in saying, because women lack the physical property that signifies desire: not having a penis (the bodily representative and support of the libido, the physical referent which in sexuality, in fantasy, becomes the signifier, or more properly the sign-vehicle, the bearer, of desire), females are effectively castrated, symbolically, in the sense that they lack – they do not have and will never have – the paternal phallus, the means of symbolic access to the first object of desire that is the mother's body. It is the potential for losing the penis, the *threat* of castration, that subjects the male to the law of the father and structures the male's relation to the paternal phallus as one of insufficiency; and it is that potential for loss which gives the penis its potential to attain the value or the stature of the paternal phallus. Having nothing to lose, the theory goes, women cannot desire; having no phallic capital to invest or speculate on, as men do, women cannot be investors in the marketplace of desire but are instead commodities that circulate in it.[8]

Feminist theorists, following Lacan, have sought to disengage the notion of castration from its reference to the penis by making it purely a condition of signification, of the entry into language, and thus the means of access to desire. Silverman, for example, states: "One of the crucial features of Lacan's redefinition of castration has been to shift it away from this obligatory anatomical referent [the penis] to the lack induced by language."[9] Yet the semiotic bond between the signification of the phallus and the "real" penis remains finally indissoluble: "No one has the phallus but the phallus is the male sign, the man's assignment. . . . The man's masculinity, his male world, is the assertion of the phallus to support his having it."[10]

In all such arguments, however, nearly everyone fails to note that the Lacanian framing of the question in terms of having or being the phallus is set in the perspective of normative heterosexuality (which both analysis and theory seek to reproduce in the subject), with the sexual difference of man and woman clearly mapped out and the act of copulation firmly in place.[11] But what if, I ask, we were to frame the question of the phallus and the fantasy of castration in the perspective provided by Freud's negative theory of perversion?

With regard to the passage from *The Well of Loneliness* (but it could be shown of other lesbian texts as well), let me emphasize that, if it does inscribe a fantasy of castration, it also, and very effectively, speaks desire, and thus is fully in the symbolic, in signification. Yet the desire it speaks is not masculine, not simply phallic. But, if the phallus is both the mark of castration and the signifier of desire, then the question is: What manner of desire is this? What acts as the phallus in this lesbian fantasy? I will propose that it is not the paternal phallus, or a phallic symbol, but something of the nature of a fetish, something which signifies at once the absence of the object of desire (the female body) and the subject's wish for it.

A model of perverse desire

In the clinical view of fetishism, the perversion is related to the subject's disavowal of the mother's castration, which occurs by a splitting of the ego as a defense from the threat of castration. Disavowal implies a contradiction, a double or split belief: on the one hand, the recognition that the mother does not have a penis as the father does; and yet, on the other hand, the refusal to acknowledge the absence of the penis in the mother. As a result of this disavowal, the subject's desire is metonymically displaced, diverted onto another object, part of the body, clothing, etc., which acts as "substitute" (Freud says) for the missing maternal penis. In this way, Freud writes, to the child who is to become a fetishist "the woman has got a penis, in spite of everything, but this penis is no longer the same as it was before. Something else *has* taken its place, has been appointed its substitute, as it were, and now inherits the interest which was formerly directed to its predecessor."[12] In this diversion consists, for Freud, the *perversion* of the sexual instinct, which is thus diverted or displaced from its legitimate object and reproductive aim. But since the whole process, the disavowal [*Verleugnung*] and the displacement [*Verschiebung*], is motivated by the subject's fear of his own possible castration, what it brings into evidence is the fundamental role in fetishism of the paternal phallus (that which is missing in the mother). And this is why, Freud states, fetishism does not apply to women: they have nothing to lose, they have no penis, and thus disavowal would not defend their ego from an already accomplished "castration."

However, argues an interesting essay by Leo Bersani and Ulysse Dutoit, Freud placed too much emphasis on the paternal phallus. "The fetishist can see the woman as she is, without a penis, because he loves her with a penis somewhere else," they say:

> The crucial point – which makes the fetishistic object different from the phallic symbol – is that the success of the fetish depends on its being seen as authentically different from the missing penis. With a phallic symbol, we may not be consciously aware of what it stands for, but it attracts us because, consciously or unconsciously, we perceive it *as* the phallus. In fetishism, however, the refusal to see the fetish as a penis-substitute may not be simply an effect of repression. The fetishist has displaced the missing penis from the woman's genitals to, say, her underclothing, but we suggest that if he doesn't care about the underclothing resembling a penis it is because: (1) he knows that it is not a penis; (2) he doesn't want it to be only a penis; and (3) he also knows that *nothing* can replace the lack to which in fact he has resigned himself.[13]

Thus, to the fetishist, the fetish does much more than *re-place* the penis, "since it signifies something which was never anywhere": it "derange[s] his *system of desiring*," even as far as "deconstructing and mobilizing the self." Unlike a phallic symbol, which stands for the perceived penis, the fetish is a "fantasy-phallus," "an inappropriate object precariously attached to a desiring fantasy, unsupported by any perceptual memory." Fetishism, they conclude, outlines a model of desire dependent on

> an ambiguous negation of the real. . . . This negation creates an interval between the new object of desire and an unidentifiable first object, and as such it may be the model for all substitutive formations in which the first term of the equation is lost, or un- locatable, and in any case ultimately unimportant.

And they suggest that

> the process which *may* result in pathological fetishism can also have a permanent psychic validity of a formal nature.
>
> (p. 71, emphasis added)

I will follow up their argument and propose that if – and admittedly it's a big if, but not a speculation alien to or unprecedented in psychoanalytic theory[14] – if the psychic process of disavowal that detaches desire from the paternal phallus in the fetishist can *also* occur in other subjects, and have enduring effects or formal validity as a psychic process, then this "formal model of desire's mobility," which I prefer to call *perverse desire*, is eminently applicable to lesbian sexuality.

The fetish as fantasy-phallus

Consider the following three statements from their essay cited above, with the word lesbian in lieu of the word fetishist: 1) the lesbian can see the woman as she is, without a penis, because she loves her with a penis somewhere else; 2) the lesbian also knows that nothing can replace the lack to which in fact she has resigned herself; 3) lesbian desire is sustained and signified by a fetish, a fantasy-phallus, an inappropriate object precariously attached to a desiring fantasy, unsupported by any perceptual memory. In other words, what the lesbian desires in a woman and in herself ("the penis somewhere else") is indeed not a penis but the whole or perhaps a part of the female body, or something metonymically related to it, such as physical, intellectual or emo- tional attributes, stance, attitude, appearance, self-presentation; and hence the importance of performance, clothing, costume, etc. She knows full well she is not a man, she doesn't have the paternal phallus, but that does not necessarily mean she has no means to signify desire: the fantasy-phallus is at once what signifies her desire and what she desires in a woman. As Joan Nestle put it,

> For me, the erotic essence of the butch–femme relationship was the external difference of women's textures and the bond of know- ledgeable caring. I loved my lover for how she stood as well as for

what she did. Dress was a part of it: the erotic signal of her hair at the nape of her neck, touching the shirt collar; how she held a cigarette; the symbolic pinky ring flashing as she waved her hand. I know this sounds superficial, but all these gestures were a style of self-presentation that made erotic competence a political statement in the 1950s. . . . Deeper than the sexual positioning was the overwhelming love I felt for [her] courage, the bravery of [her] erotic independence.[15]

The object and the signifier of desire are not anatomical entities, such as the female body or womb and the penis respectively; they are fantasy entities, objects or signs that have somehow become "attached to a desiring fantasy" and for that very reason may be "inappropriate" (to signify those anatomical entities) and precarious, not fixed or the same for every subject, and even unstable in one subject. But if there is no privileged, founding object of desire, if "the objects of our desires are always substitutes for the objects of our desires" (as Bersani and Dutoit put it), nevertheless desire itself, with its movement between subject and object, between the self and an other, is founded on difference and dependent on "the sign which describes both the object and its absence" (Laplanche and Pontalis).

This is why a notion of castration and a notion of phallus as signifier of desire are necessary to signify lesbian desire and subjectivity, although they must be redefined in reference to the female body, and not the penis. It is not just that fantasies of castration have a central place in lesbian texts, subjectivity and desire. It is also that what I have called the fetish or fantasy-phallus, in contradistinction to the paternal penis-phallus, serves as the bearer, the signifier, of difference and desire. Without it, the lesbian lovers would be merely two women in the same bed. The lesbian fetish, in other words, is any object, any "inappropriate object precariously attached to a desiring fantasy," any sign whatsoever, that marks the difference and the desire between the lovers – for instance, again in Nestle's words, "the erotic signal of her hair at the nape of her neck, touching the shirt collar," or "big-hipped, wide-assed women's bodies."

The wound and the scar

Returning, then, to the text started from, it may now be possible to see its fantasy of bodily dispossession as related to a somewhat different notion of castration. Let me recall for you the passage in The Well of Loneliness where, in describing Stephen's purchase of clothes intended to impress Angela – and they are, as we know, masculine-cut or mannish clothes – the narrator tells us: "Nor could she resist the lure of pyjamas made of crêpe de Chine [which] led to a man's dressing-gown of brocade – an amazingly ornate garment." Now we can be almost sure that Angela would never see those pyjamas and dressing gown. And yet Stephen *could not resist* their *lure*. Just as she hates her masculine body naked, so does she respond to the lure of masculine clothes; and we may remember, as well, the intensity with which both Stephen Gordon and her author Radclyffe Hall yearned to cut their hair quite short, against all the contemporary appearance codes. What I am driving at, is that masculine clothes, the insistence on riding astride, and all the other accoutrements and signs of masculinity, up to the war scar on her face, are Stephen's fetish, her fantasy-phallus. This does explain the paradox of the scene at the mirror, in which she hates her *naked* body and wants to "maim" it (to inscribe it with the mark of castration) precisely because it is masculine, "ardent and sterile . . . so strong and so self-sufficient," so phallic, whereas the body she desires and wants to make love to, another's or her own, is a feminine, female body.

Consider, if you will, this scene at the mirror as the textual reenactment of the Lacanian mirror stage which, according to Laplanche and Pontalis, constitutes the matrix or first outline of the ego.

> The establishment of the ego can be conceived of as the formation of a psychical unit paralleling the constitution of the bodily schema. One may further suppose that this unification is precipitated by the subject's acquisition of an image of himself founded on the model furnished by the other person – this image being the ego itself. Narcissism then appears as the amorous captivation of the subject by this image. Jacques Lacan has related this first moment in the ego's formation to that fundamentally narcissistic experience which he calls the *mirror stage*.[16]

What Stephen sees in the mirror (the image which establishes the ego) is the image of a phallic body, which the narrator has taken pains to tell us was so from a very young age, a body Stephen's mother found "repulsive." This image which Stephen sees in the mirror does not accomplish "the amorous captivation of the subject" or offer her a "fundamentally narcissistic experience", but on the contrary inflicts a narcissistic wound, for that phallic body, and thus the ego, cannot be narcissistically loved.[17]

The fantasy of castration, here, is explicitly associated with a failure of narcissism, the lack or threatened loss of a *female* body, from which would derive in consequence the defense of disavowal, the splitting of the ego, the ambiguous negation of the real. What is formed in the process of disavowal, then, is not a phallic symbol, a penis-substitute (indeed Stephen hates her masculine body), but a fetish – something that would cover over or disguise the narcissistic wound, and yet leave a scar, a trace of its enduring threat. Thus Stephen's fetish, the signifier of her desire, is the sign of both an absence and a presence: the denied and wished-for female body is both displaced and represented in the fetish, the visible signifiers and accoutrements of masculinity, or what Esther Newton has called a "male body drag." That is the lure of the mannish lesbian – a lure for her and for her lover. The fetish of masculinity is what lures and signifies her desire, and what in her lures her lover, what her lover desires in her. Unlike the masculinity complex, the lesbian fetish of masculinity does not refuse castration but disavows it; the threat it holds at bay is not the loss of the penis in women but the loss of the female body itself, and the prohibition of access to it.

To conclude, in this lesbian text, the subject's body is inscribed in a fantasy of castration, which speaks a failure of narcissism. I cannot love myself, says the subject of the fantasy, I need another woman to love me (Anna Gordon was repulsed by her daughter) and to love me sexually, bodily (the sexual emphasis is remarked by the masturbation scene barely disguised in the passage). This lover must be a woman, not a man, and not a faulty woman, dispossessed of her body (such as I am) but a woman-woman, a woman embodied and self-possessed, as I would want to be and as I can only become by her love.

> But in fact we were always like this,
> rootless, dismembered: knowing it makes the difference.
> Birth stripped our birthright from us,
> tore us from a woman, from women, from ourselves
> so early on
> and the whole chorus throbbing at our ears
> like midges, told us nothing, nothing
> of origins, nothing we needed
> to know, nothing that could re-member us.
>
> Only that it is unnatural,
> the homesickness for a woman, for ourselves,
> for that acute joy at the shadow her head and arms
> cast on a wall, her heavy or slender
> thighs on which we lay, flesh against flesh,
> eyes steady on the face of love; smell of her milk, her sweat,
> terror of her disappearance, all fused in this hunger
> for the element they have called most dangerous, to be
> lifted breathtaken on her breast, to rock within her
>
> – even if beaten back, stranded again, to apprehend
> in a sudden brine-clear thought
> trembling like the tiny, orbed, endangered
> egg-sac of a new world:
> *This is what she was to me, and this*
> *is how I can love myself –*
> *as only a woman can love me.*
>   (Adrienne Rich, from "Transcendental Etude")[18]

Nevertheless, the fantasy of dispossession is so strong in the text that Stephen ends up still dispossessed, in spite of having had (and given up) a woman lover. Although the sense of belonging to "one's own kind," the political presence of a community – the "thousands" and "millions" like her for whom Stephen writes and implores God at the close of the novel, mirroring the author's purpose in writing it and predicting its enormous success and impact on its readers – can soothe the pain and provide what Radclyffe Hall calls "that steel-bright courage . . . forged in the furnace of affliction," nevertheless the narcissistic wound remains, unhealed under the scar that both acknowledges and denies it. The wound and the scar, castration and the fetish, constitute an original fantasy that is repeated, re-enacted in different scenarios, in lesbian writing and in lesbian eros.

Notes

1 A shorter version of this chapter was presented as a paper at the 1990 MLA convention in Chicago.
2 Havelock Ellis, "Sexual Inversion in Women," *Alienist and Neurologist*, vol. 16 (1895), pp. 141–58. A similar view of female homosexuality is expressed in Ernest Jones, "The Early Development of Female Sexuality," *International Journal of Psycho-Analysis*, vol. 8 (1927), pp. 459–72, and later repeated by Jacques Lacan, "Guiding Remarks for Congress on Feminine Sexuality," in *Feminine Sexuality*, ed. Juliet Mitchell and Jacqueline Rose, New York: W. W. Norton, 1983, pp. 96–7.
3 This reconceptualization of perversion is made possible by Freud's notion of the sexual drive as independent of its object. See Arnold Davidson, "How To Do the History of Psychoanalysis: A Reading of Freud's *Three Essays on the Theory of Sexuality*," in *The Trial(s) of Psychoanalysis*, ed. Françoise Meltzer, Chicago: University of Chicago Press, 1987–8, pp. 39–64.
4 Diane Hamer, "Significant Others: Lesbianism and Psychoanalytic Theory," *Feminist Review*, no. 34 (Spring 1990), pp. 143–5.
5 Radclyffe Hall, *The Well of Loneliness*, New York: Avon Books, 1981. See Esther Newton, "The Mythic Mannish Lesbian: Radclyffe Hall and the New Woman," *Signs*, vol. 9, no. 4 (Summer 1984), pp. 557–75, and Catharine R. Stimpson, "Zero Degree Deviancy: The Lesbian Novel in English," in *Writing and Sexual Difference*, ed. Elizabeth Abel, Chicago: University of Chicago Press, 1982, pp. 243–60. Rebecca O'Rourke, *Reflecting on THE WELL OF LONELINESS* Routledge London: 1989 contains an interesting, if partial, study of the novel's reception.
6 The distinction between repression (*Verdrangung*) and repudiation or foreclosure (*Verwerfung*) is that, while the repressed contents are accessible to consciousness and to be worked over, for example in analysis, what is repudiated or foreclosed is permanently repressed, for ever lost to memory.
7 J. Laplanche and J.-B. Pontalis, "Fantasy and the Origins of Sexuality," in *Formations of Fantasy*, ed. Victor Burgin, James Donald and Cora Kaplan, London: Methuen, 1986, pp. 5–34.

8 See Luce Irigaray's critique in "Commodities Among Themselves," in *This Sex Which Is Not One*, trans. Catherine Porter Ithaca: Cornell University Press, 1985.

9 Kaja Silverman, "Fassbinder and Lacan," *Camera Obscura*, no. 19 (1989), p. 79.

10 Stephen Heath, "Joan Riviere and the Masquerade," in *Formations of Fantasy*, p. 55.

11 As Lacan himself puts it, "The phallus is the privileged signifier of that mark where the share of the logos is wedded to the advent of desire. One might say that this signifier is chosen as what stands out as most easily seized upon in the real of sexual copulation, and also as the most symbolic in the literal (typographical) sense of the term, since it is the equivalent in that relation of the (logical) copula. One might also say that by virtue of its turgidity, it is the image of the vital flow as it is transmitted in generation" (*Feminine Sexuality*, p. 82).

12 Sigmund Freud, "Fetishism," in *The Standard Edition of the Complete Psychological Works of Sigmund Freud*, trans. and ed. James Strachey, London: Hogarth Press, 1953–74, vol. 21, pp. 155–6.

13 Leo Bersani and Ulysse Dutoit, *The Forms of Violence: Narrative in Assyrian Art and Modern Culture*, New York: Schocken Books, 1985, pp. 68–9.

14 Juliet Mitchell also extrapolates from disavowal and fetishism a more general, formal model of the constitution of the subject in her "Introduction I" to Lacan, *Feminine Sexuality*, p. 25.

15 Joan Nestle, *A Restricted Country*, Ithaca: Firebrand Books, 1987, pp. 104–5.

16 Laplanche and Pontalis, "Narcissism," in *The Language of Psychoanalysis*, New York: W. W. Norton, p. 256.

17 "A second theoretical characteristic of the castration complex is its impact upon narcissism: the phallus is an essential component of the child's self-image, so any threat to the phallus is a real danger to this image; this explains the efficacity of the threat, which derives from the conjunction of two factors, namely, the primacy of the phallus and the narcissistic wound" (Laplanche and Pontalis, *The Language of Psychoanalysis*, p. 57).

18 Adrienne Rich, *The Dream of a Common Language: Poems 1974–1977*, New York: W. W. Worton, 1978, pp. 75–6.

14

READING THE SEXED BODIES AND SPACES OF GYMS

Lynda Johnston

Introduction

The body builder . . . is involved in actively reinscribing the body's skeletal frame through the inscription of muscles (the calculated tearing and rebuilding of selected muscle according to the exercise chosen) and of posture and internal organs.

(Grosz 1994: 143)

Female body building disrupts binary notions of femininity/masculinity. The specific materiality of women's[1] muscled (built) bodies provides the ground for contesting hierarchical dualisms such as femininity/masculinity, nature/culture, body/mind and sex/gender. Not only does it disrupt notions of the fixed biology of "the body"[2] but also provides a new space for "thinking through the body" (Gallop 1988) in geography. Bodies become sexed according to a particular place and time. I argue that both female "built" bodies and their training environments are politicized sites of change. Further, I argue (female) *body* (builders) and *environments* (of training) are mutually constitutive. The actual materiality of bodies is constructed and inscribed by the environment. The specificity of this relationship of *bodies* and *environment* provides a window through which to examine contradictory aspects of performative corporeality of female body builders.

The matter of bodies has become a recurrent theme in recent, diverse feminist writings.[3] Reconsideration of bodies has also been the task of some geographers.[4] Epistemological implications are that geographers are no longer able to start with "the body" as a natural given. Rather, attention needs to be directed towards the theorizing of bodies as "made" and "remade" in specific contexts of place, time and identity politics. Biological, social and sexual "truths" of "the body" are contested by built female bodies.

My own gym training and interest in bodies has prompted three questions which guide this chapter. My first question is: why and in what ways are the environments of fitness centres sexed spaces? To an extent this question is anchored in geographical analyses of gendered space. This leads on to my second question: how and in what ways do the sexed spaces of fitness centres construct bodies feminine and/or masculine? This question requires a more detailed recognition and analysis of contemporary debates about nature/ culture dualisms. My third and final question is: how and in what ways do female body builders disrupt the feminine/masculine spaces of gyms? Three readings of female built bodies are offered in response to this question.

This chapter draws from contemporary feminist and poststructuralist challenges to notions of what constitutes the identity of human bodies and their capacity to change. In what follows I discuss fitness centres, specifically, the hard core gym[5] to explore the claim that the environments of gyms confirm traditional masculine and feminine body stereotypes. Political struggles occur in specific gym spaces. Hegemonic masculinist definitions of femininity are both reinforced and resisted in these gym spaces. It is this terrain that the specific and transgressive corporeality of female body builders are providing challenges to stereotypical constructions of femininity and masculinity.

Professional body builders Lenda Murray (seven times winner of the Ms Olympia title and regularly featured in body building magazines) and Bev Francis (see Figure 14.1) are, in part, the reason I am motivated to "gaze"[6] at female body builders. I suggest that there are a number of possible readings of female built bodies. In this chapter, I focus on three readings: first, bodies are "built" in order to become more feminine – "docile" and self disciplining bodies; second, bodies are "built" as a transgressive gesture whereby feminine bodies obtain those privileges usually associated with men; third, built bodies are driven by the desire for corporeal erotic sensations, such as the pain/pleasure of feeling and watching the "pump." The erotic gaze also produces/personifies female body builders as abject. Ironically, some of the body builders could be read all three ways – feminine, transgressive and erotic.

These three non-exclusive readings and the examination of the produced spaces of gyms are provided by the record of empirical work I have conducted over eighteen months.[7] I critically examine sexed "built" bodies and ask the question posed by Cream (1995b: 31): "What, then, is this thing we call the sexed body?"

Western philosophical dualities

Debates which question the construction of knowledge and discourse in geography have elaborated on the existing critique of Western "scientific" methods (Bondi 1990; Johnson 1994; McDowell 1991, 1992). Contemporary deconstructive debates on modernism/humanism have identified geography as a masculinist discourse. It has been argued that Western philosophical traditions of dualities shape "our" conceptualization of people and places

Figure 14.1 Bev Francis
Source: Teper and Neveux, 1994: 43. Reprinted with
permission from *Ironman* magazine

(Grosz 1993, 1995). Feminists such as Moria Gatens (1988) work to make explicit what remains intolerable in "malestream knowledges" (Grosz 1988: 56) and to detail the exclusionary practices of Western thought.

Western rationalist tradition entails a radical separation of body and mind that accords primacy to the mind. Johnson (1989) argues that geography, as well as the rest of the social sciences, has been built on the mind/body dualism. She insists that the mind and body have been conceived as separate and acting on each other; the mind traditionally connected to "positive" terms such as reason, rationality, subject, culture, consciousness and masculinity. "The body," however, has been associated with "negative" terms such as passion, irrationality, object, nature, non-consciousness and femininity.

Treatment of "the body" in geographical literature has not lead to the body simply being absent and the mind being present in geographical discourse:

> Rather, it is as though the body has acted as geography's Other; it has been both denied and desired depending on the particular school of geographical thought under consideration.
>
> (Longhurst 1995: 99)

Some feminist geographers have engaged in deconstructing and reconstructing these binaries in an explicitly poststructural, corporeal, feminist politics. Robyn Longhurst (1995: 97) suggests an "upheaval" of these dualisms is needed for a transformed geographical discipline.

My argument is that bodies cannot be understood simply as "raw material," that is, non- or pre-social. On the other hand, I do not wish to imply that bodies can be regarded as purely a social, cultural and signifying effect lacking in their own "weighty materiality" (Grosz 1994: 21).

In the following section I present a case study of a Hamilton, Aotearoa/ New Zealand gym. I highlight the "feminine" and "masculine" gym spaces which help constitute traditionally sexed bodies. I suggest that the gym environment provides an order and organization that links bodies. It is in the condition and milieu of body building environments that corporeality is socially, sexually, politically and discursively produced. Female body builders are, in some instances, positioned, aligned and compliant within these traditional sexed spaces. Ironically, however, their transgressive corporeality and activities provide challenges to these gym spaces.

Sexing the gym spaces

Space is bound into power/knowledge relationships and therefore the spaces of gyms are central to the subjectivity of gym users. Gillian Rose (1995: 335) asserts that "particular imagined spatialities are constitutive of specific subjectivities. Identities are in part constituted by the kind of space through which they imagine themselves." Elsewhere I have argued (Johnston 1995:

16) "that the sexed space of the gym environment is an integral aspect of the construction of sexed bodies in particular ways." Elizabeth Grosz (1994: 142) pushes the constitutive relationship between bodies and environments further, arguing:

> it is crucial to note that these corporeal inscriptions do not simply adorn or add to a body that is basically given through biology; they help constitute the very biological organization of the subject – the subject's height, weight, coloring, even eye color, are constituted as such by a constitutive interweaving of genetic and environmental factors.

The socio-political structures of the hard core gym environment constructs particular kinds of bodies. These bodies confirm some of the most blatant feminine and masculine body stereotypes. Bodies are often reworked in this space to an explicit female/male binary. Hegemonic powers operate within gym environments actively producing and reproducing difference as a key strategy to create and maintain spatial divisions. Female body builders and their training regimes, which I discuss later in this chapter, disrupt the produced female/male spaces.

The gym I studied is divided into five distinct areas. These are the aerobics and circuit room; the room containing free weights and weight training machines; squash courts; exercycles/rowing machines/steppers; and an arena for team sports. Each area is central to the gym user's corporeality. Various users occupy different areas of the gym. Most noticeable is the use of the free weights facilities which is dominated by males. During the course of my participant observation this room usually contained between 20–40 men and 0–4 women at peak use time.

Images representing sexual differences are made meaningful in each different gym space. Step Reebok, a type of exercise to music, is an activity that is dominated by females. Step Reebok posters promoting the gyms activities are displayed on the fitness centre walls. The body in the poster, toned and muscular with very little body fat, is presented as a goal for those participating in the Step Reebok programme. The female in the poster is definitely a "female." The muscle bulk does not transgress the "allowed" amount for women, therefore she cannot be confused with a muscular male. The hair is long in an attempt to stay firmly on the feminine side of the feminine/masculine divide. Hegemonic constructions of masculinity and femininity are understood in relation to these stereotypes. The Reebok woman "fits": she is not out of place. The aerobics and circuit room is painted a pastel shade and is brightly lit. However, existing at the other end of the decor/sex space binary is the room containing the free weights which is painted black and blue. Until recently the name of the room was "The Black and Blue Room". The Black and Blue room, its name and decor, links weight training with (potential) masculine violence.

Members I have talked to (both female and male) communicate some dis-ease about being in a strongly male/female segregated environment. This effects the time of day they go to the gym, since at different times, different ratios of females/males are present. It also effects the places within the gym that they occupy, or move through. The following remark came from a woman waiting for a male friend who was weight training in "The Black and Blue Room:" "That's the male bonding room in there. I'm not going in to find him" (participant observation notes, 6 June 1994).

Space, therefore, becomes something to be negotiated, a territory owned by someone else. Most women do not train in the weight training room, but choose to train in the feminine space of the aerobics and circuit room. Iris Young (1990b) theorizes the spatiality of embodiment. She links an awareness of embodiment to women's sense of space not being their own. The following dialogue highlights the strongly sexed environment – segregated masculine and feminine spaces – of the hard core gym:

> On one of my evening work shifts at the gym as receptionist, two young secondary school women claimed, "Oh, we went into the men's room by mistake." I replied, "What, the changing room?" They said, "No, the weights' room with men in it." I then said, "Oh no, that's for everyone, not just men." They then told me about the reaction they received when using the facilities of the room. One male, who was waiting to use the equipment said, "This is for men – this room – women can use the other [aerobics and circuit] room.
> (Participant observation notes, 1 May 1994)

The two young women's experience of the hard core gym environment are similar to Rose's (1993b: 144) illustrations of "confinement in space." She argues that "we often do not gesture and stride, stretch and push the limits of our physical capabilities." The corporeal consequences of only doing aerobics and circuit training, without the free weights of "The Black and Blue" room is that women's muscle bulk is unlikely to become excessive (read unacceptable).

Female body builders disrupt the masculine weight training (male bonding) spaces. Sarah highlights the response to her training routines:

> *Lynda:* And when you started body building did anyone help you in the gym?
> *Sarah:* No they didn't help me um no. It was, um, it was more like a *man's gym*. It was out there on your own and your experience and you train on your own.
> (Interview, 22 June 1994, original emphasis)

Sarah reflects here on the masculine space of a "man's gym." The masculine environment she refers to is unsupportive, isolating and discouraging for her weight training. Although she was not asked to leave, she was not made to feel welcome. Historically, entry by women into gyms – an exclusive male environment – was not easily achieved.

> Back home I was finding that training facilities for women were few and far between, a long way from today's explosion of gyms and health studios. In fact, it was a real struggle to find places to train each week day. Finally at the eleventh hour my luck changed. . . . St. George's was a gym of the old school – old, damp, tough, rough and ready. After applying for membership for several months with little response I finally received a reply. It was positive. I had been accepted. . . . But all was not over yet. The first day I walked in I was greeted by a petition. On the notice board was a list of names supporting the motion that women members should be debarred. Fortunately good sense prevailed and the petition died a death.
>
> (Cheshire and Lewis 1985: 9)

Today women are not barred from entering gyms. However, once inside the gym, female body builders are subject to comments and innuendos made by some of the male users:

> *Anja:* Oh, you shouldn't be lifting that heavier weight, you might get muscley.
>
> *Jenny:* One guy said "Oh, do you need a hand with that?" And I was carrying it [the bar bell] over to prepare to set up and to do an exercise and I said, "No, I'm alright." He said "Oh, are *you* going to use it?" and I said "Yeah," and he had his mouth wide open.
>
> *Carmen:* And, like, quite often guys ask you, like, "What are you planning on doing with those weights?" as if you shouldn't be, you know, lifting those, you know. You should put them back down. Use something a bit lighter.
>
> *Jenny:* I often find that people will look, especially at a small woman, and they may not say anything but there is a constant look at like, "What is *she* doing with those weights?"
>
> (Focus group, 12 June 1994, original emphasis)

The male gym user may experience a double disruption in his conception of the "natural" order of femininity and masculinity. First, because femininity and "nature" are often considered to be so closely allied, any attempt to reconstruct the body is transgressive against the "natural" identity of the female body. Second, when female athletes use the technology of the hard core

250

gym to achieve physical muscularity – the male prerogative – they transgress the "natural" order of sexed identity.

In Western traditions, muscularity has belonged to the male sex. Body building for men can be seen as the fulfilment of a certain notion of masculinity and/or virility. Male body building can be interpreted as an attempt to render the whole body into the phallus, "creating the male body as hard, impenetrable, pure muscle" (Grosz 1994: 224). Ironically, it could also be argued that the display of hypermasculinity with the narcissistic reinvestment of male bodies, is a feminine activity.[8] The female body builder works her body and creates a differently sexed body which confuses and confounds traditional notions of "femininity." Muscles, supposedly "natural," are dangerous and out of place on women.

In the following section I present and discuss the specific materiality of female built bodies. I argue that the built female body disrupts the nature/ culture dichotomy and can be read in multiple ways. The first reading focuses on bodies which become more feminine or docile.

Becoming more feminine: docile bodies

One of the participants, Sarah explains her self monitoring observations concerning femininity:

> *Sarah:* If anything you create more of a femininity because you are constantly aware of how your body is looking, how your body is changing, um, yes, I think you . . . concentrate more on femininity.
>
> (Focus group, 12 June 1994)

In an attempt to understand the body builders' training environments, I draw on Foucault's (1976) *Discipline and Punishment*. In this text Foucault uses the notion of the panoptican, referring to Bentham's design for a prison that leaves prisoners perpetually exposed to a one way viewing tower, therefore likely to police themselves. The following description of power by Foucault (1980: 105) requires "minimum expenditure for maximum return" and its central organizing principle is that of discipline:

> There is no need for arms, physical violence, material constraints. Just a gaze. An inspecting gaze, a gaze which each individual under its weight will end by interiorizing to the point that s/he is her/his own overseer, each individual thus exercizing this surveillance over and against, her/himself. A superb formula: power exercised continuously and for what turns out to be at minimal cost.

Jenny explains her, and other female body builders', constant surveillance of femininity:

Jenny: I think it enhances it [femininity] a lot, you know. . . . You know [female body builders are] really beautiful. Can look after themselves and you know, I think it really enhances your femininity 'cause of all the posing routines especially. You have to be like very, very graceful.

(Interview, 10 July 1994)

Female body building accents the contradictory demands of contemporary femininity. The double bind of achieving a muscular body and remaining feminine demands a high level of body maintenance. The systems of surveillance in the gym (such as the gaze, reflections via mirrors, and comments) provide an atmosphere that (self) regulates body builders, forms of femininity and heterosexuality. Powerful panoptic technologies produce self monitored "docile" bodies (Foucault 1976: 135). Eating, like training, for body builders is a constant process of self monitoring. Fiona gives an example of social power expressed through what Foucault (1976) called "technologies of the self."

Fiona: I've had to start writing timetables. Not just routine timetables for the gym but whole day timetables. I make my lunch the night before, when I'm going to cook the steak and boil the rice, and stuff like that or what ever it is I'm having, and make sure I do it at the right time. I plan. I just, I've just starting doing that because there's a lot to plan, and I take my lunch to work and eat at odd times during the day compared to everyone else.

(Interview, 18 June 1994)

Timetables, schedules, a sedentary occupation to complement the muscle development and carefully prepared foods, in part constitute Fiona's built body. Fiona's notion of discipline, Foucault (1976: 139, original emphasis) would argue, is intimately linked to power. Biopower constitutes:

> the body as a machine: its disciplining, the optimization of its capacities, the extortion of its forces, the parallel increase of its usefulness and its docility . . . all of this was ensured by the procedures of power that characterized the *disciplines: an anatomo-politics of the human body.*

Fiona (and other body builders) are caught up in, and constituted by, other (institutional) biopower discourses relating to the sexed spaces of gyms. As body builders use power through the medium of their bodies, certain "contradictions" emerge. Some female body builders could be understood as docile bodies, and others transgress docile bodies. Female body builders can be partly understood within each regime. Sarah reinforces this double bind.

Whether built female bodies are admired or not, they are always subject to the gaze.

> *Sarah:* But some guys love it but some guys say just, "No way." Women shouldn't look like that no matter whether she got a lot of muscle or not much muscle. She shouldn't have any muscle at all. Some men are like that.
>
> (Focus group, 12 June 1994)

Sarah illustrates the ambiguity of being a female body builder. The feminine/masculine debate of weakness (female) and muscle (male) is played out consistently on and through corporeality. The next section teases out some of the transgressive notions of female body building.

Transgressive bodies

> New images of women are produced when some women develop strong, muscular bodies. And as female body builders defy canons of the feminine aesthetic, building their bodies beyond traditional limits, they destabilize feminine bodily identity and confuse gender.
>
> (Sawicki 1991: 64)

Many of the participants established themselves as transgressive of hegemonic constructions of femininity, referring to their bodies in terms of the natural and unnatural:

> *Fiona:* I really quite like the idea that it's unnatural. I don't know whether I should admit that or not? There's been massive changes in me and yes, I think I quite like the fact that it's a bit unnatural. In fact, people make comments and it just spurs me on all the more.
>
> (Focus group, 12 June 1994)

Transgressive bodies become unacceptable:

> *Jenny:* It's still not generally accepted for women to have big muscles you know.
>
> (Interview, 10 July 1994)

As a consequence of transgressive bodies being unacceptable, Fiona reminds us that: "There aren't many women who do it [body build]." (Focus group, 12 June 1994).

Female body builders also become transgressive, "unnatural" and dangerous when they reduce their body fat to accentuate muscles, especially before competing. As a consequence their breasts are reduced in size and in some cases eliminated, and menstruation ceases. Without recognizable signifiers of femininity (breasts), the female body builder becomes "dangerous" and transgressive. Breast implants are being used by top female body builders to maintain femininity. In an article by feminist Gloria Steinem (1994: 71) the once professional body builder, Bev Francis (see Figure 14.1), offers her opinion on the breast implant issue.

> All the top women body builders have implants now. That's one of the things that annoys me about body building. We're not supposed to be what conventional women look like, because we've built our bodies. How can we have low body fat and still have breasts? My sexuality isn't threatened enough to stuff things in my chest.

Particular bodily zones serve to emphasize women's difference from, and otherness to, men. Women's breasts (and genitals) are the loci of (potential) flows. Grosz (1994) drawing on Julia Kristeva's (1982) notion of the abject, develops an argument on sexed bodies, focusing on the elision of fluids in the male body and the derogation of the female body in terms of various forms of uncontrollable flow. Grosz (1994: 203) asks the question:

> Can it be that in the West, in our time, the female body has been constructed not only as a lack or absence but with more complexity, as a leaking, uncontrollable, seeping liquid; as a formless flow; as viscosity, entrapping, secreting; as lacking not so much or simply the phallus but self-containment – not a cracked or porous vessel, like a leaking ship, but a formlessness that engulfs all form, a disorder that threatens all order?

Traditionally, women's corporeality has been inscribed as a mode of seepage, as lived liquidity. A hard female body builder without breasts and temporarily without menstruation, subverts this representation. In order to establish the solid male body, there must be a contrast with the non-solid or liquid female body. Implicit in this categorization of bodies are the dualistic organizing mechanisms of Western thought. The Same or Self requires an Other against which to identify itself. For example, representations of male body builders in popular body building magazines often portray the hard male body builder (Self) with a soft female non-muscular (Other) body.

Idealized notions of femininity and compulsory heterosexual attractiveness are disrupted by female body builders. Sarah offers an example of being both constrained by heterosexual hegemony and encouraged by it. Her motivation for beginning body building came from her fiancé's input and encouragement. However, the encouragement for her to build muscles also has a ceiling:

Sarah: And, um, I thought, well, I can only give it a try and, um, so that's good having him there because *he likes how the female looks with muscles, without too much muscle.*

(Interview, 22 June 1994, my emphasis)

Sarah qualifies her fiancé's acceptance of and limits to female built bodies. However, the confusion of "how far to go" with her muscles seems not to be an issue when she discovers the pleasures of building her body bigger. Sarah neatly encapsulates the initial unnatural "horror" of women with muscles. This is turned around once she begins to see her own (erotic) muscled body.

Sarah: I used to think *no way.* I'm not going to be a woman with muscles because again I envisaged this big huge thing and until you start to push a little bit of weights and notice the change in your body you start to appreciate that, you know, it looks nice.

(Focus group, 12 June 1994, original emphasis)

The initial horror disappears I suggest, because a feeling of empowerment is entailed in transgressing the traditionally masculine domain of body building and granting herself the attributes usually associated with men: strength, stamina, muscularity and control. Desire to have the hardness and solidity that represents and defines men, means women forego the formlessness which has traditionally represented them.

Grosz (1994: 224) writes:

On [the] one hand, it [female body building] may, depending on the woman's goals, be part of an attempt to conform to stereotyped images of femininity, a form of narcissistic investment in maintaining her attractiveness to others and herself. On the other hand, it can be seen as an attempt on the part of the woman to take on for herself many of the attributes usually granted to men.

I suggest there is (at least) another reading of female built bodies.

Pumped/eroticized Bodies

The third and final reading of female body builders is through the pleasure/pain experience of the pump. The pump is the result of high intensity training and muscle stimulation. Muscles become engorged with blood and short of oxygen. The skin stretches tight over the muscles. The pump normally means muscle growth because more blood flushes through the muscles. A pumped muscle is larger than a non-pumped muscle. "Many advanced bodybuilders can pump their arms 2 inches bigger than their cold arm measurement" (Zulak 1994: 40). There is an associated thrill of watching the pump transform the body. One professional female body builder explains the pump as a pleasure/pain feeling. "Even the pain has its reward: it is invariably followed by the most desired effect of all – the pump" (Parker 1989: 4).

Fiona elaborates on the pleasure/pain experience:

> *Fiona:* It's not just the way you end up looking um, it's the feel of lifting the heavy weights, I really enjoy that. There's something about it. Like this morning – I did chest this morning – just chest alone and I left and I felt all tight across here [points to her chest]. It's a really good feeling and also you get to *like* the pain you get the next day. For legs you usually get sort [of] two days of pain. . . . I'm still feeling it from the other day on me.
>
> (Interview, 18 July 1994, original emphasis)

The effects of the pleasure/pain experience stay with Fiona in her everyday activities:

> *Fiona:* Yeah I feel good when I'm walking along and even covered up um, there is some sense of feeling good and holding your head up, walking square shoulders and knowing underneath you've got this changed body that you're working on.
>
> (Interview, 18 July 1994)

The corporeal sensations of the pump could be further explained by psychoanalysis, using the concept of the erotic body.

> The erotic is a very slippery notion, encompassing sexual marginality, excesses of flesh, and perverse pleasures of the body. The erotic is tied up with notions of pleasure and pain.
>
> (Worth 1994: 37)

Psychoanalytic theory provides various ways to rethink bodies and the erotic. (See also Pile 1996, for a discussion on the ways in which the space of the body might be psychoanalysed.) This theory claims that bodies are literally written on and inscribed by desire and signification, at the anatomical, physiological and neurological levels. Psychoanalytic theory has enabled feminists to reclaim bodies from the realms of biology/the natural/essential given, in order to see it as a psycho-social product. Bodies can be read as open to transformations in meaning and functioning, capable of being contested and re-signified (Wright 1992).

To understand how female body builders become erotically constructed by the gaze, I have drawn from the work of French philosopher and psycho-analyst Kristeva. She develops the notion of abjection in *The Powers of Horror* (1982). The feeling of abjection is one of disgust, often evoking nausea, as well as fascination. Abjection

> is an extremely strong feeling which is at once somatic and symbolic, and which is above all a revolt of the person against an external menace from which one has the impression that it is not only an external menace but that it may menace us from the inside. So it is a desire for separation, for becoming autonomous and also the feeling of impossibility of doing so.
>
> (Kristeva 1980: 135)

The abject exists on the border, but does not respect the border. It is "ambiguous," "inbetween," "composite" (Kristeva 1982: 4). The abject is what threatens identity. It is neither good nor evil, subject nor object, but something that threatens the distinctions themselves. Kelly Oliver (1993: 56) claims "Every society is founded on the abject – constructing boundaries and jettisoning the antisocial – every society may have its own abject." Iris Young (1990a) uses the concept of the abject to refer to societies' "ugly bodies." Young analyses the political importance of feelings of beauty and ugliness, cleanliness and filth, in the interactive dynamic and cultural stereotyping of racism, sexism, homophobia, ageism and ableism. I extend this deployment of the concept of abject to the (extremely) built female body which, I suggest, in the following example, personifies the abject. The female body builder threatens the border between female and male sexed bodies. "Looks like a man's body with a female head on it" (questionnaire response). She represents a loss of traditional female identity. I suggest these responses were not surprising when placed in the context of Kristeva's abjection theory. The abject threatens the unity and identity of both society and the subject. It questions the boundaries upon which they are constructed.

Back within the spaces of the gym, built female bodies become the abject. The (heavily muscled) chief fitness instructor reacted in the following fashion when he looked through my *"Female Bodybuilder"* magazine:

> Most of these women are bigger than me. She is, she's *massive*! Too many drugs for me. You don't aspire to that do you?
>
> (Participant observation notes, 9 October 1994, original emphasis)

This reaction is interesting because it comes from within the (biopower) institution of "muscle culture." The instructor expresses abhorrence towards performance enhancing drugs that help create the built female body. But he does not question the pictures of the many other (non-body builder) athletes displayed on the walls of the gym. Whether the bodies on display are enhanced by drugs is unknown. The instructor conflates the excessive musculature of the female body builders with performance enhancing drugs. I argue this conflation is the instructor's inability to accept the female "built" body thus establishing the female body builder as the abject.

The abject becomes a personalized horror: "Most of these women are bigger than me." To echo Kristeva (1980: 135) "[The abject] is not only an external menace but that it may menace us from the inside." Body building associations stipulate that the use of performance enhancing drugs is forbidden for those competing. However, the discourses around steroid use are sexed. The women are discouraged from using steroids based on the fact that they will become non-women. Men using steroids can be read as enhancing their "natural" testosterone levels. What is not clearly voiced in the debates over the use of steroids is the possibility of life threatening damage to the liver for both men and women. I suggest that the woman who takes steroids becomes too dangerous to accept. She disrupts the sexed body dichotomy.

These three readings of built female bodies are not exclusive, but they do serve to contest what "we" have traditionally understood sexed bodies to be.

Conclusions

The gym in my case study provides socio-political spaces which confirm masculine and feminine stereotypes. Not only do bodies shape the hard core gym environment, but they are shaped by it. Socio-political spatialities within the hard core gym environment help constitute the corporeality of gym users. Sexual difference creates masculine and feminine spaces, and these sexed spaces help create masculine and feminine bodies. The body becomes invested with relations of power and domination. Women are encouraged to participate in aerobics and circuit training, away from the masculine sexed space of the potentially violent "Black and Blue Room." Women dominate the aerobics and circuit room, while the men dominate and actively dis-courage women from participating in the free weights "Black and Blue Room." This particular gym confirms the construction of masculine and feminine spaces. Within the terrain of the hard core gym environment, female body builders provide challenges to hegemonic notions of sex and sexuality, as well as participating in hegemonic discourses which shape female bodies.

Docile, transgressive and eroticized built female bodies all sexually refigure "the body." Attention given to bodies and places contest the Western hierarchical dualisms which have traditionally shaped the construction of knowledge. Female body building disrupts the Western Othering of female bodies. The specificity of flows in and through sexed bodies is another way of rethinking sexual difference in corporeal terms.

Julia Cream (1992: 4) asks: "If the sexed body is no longer stable and fixed, no longer that biological bedrock of truth upon which we can build our

258

theories, then what is left?" I suggest what is left is a myriad of non-dichotomous human subject identities which resist the normative structures inherent in geographers' treatment of "the body," sex, sexuality and place.

Acknowledgements

An earlier version of this chapter was published in *Gender, Place and Culture: A Journal of Feminist Geography*. I am especially grateful to Robyn Longhurst and Robin Peace for support, suggestions and critical comments. Thanks also to Steve Pile and Heidi J. Nast for the comments on earlier drafts.

Notes

1 Throughout this chapter the categories "women," "woman" and "female" are used to "sex" body builders. While retaining "women," "woman" and "female" as meaningful categories, I have aimed to work against the reification of "women" with a single political meaning and single essence, outside of time and place but as occupying critical space within social structures.

2 Grosz (1992: 242) offers a useful definition of "the body:"

> a concrete, material, animate organization of flesh, nerves, muscles, and skeletal structure which are given unity, cohesiveness, and organization only through their psychical and social inscription as the surface and raw materials of an integrated and cohesive totality. The body is, so to speak, organically/biologically/naturally "incomplete;" it is indeterminate, amorphous, a series of uncoordinated potentialities which require social triggering, ordering, and long-term "administration," regulated in each culture and epoch by what Foucault has called "the micro-technologies of power."
>
> Where possible, I refer to multiple bodies, by insisting on pluralizing the term "body." "The body" is used in quotation marks to signify its traditional references as fixed/singular and static. I argue that bodies maintain multiple subject positions.

3 See for example Bordo 1989; Butler 1989, 1990, 1993; Douglas 1982; Gatens 1991a, 1991b, 1991c; Grosz 1987, 1988, 1989, 1992, 1993; Irigaray 1985; Kirby 1991; Kristeva 1982.

4 See Cream 1992, 1995a, 1995b; Dorn and Laws 1994; Dyck 1992; Johnson 1989; Kearns 1993; Longhurst 1994, 1995, 1996; Pile 1996; Rose 1991, 1993a, 1993b.

5 The "hard core" gym needs to be distinguished from the health spa or fitness studio environment. Over the last ten to fifteen years there has been a rapid increase in the United States of the number of health clubs commonly associated with meeting the exercise needs of middle class professionals. A similar trend exists in New Zealand. Body builders are actively discouraged from using this type of institution by the deliberate exclusion of certain types of the technology necessary to carry out the activity. The ambience created by the body builder is not one that is valued in health clubs or fitness studios (Mansfield and McGinn 1993).

6 The pleasure in the "gaze" for social scientists is often omitted from their accounts. However, Rose (1993b) is involved in theorizing the gaze of geographers using psychoanalytic techniques. Rose argues that geographers privilege the visual, the gaze. Looking in order to know without expressing the involvement of the researcher, makes for what Rose calls masculinist knowledge. Popular body building magazines frequently feature professional female body builders, such as Lenda Murray and Bev Francis. I became a regular "reader" of these magazines.

7 Research methods I used were focus groups, individual in-depth interviews, participant observation and discourse analysis of popular body building magazines. Focus groups enabled similar people to come together and discuss their gyms and their bodies. Individual in-depth interviews enabled participants to elaborate on previous focus group topics. Participant observation was carried out while I both worked as a receptionist and "worked out" in the gym of the case study. Pseudonyms have been used for all participants.

8 Several points could be added here on the subject of feminization of male body builders. For example: men's small waists, curvaceous pectoral muscles, hairless and oiled flesh are all corporeal indicators of femininity. The act of wearing high heel shoes by men to help develop calf muscles would be worth theoretical pursuit. Unfortunately there is not space in this chapter to further the argument that male body builders feminize themselves.

References

Bartky, S. (1988) Foucault, femininity, and the modernization of patriarchal power, in I. Diamond and L. Quinby (eds) *Feminism and Foucault: Reflections in Resistance* (Boston: Northeastern University Press).

Bondi, L. (1990) Feminism, postmodernism and geography: space for women?, *Antipode*, 22: 156–67.

Bordo, S. (1989) The body and the reproduction of femininity: a feminist appropriation of Foucault, in A. Jagger and S. Bordo (eds) *Gender/Body/Knowledge: Feminist Reconstructions of Being and Knowing* (London: Rutgers University Press).

Butler, J. (1989) Foucault and the paradox of bodily inscriptions, *The Journal of Philosophy*, 86 (11): 601–7.

Butler, J. (1990) *Gender Trouble: Feminism and the Subversion of Identity* (London: Routledge).

Butler, J. (1993) *Bodies That Matter: On the Discursive Limits of Sex* (London: Routledge).

Cheshire, C. and J. Lewis (1985) *Body Chic* (London: Pelham Books).

Cream, J. (1992) Sexed Shapes. Proceedings from the *Sexuality and Space Network Conference, September 1992*.

Cream, J. (1995a) Women on trial, in S. Pile and N. Thrift (eds) *Mapping the Subject: Geographies of Cultural Transformation* (London: Routledge).

Cream, J. (1995b) Re-solving riddles: the sexed body, in: D. Bell and G. Valentine (eds) *Mapping Desires: Geographies of Sexualities* (London: Routledge).

Deveaux, M. (1994) Feminism and empowerment: a critical reading of Foucault, *Feminist Studies*, 20(2): 223–47.

Diamond, I. and L. Quinby (eds) (1988) *Feminism and Foucault: Reflections on Resistance* (Boston, Northeastern University Press).

Dorn, M. and G. Laws (1994) Social theory, body politics and medical geography: extending Kearn's invitation, *Professional Geographer*, 46(1): 106–10.

Douglas, M. (1982) *Natural Symbols* (New York: Pantheon).

Dyck, I. (1992) Health and healthcare experiences of immigrant women: questions of culture, context and gender, in M. Hayes, L. Foster and H. Foster (eds) *Community, Environment and Health: Geographic Perspectives: Western Geographic Series*, 27 (Victoria: University of Victoria Press).

Foucault, M. (1976) *Discipline and Punishment: The Birth of the Prison* (New York: Vintage).

Foucault, M. (1980) *Power/Knowledge* (New York: Pantheon Books).

Gallop, J. (1988) *Thinking Through the Body* (New York: Columbia University Press).

Gatens, M. (1988) Towards a feminist philosophy of the body, in B. Caine, E. Grosz and M. de Lepevanche (eds) *Crossing Boundaries: Social and Political Theory* (Sydney: Allen and Unwin).

Gatens, M. (1991a) *Feminism and Philosophy: Perspectives of Difference and Equality* (Cambridge, Polity Press).

Gatens, M. (1991b) A critique of the sex/gender distinction, in S. Gunew (ed.) *A Reader in Feminist Knowledges* (New York: Routledge).

Gatens, M. (1991c) Embodiment, ethics and difference, *Occasional Paper Series 1* (University of Waikato, Department of Women's Studies).

Grosz, E. (1987) Notes towards a corporeal feminism, *Australian Feminist Studies*, 5: 1–16.

Grosz, E. (1988) Desire, the body and recent French feminisms, *Intervention*, 21/22: 28–33.

Grosz, E. (1989) *Sexual Subversions: Three French Feminists* (Sydney: Allen and Unwin).

Grosz, E. (1992) Bodies-cities, in B. Colomina (ed.) *Sexuality and Space* (London: Routledge).

Grosz, E. (1993) Bodies and knowledges: feminism and the crisis of reason, in Alcoff and E. Potter (eds) *Feminist Epistemologies* (New York: Routledge).

Grosz, E. (1994) *Volatile Bodies: Toward A Corporeal Feminism* (St Leonards, NSW: Allen and Unwin).

Grosz, E. (1995) *Space, Time and Perversion* (St Leonards, NSW: Allen and Unwin).

Irigaray, L. (1985) *This Sex Which is Not One*, trans. C. Porter and C. Burke (Ithaca: Cornell University Press).

Johnson, L. (1989) Embodying geography: some implications for the sexed body in space, *Proceedings of the 13th Geography Conference*: 134–8 (Dunedin, New Zealand Geographical Society).

Johnson, L. (1994) What future for feminist geography? *Gender, Place and Culture: A Journal of Feminist Geography*, 1(1): 103–13.

Johnston, L. (1995) The Politics of the Pump: Hard Core Gyms and Women Body Builders, *New Zealand Geographer* 51(1): 16–18.

Kearns, R. (1993) Place and health: towards a reformed medical geography, *Professional Geographer*, 45(2): 139–47.

Kirby, V. (1991) Corporeal habits: addressing essentialism differently, *Hypatia*, 6(3): 4–24.

Kristeva, J. (1980) Interview with Kristeva, in E. Baruch and L. Serrano (eds) (1988) *Women Analyze Women*. (New York: New York University Press).

Kristeva, J. (1982) *The Powers of Horror: An Essay in Abjection* (New York: Columbia University Press).

Longhurst, R. (1994) The geography closet in – the body, *Australian Geographical Studies*, 32(2): 214–23.

Longhurst, R. (1995) The body and geography, *Gender, Place and Culture: A Journal of Feminist Geography*, 2(1): 97–105.

Longhurst, R. (1996) Refocusing groups: pregnant women's geographical experiences of Hamilton, New Zealand/Aotearoa, *Area*, 28(2): 143–9.

Mansfield, A. and B. McGinn (1993) Pumping irony: the muscular and the feminine, in S. Scott, and D. Morgan (eds) *Body Matters: Essays on the Sociology of the Body* (London: The Farmer Press).

McDowell, L. (1991) The baby and the bath water: diversity, deconstruction and feminist theory in geography, *Geoforum*, 22(2): 123–33.

McDowell, L. (1992) Multiple voices: speaking from inside and outside "the project," *Antipode*, 24: 56–72.

Oliver, K. (1993) *Reading Kristeva: Unravelling the Double Bind* (Bloomington and Indianapolis: Indiana University Press).

Parker, R. (1989) *Female Bodybuilding*, July: 2–4.

Pile, S. (1996) *The Body and the City: Psychoanalysis, Space, and Subjectivity* (New York: Routledge).

Rose, G. (1991) On being ambivalent: women and feminisms in geography, in C. Philo (ed.) *New Words, New Worlds: Reconceptualising Social and Cultural Geography*, Conference Proceedings, Department of Geography, University of Edinburgh, 10–12 September.

Rose, G. (1993a) Some notes towards thinking about spaces of the future, in J. Bird, T. Curtis, T. Putman, G. Robertson, and L. Tickner (eds) *Mapping Futures: Local Culture, Global Change* (New York: Routledge).

Rose, G. (1993b) *Feminism and Geography: The Limits of Geographical Knowledge* (Cambridge: Polity Press).

Rose, G. (1995) Making space for the female subject of feminism: the spatial subversions of Holzer, Kruger and Sherman, in S. Pile and N. Thrift (eds) *Mapping the Subject: Geographies of Cultural Transformation* (London: Routledge).

Sawicki, J. (1991) *Disciplining Foucault: Feminism, Power and the Body* (London: Routledge).

Steinem, G. (1994) The strongest woman in the world, *New Women*, July: 69–73.

Teper, L. and M. Neveux (1994) Championship Tips, *Ironman* Magazine, February: 42–3.

Worth, H. (1994) Erotic bodies, *Broadsheet*, 202: 36–9.

Wright, E. (ed.) (1992). *Feminism and Psychoanalysis: A Critical Dictionary* (Oxford: Blackwell).

Young, I. (1990a). *Justice and The Politics of Difference* (Princeton N.J., Princeton University Press).

Young, I. (1990b). *Throwing Like a Girl and Other Essays in Feminist Philosophy* (Bloomington: Indiana University Press).

Zulak, G. (1994). Priming the pump, *Musclemag International*, 146 (July): 38–40.

15

LADIES AND GENTLEMEN

Train rides and other Oedipal stories

Virginia L. Blum

In Freud's famous case history of the five-year-old boy, Hans, we see explicitly how Hans's emergent gender identity happens in relation to his experience of space. Masculinity, for Hans, will be construed as that which leaves the house/space of the mother and enters the street, that which forces itself through openings (doors or gates) and along passageways (streets). As a result of his struggle between the discursive orders and the domains of his mother and father, the child produces a spatial phobia (specifically agoraphobia)[1] that symbolizes an attempt to reconcile the difference between the maternal and paternal accounts of bodies, spaces, and genders.

The following are some preliminary remarks on how psychoanalysis represents the emergence of gender identity as a spatial event. Depending on a theory of sexual developmental stages, whereby the subject has different desires, fantasies and psychical motivations according to age, psychoanalytic discourse in the twentieth century has elaborated as though fundamental a particular metafiction – the train ride to gender. Looking at Lacan's parable of the train station and Melanie Klein's treatment of the child-analyzand, Dick, and then focusing at length on Freud's case history of a five-year-old, I will consider the degree to which all three writers engage space at the same time that they neutralize its pivotal effects by ignoring it as a precondition. While the material spatial aspects of each case are clearly central and in many ways formative, such aspects must be suppressed in order for the psychoanalytic economy of gender to be sustained as the unimpeachable and overarching account of human sexual development. Indeed, as we will see in the case of Melanie Klein's autistic four-year-old patient, the child's spatial discourse is reread/misread through the prestructure of Oedipal logic.

By way of introducing the subject of gender and space in psychoanalysis, let me begin with two related passages. In both passages the process of becoming gendered is represented as a spatial event; specifically, gender happens through the selection of a bathroom door. Note in particular that the choice will be not between male and female but between Ladies and

Gentlemen or Gents, emphasizing how anatomical difference is processed through language and space. The anatomical, then, appears to be an effect of spatial significations whereby the selection of the door tells you what you *have* (or lack, as the case may be). The first passage is from Jacques Lacan's "The Agency of the letter in the unconscious or reason since Freud."

> A train arrives at a station. A little boy and a little girl, brother and sister, are seated in a compartment face to face next to the window through which the buildings along the station platform can be seen passing as the train pulls to a stop. "Look," says the brother, "we're at Ladies!"; "Idiot!" replies his sister, "Can't you see we're at Gentlemen." . . . For these children, Ladies and Gentlemen will be henceforth two countries towards which each of their souls will strive on divergent wings, and between which a truce will be the more impossible since they are actually the same country and neither can compromise on its own superiority without detracting from the glory of the other.
>
> (Lacan 1977: 152)

This second passage is from the opening of an interactive text adventure computer game entitled *The Leather Goddesses of Phobos*:

> The place is Joe's Bar
> An undistinguished bar, yet the social center of Upper Sandusky. Doors marked "Ladies" and "Gents" lead, respectively, northeast and northwest. You feel an urge.
>
> (Mazursky 1986)

You cannot play the game until you choose a door. Once you've chosen, you will become a man or a woman for the duration of the game. There is nothing behind the doors save more text – about the spaces that shape gender, the places where gender should and can happen. Like Lacan's "Ladies" and "Gentlemen" it was the choice of the door, the signifier itself, that made "you" who "you" are – a boy or a girl. While *Leather Goddesses* asks you to choose the door of your own (assumed) gender, Lacan's doors signify at once the object of desire and the place from which you will forever after be banned. Once "gentleman" becomes the endpoint of your journey, you can only stop outside his door. You cannot enter his domain – you cannot *be* him. Your experience of gender happens *as a result* of your sexual object – which is no more than a "door" you recognize as a sign of prohibition – the door whose very signifier marks a refusal of entry to your "opposite" body. Not only, then, do you long for the object represented by the "other" signifier, you also long to *be* signified by the other signifier.[2] You long to be in its *place*.

Lacan's train station

Lacan's train ride to a gendered destination strands the boy and girl in a "country" built according to the logic of two dimensions – the length and breadth of doors flatten one's potentially three-dimensional (polymorphous) sexual subjectivity upon their entrance into the ostensibly stable "space" of the signifier.[3] Even though you can't leave it, once you've committed yourself (been committed) to living there, you can dwell in it indefinitely, depending on the unwavering integrity of its zoning laws. You are on protected property ... this is part of its structural guarantee. It is why the outrage over contractual transgression (of gender) is so extreme. You were promised fences for not just *your* property but *theirs* – in relation to which the parameters of your own sexuality are at once forged and maintained.

This train ride simultaneously fashions boy and girl as self-consciously gendered and prescribes an object and an aim – the tracks to heterosexuality. Indeed, for Lacan, the developmental transition into gender and hetero-sexuality are one and the same. It is structurally necessary, in other words, that they recognize only the bathroom doors of the opposite sex. But what if the train gets derailed or, more significantly, what if its arrival is slightly delayed or what if the little girl and little boy are facing in different directions? The girl might see "Ladies" and the boy might see "Gentlemen." This is the implicit threat of the train ride, that it will stop at the wrong station. This is the problem of time as well, that has come increasingly, between the late nineteenth and early twentieth centuries, to depend on standardized, atomizable clock time in relation to precisely measured distances. Like what Stephen Kern has called "the dominating power of clock time" that sustains, measures, and rules over a whole world of synchrony, psychoanalytic theory elaborates developmental stages that are construed as cross-cultural, trans-historical measurements of the burgeoning psycho-sexual life of the human subject (Kern 1983: 23).[4]

Tellingly, Lacan's train ride links up with a certain brand of homophobic discourse in psychoanalytic theory. Popular among the psychoanalytic community from the 1930s through the 1950s was the theory that a traumatic event or situation would derail the young subject from the "anatomically inborn and inevitable fate of both sexes" – heterosexuality (Feldman 1956: 74). Homosexuality, the theory held, was a kind of "detour" (Feldman 1956: 75). Lacan represents female homosexuality as disappointed heterosexuality (Lacan 1982: 96–7). The train got off course as it were. Mas-culinity and femininity as such are not really the issue for Lacan; or rather, they are secondary phenomena, the residue of the desiring trajectory. Thus, it is where the little boy and little girl are heading (towards Ladies or Gentlemen – the object of desire, an object that is a place, a place that is two dimensions) that marks them retrospectively as boy and girl.

It is worth considering in this context the distinction Freud makes between the object and the aim of desire in order to see how the train functions heuristically for psychoanalytic theory as a device for "training" twentieth-century desire to progress unidirectionally – in sequence, on schedule. "Let us call the person from whom sexual attraction proceeds," writes Freud, "the *sexual object* and the act towards which the instinct tends the *sexual aim*" (Freud 1905: 1–2). "Perversions," he explains, "are sexual activities which either (a) extend, in an anatomical sense, beyond the regions of the body that are designed for sexual union, or (b) linger over the intermediate relations to the sexual object which should normally be traversed rapidly on the path towards the sexual aim" (Freud 1905: 16). The parable of the train ride gives us two objects and two aims: first, the train itself has an object – the motion itself, the movement through space and time – while its aim is the arrival at the station. If the train went down the wrong track, or did not make it all the way to the station, we might say that this constituted a "perversion" of the normal aim – the arrival at its proper (predestined) destination. For the little boy and the little girl, their objects are respectively Ladies and Gentlemen; the aim in both cases is the assumption of heterosexuality *via* the selection of the hetero-appropriate object.

Luce Irigaray writes that "The change of the body and the modification of the interval represent an important issue in the economy of desire. The *locomotion toward* and *reduction in interval* are the movements of desire." (Irigaray 1993: 48). This locomotion, this train ride of desire (in which desire functions as cause and effect of the movement to place, the movement in space, the urge to collapse the interval, the ultimate implacability of the interval that sustains desire as such) takes one to the doors of gender which consequently become the "toward which" that desire pursues; this is potentially unharnessed movement that must be reconceived as a movement toward a destination, movement that is framed historically, chronologically, and with reference to gendered subjectivities. Equally dangerous is no movement at all – the ultimate possibility that the train never gets started – that desire short-circuits or, even more importantly, doesn't happen at all, never gets off the ground as it were. So – our opposing and dangerous possibilities are utter stasis and pure movement – what might they have in common? It seems to be that both alternatives eliminate time as the overarching modality of existence. If time happens in relationship to beginnings and endings, origins and destinations, if time depends for its measurement on teleology, then stasis of course is outside time. Stasis is only place, a place that remains in place, no leave-taking of place. In psychoanalytic theory, the place of places, the origin of place, the place of origin, the place to which one's whole life marks the urgency of return is the mother's body. Thus (on the level of fantasy) there is only one place, no interval between places, the place one began, mother, and her eventual substitute, the object of desire. Pure movement likewise happens without reference to origin and destination and in this sense has no place, taking place as it does only among places without ever being emplaced. Pure movement threatens the order of the mother's body that drives psychoanalytic theory.

Why might characterizing the assumption of gender identity as a train ride be significant? The use of the train by psychoanalytic theory as a trope (for developmental stages as well as a penis-equivalent) is rooted in the train's role both materially and imaginatively in industrial and post-industrial society. David Harvey writes that, "space relations may be revolutionized through technological and organizational shifts that (quoting Marx) 'annihilate space through time.' Such revolutions (the impact of turnpikes, canals, railways, automobiles, containerization, air transport and telecommunications) alter the character of places (if only in relation to each other) and thereby interact with the activities of place construction" (Harvey 1993: 6). Trains are the major route whereby capital interlocks places whose very placiality is reconfigured through both capital and the railway system. In 1883 local church time was superseded by standardized national time specifically in order to regulate the movement of a national railway.[5] The reliable schedule of the railroad system intersected with the developmental achievements of the bourgeois subject. Childhood itself was heavily subdivided into stages of expectation – infants crawl, toddlers walk, you learn to read and write at age six and so on down the chain of progress toward a self-sufficient and culturally normative adulthood.[6] Edward Casey isolates an earlier but no less significant transformation in the reconceptualization of the relationship between place and time, the marine chronometer, perfected in the early nineteenth century. Casey writes: "The winning logic was this: when lost in space, turn to time. In other words, *where* one was became equivalent to *when* one was. The determination of the latter allowed the specification of the former. The measure of space was taken by time" (Casey 1993: 6).

So gender is not just movement but timed movement that at the same time creates the subject of, to, and in time. It is not surprising that in concert with increasing industrialization another preeminent twentieth century discourse, psychoanalysis, ushered in a theory of developmental stages that depend on a newly forged notion of space–time. Significantly, post-Freudian twentieth-century debates in psychoanalysis have centered on the timing of the subject's emergence into human subjectivity. The events, in other words, such as oral, anal, and genital stages are chronologically timed. Melanie Klein's emphasis on pre-Oedipal over Oedipal-level experience occurs in light of this linear account of human development. While it has been argued that Klein subverts the linear Oedipal model with her notion of "positions" that are always available to the subject, she too is committed to a narrative of train rides (Mitchell 1987: 25–30).

Out of the mummy-station, into the Oedipus

Klein's case history of the autistic child, Dick, is clearly the precursor for Lacan's parable of the train station. Klein claims that Dick hasn't achieved symbol-formation: "The ego had ceased to develop phantasy life and to establish a relation with reality. After a feeble beginning, symbol formation in this child had come to a standstill. . . . The child was indifferent to most of the objects and playthings around him, and did not even grasp their purpose or meaning. But he was interested in trains and stations and also in door-handles, doors and the opening and shutting of them" (Klein 1987: 101). Lacan, in commenting on this case, observes that Dick lives in the real by which he means that he is still immersed *in* the experience instead of representing it, the developmental and necessary accomplishment that makes us subjects of and to representation. "This young subject is completely in reality, in the pure state, unconstituted. He is entirely in the undifferentiated" (Lacan 1988: 68). Klein describes the child repeatedly rushing to the "space between the [inner and outer] doors' of her office, a space that Klein tells us he "identifies . . . with the mother's body." I would argue that Dick's chronic return to the space between the doors represents his refusal to enter the adult Oedipal system – the only symbolic system Klein recognizes. He *performs* this refusal of the Oedipal through his refusal of the rest of Klein's office. The space between the doors could be seen as a threshold between the psychical world, the world of internal objects and the material world, the outside. Or, even more importantly, this space between could be the place where internal and external objects are not yet differentiated – where they don't have to be. To the degree that such spatial arenas are culturally gendered (the inside world/the world of the home being associated with mother and the outside world being associated with the reality-bearing father), Dick is also on the threshold of gender identification – which door shall he choose? The point is that a door must be chosen – he will not be allowed to linger in the "space between" as Klein actively beckons him into the country of Oedipus.

Klein takes Dick on a train ride:

> I took a big train and put it beside a smaller one and called them "Daddy-train" and "Dick-train." Thereupon he picked up the train I called "Dick" and made it roll to the window and said "Station." I explained: "The station is mummy; Dick is going into mummy." He left the train, ran into the space between the outer and inner doors of the room, shut himself in, saying "dark" and ran out again directly.
>
> (p. 102)

In Klein's hands the train becomes the phallic route back into the interior of the mother's body.[7] Klein takes Dick there in order to start him from scratch as it were, at the originary moment of the Oedipus complex. She believes that she must symbolically lead him back into the mother's body in order to lead him *out*. It is at the intersection between the fantasy of penetrating the mother's body and *symbolizing* it that Klein locates the origin of the Oedipus complex. In her gender economy, then, the mother's body is the passive station in relation to the boy's active locomotion – his train-identity.

Klein's therapeutic strategy here should make us rethink Lacan's parable. Is there any real place for femininity in this story of thrusting masculine train-speed hurtling to the station of gender? If we superimpose Klein's story onto Lacan's, then we see a boy-train carrying off both boy and girl to the passive mummy-station where the train's phallic thrust in contrast to the station's waiting *doors* (places to be entered) becomes a dramatization of how gender happens through phallic pressure against and through time and space. The phallus-train, then, is what puts everything in order, aligns bodies and spaces, introduces them, matches them up.[8]

Deleuze and Guattari claim that Klein has "no understanding of [Dick's] cartography . . . and is content to make ready-made tracings – Oedipus, the good daddy and the bad daddy, the bad mommy and the good mommy – while the child makes a desperate attempt to carry out a performance that the psychoanalyst totally misconstrues' (1987: 13). This is the only map Klein has mastered (and is mastered by) – it is the psychoanalytic cartography of a heterosexualized and Oedipalized gender. Dick must choose a door and there's no going back – which Dick on some level seems to recognize as he clings to the space between. This space between is not some neutral, amorphous space in contradistinction to the material and steadfast reality of the doors. Rather, it is "between" the doors, between the inside and the outside, the not-doors, the space where one can say "no" to doors that are (from Dick's perspective) considerably more terrifying than the space between them. Melanie Klein's office is where Dick is socially spatialized. He is taught about what bodies enter what spaces, about who plays the train and who plays the station, about, ultimately, how to inhabit social spaces that up to now, he has refused. This is not to suggest that we should celebrate autism or that Dick (whose condition improved through Klein's treatment) should have been left alone to enjoy some fantasmatic gender-free state. Rather, I am illustrating that Klein's cure from the very beginning took account (unwittingly) of the degree to which the assumption of gender is linked to the regulation of the body in space.

These doors that loomed before Dick, that circumscribed his pathway and his options en route to becoming subject to gender, suggest what Lacan left out of his parable about restrooms – that the boy and the girl will experience these doors differently. The door taken by the boy could be to a life of power and privilege while the girl could be entering the prison house of her options. If indeed they were identical then why are experiences of femininity and masculinity so very different? Further, if as Freud maintains, femininity and masculinity are "a phase apart," how did Lacan's children manage to arrive at the station at the same time? As any psychoanalytic account emphasizes, their Oedipal trajectories in space and time are significantly different.

Further, what kinds of subjects are these? What, for example, is their class and cultural affiliation? Who rides on trains? There are diverse space–time compressions taken for granted by Western bourgeois subjects. We need to consider the variations among social circumstances; psychoanalytic train rides are for very specific subjects whose identities are forged and enacted in relation to particular sets of economic and social privilege. Who is driving the train? What kind of people ride on trains? Is everyone ideologically and psychologically "trained" according to the paradigm of developmental stages? In other words, along with regulating and taking for granted the time–space continuum of the progress of gender, psychoanalytic accounts of developmental stages assume that subjects who in many ways are conditioned by their bourgeois circumstances come into the world *pre-conditioned* for them.

Becoming a boy: little Hans and big horses

In 1909 Freud published his case history of a phobia in a five-year-old boy. This was Freud's only "analysis" of a child and even in this instance the case was diagnosed and treated largely by way of the child's father who would report back to Freud. Hans developed a rather complex phobia involving horses that led to a fear of leaving his house. In light of the foregoing accounts of trains and gender, I am going to read this case as an illustration par excellence of how the accomplishment of gender is a spatial event and how Hans's phobia acts out in response to, as a commentary on, and ultimately as a resistance to the implicit spatial directives of the Oedipal trajectory.

Freud and Hans's father, Max Graf, trace Hans's phobia to his relationship with his father, and offer a classic Oedipal analysis in which the small child projects onto horses the dangerous Oedipal father who might punish him for desiring his mother. I will suggest, in contrast, that Hans's phobia is his "solution" to the dilemma posed by his simultaneous conflictual identifications with the mother's and the father's bodies, identifications that are very spatial in their content. Only after mapping effectively the mother's body can the son find his "way out" of feminine sexuality. It is the mother's body that the father's narrative/analysis/law (they are all three combined here) is insisting Hans forfeit and ultimately revile – not just because she is the father's possession, the taboo object; it is also because Hans must reject the very shape of the mother's body as insufficient, as lacking the "wiwimacher" the father repeatedly claims he and Hans *have*.

Hans suffers from a spatial phobia. Originally, the reason he is afraid to leave his house is unclear. Freud points out that prior to locating a phobic object, what Hans experiences is diffuse anxiety and that his phobia has the effect of containing his anxiety through naming/locating it (Freud 1909: 66).

While Freud and the child's father never imagine that his fear of leaving the house may be an expression of separation anxiety from his mother, most twentieth-century psychoanalysts (after the development of "attachment theory") would at least consider this interpretation. One might argue, reasonably, that he is afraid of being separated from his mother and that "going out in the street" symbolizes for him the route away from her house/body. The child's attachment to his mother, then, would act out spatially in the amount of distance he can tolerate between his body and hers.

Horses assume the role of phobic object. Fearful of seeing horses leading carts in the street, Hans refuses to go out in the street at all, so terrified is he of seeing the horses fall down and "make a row with their legs" as he terms it. Freud suspects that this "row with their legs" refers to Hans's understanding of parental intercourse. Most importantly, Hans is afraid of the horses drawing the loaded carts through a gate. This gate, I will argue, is the gate of gender that his dawning recognition of his role in relation to parental bodies compels the child to enter. His phobia signals a moment of resistance to entering the "outside" world of adult gender laws – an outside world he must inhabit once he gives up his mother's house/body.

It is in the transition from inside the home (where girls and boys are just undifferentiated babies and the mother's love reigns supreme) to the outside social world of the father's law that Hans's phobia takes place – represented by the horses passing through the gate. Indeed, his phobia emerges as a result of multiple spatial disturbances in his life and consciousness: his mother's pregnancy; his simultaneous awareness of sexual difference and death. As Lacan describes Hans: "the world appears to him as punctuated throughout by a series of dangerous places, alarming places, that restructure him" (Lacan 1994: 245). Edoardo Weiss observes that agoraphobia seems to emblematize a psychical spatial confusion between outside dangers and internal enemies. "In traumatic neurosis . . . *internal*, destructive, instinctual stimuli are diverted from the channels in which they have hitherto been employed and continue the effects of the external stimuli which had formerly been operative" (Weiss 1935: 79). In Hans's case it is as though the emergent cartography of internal psychical space is traumatically projected onto the division between domestic and outdoor space. At the same time, such spatial articulations are closely allied to a gender identity as it shapes body-space.

Hans's own body has become for him a landscape of danger and possibility. First, he has a growing repugnance to anything that reminds him of excrement – what he terms "lumf." Soon, it turns out that he worries about the "loud row" made by horses "doing lumf" in the street. Hans's father concludes: "The bus-horse that falls down and makes a row with its feet is no doubt – a lumf falling and making a noise. His fear of defecation and his fear of heavily loaded carts is equivalent to the fear of the heavily loaded stomach" of his mother (Freud 1909: 105). His sister Hanna is herself equated with a lumf. Based on the story of the stork-brought baby told to him by his parents, Hans (although highly suspicious of this fable) invents a variation on the theme: he posits a "stork-baby-box" that holds the babies. In light of this fantasy of the big box, Hans insists that his sister Hanna was with his parents and him on a holiday they took prior to Hanna's birth (when his mother was pregnant with her). It is through this misrecollection in fact that Freud and Graf learn that Hans was perfectly aware of the relationship between his mother's pregnant body and the subsequent birth of a child. It is his mother's body then that figures as the "box" in Hans's memory. He tells his father: "Hanna traveled with us to Gmunden in a box. . . . We got a big box and it was full of babies" (p. 109). Elsewhere he describes the unborn Hanna's enormous appetite: "'she kept on eating all the time. . . . She gobbled everything up like a hare.'" (p. 116). Babies are confused with "lumfs" just as food might be babies who are voraciously hungry (for the mother's body) at the same time that they are contained by (eaten by) the mother's body. At times he remembers sharing the box with Hanna. "'Hanna and I really traveled together in the box; I slept the whole night in the box'" (p. 116). Hans isn't entirely certain where to locate his own body (inside or outside the box) in these various versions of the train ride.

Importantly, it is in his account of the train ride to Gmunden that Hans distorts developmental stages – specifically, spatial logic is privileged over temporal progression as though in utter defiance of train-time. The interior and exterior of the mother's body are undifferentiated in Hans's recollection (his sister who was inside the pregnant mother at the time of the journey is represented as exterior just as Hans represents himself as interior – in the box with Hanna). Hans's spatial reassignments complement his temporal reversals. Thus, Hanna is "there" with them in the train to Gmunden even though she is in fact unborn – in Hans's mind, the produced Hanna precedes the unborn Hanna. Hans has Hanna already walking and talking, even though he concedes that thereafter she had to be carried around. Such a reversal offers an almost intact recapitulation of the formative sequence of Hans's understanding; certainly, it is after Hanna is born that the mother's pregnant body assumes significance.

Both his reaction to his mother's pregnancy and his consequent awareness of the spatial characteristics of his own body lead to Hans's spatial disturbance. He contains his growing fear by dis-placing the dangerous structure of his body-map onto the world. His inside/outside dichotomy is based on the mother's body; she is simultaneously the blueprint for his spatial organization of the world and that which imperils spatial difference as such (through the ontological confusions of pregnancy). Lacan writes that the phobia

> introduces into the world of the child a structure, it precisely lays out in the form of a first map the function of an interior and an exterior. Until now, the child was on the whole in the inside of the mother, from which he was rejected, or from which he imagined himself

rejected, which makes him anxious, and thus, through the aid of a phobia, installs a new order of the inside and the outside, a series of thresholds that impose a structure on his world.

(Lacan 1994: 246)

Bodies harbor babies and feces and the surface of the earth only thinly veils dead bodies whose "death" in Hans's understanding is signified by their invisibility. Max Graf notes of his son that, "Once he knocked on the pavement with his stick and said: 'I say, is there a man underneath? – some one buried? – or is that only in the cemetery?' " So he is occupied not only with the riddle of life but with the riddle of death" (Freud 1909: 108). Hans shows here his concern with the containment of the concealed – about how the concealed (babies, feces, dead bodies) might be revealed or marked. The mother's pregnant body, the gravestones in a cemetery – these are tokens of the concealed; to the extent that the concealed is marked, it is revealed. It is located (the dead bodies are *here* in their designated space, not just underground) and it has a shape. The shape of Hans's sister, Hanna, is his mother's pregnant body.

The fact that his parents *concealed* the origin of babies from Hans coalesces with the pregnancy inasmuch as the secret of parental sexuality becomes the secreted body of the sister. Perhaps this convergence of concealments founds the sexual as such: the polymorphously perverse child gets indoctrinated into the adult sexual arena through learning (and internalizing) where the sexual may and may not appear – adult sexuality itself is spatially orchestrated. The secret of the unborn baby, then, is twofold: it is there, signified by the mother's growing body, yet it remains spatially dislocated. It occupies the space of the mother's body and to the extent that her body occupies space in the world, so does the unborn child. But its space is not its own – it inhabits the mother's space and before it is born the child does not have an ontological space in the visible world. The sexual secret of the child's origins (the story of the stork) doubly dislocates the unborn child. In part, Hans assumes his own spatially registered subjectivity through an identification with these two (dis)locations of his little sister's body.

For Hans, sexuality as it gets processed through language is founded upon an original reversal in the parental accounts of sexual difference. When Hans asks his mother if she has a "widdler," she responds " 'Of course' " (Freud 1909: 49), an assertion that later will be called into question – for Hans, by his father (p. 72) as well as for the reader, by Freud (pp. 73 and 76). Neither Freud nor Hans's father offer in place of the mother's "widdler" another anatomical marker; rather, the effect of their disavowals (what else can we call it?) is not only to invalidate the mother's sexuality and her own account of her sexuality; it is to mark sexuality as a place of parental indeterminacy, an arena where the mother's story and the father's diverge. Sexual difference, as a result, emerges for the child in the incompatibility of parental versions. These "lies" cancel each other out and make of gender difference two irreconcilable perspectives; one can take the word of the mother *or* the father, never both. In this sense, gender difference as lived for each sex is transacted according to the child's affiliation with the word of the mother or the father – a word, like Ladies and Gentlemen, that is a position. Northwest or northeast, as *The Leather Goddesses* illustrates, this is one's destiny.

Sexual secrets become sexual zones. This secret of adult sexuality – which ultimately involves confining the spaces of sexuality, in other words, putting it behind doors of various sizes and shapes – entails another kind of spatial dislocation, the divergence of the signifier from the signified, the felt loss of presence – of desire from its object, of words from meaning, of form from content. The mother's body poses the problem for Hans of the relationship between form and content. About Hans's understanding of the sexual relation, Freud writes:

> it was a question of some act of violence performed upon his mother, of smashing something, of making an opening into something, or forcing a way into an enclosed space – such were the impulses that he felt stirring within him. But although the sensations in his penis had put him on the road to postulating a vagina, yet he could not solve the problem, for within his experience no such thing existed as his widdler required.
>
> (Freud, 1905: 169–70)

The vagina, then, is figured as an invisible space for the penis. Its imaginary contours derive from the penis it encloses (Irigaray 1993). Yet enclosure is precisely what terrifies Hans – even as he longs to fix its parameters. Similarly, the coordinates of the maternal anatomy derive from the child she produces. In this way, the mother's body is configured as ontologically secondary to the child whose body is the shape of her interior dimensions. The mother's body is reconfigured as a container (or box) for Hans subsequent to his confrontation with its *contents*.

Hans's body as well is now the site of possibility, a possible container of children. Like the fecal matter that reveals the shape of the inside, like the transition from inside to outside, like horses passing through gates, like babies that emerge from bodies, it becomes crucial to know whether *you* go in and out or things come in and out of *you*. If masculinity penetrates and femininity is that which is entered, what is a child supposed to do with the bodily and spatial inconsistency of these roles? Does he identify with the horse or the gate, the baby or the mother, the father or the mother? Of course, such questions have the effect of collapsing the distinctions among roles that the whole social order is prodding Hans to navigate. These mutable spatial experiences of gender that are developmentally appropriate for the five-year-old, are pathologized in a twenty-eight-year old patient of analyst Paul

Schilder: "'I have a subconscious feeling when I see a woman that I am entering that woman. It is a mental state only. When a man talks and his organ flies in my mouth it makes me feel I am a woman'" (Schilder 1935: 20). Femininity and masculinity are assumed as positions of penetrating/being penetrated. The pathology of Schilder's patient is in his continued unmediated (unacculturated) experience of the spatial foundations of masculinity and femininity.

Hans's expressed desire to have children (to be himself a container) points to an identification with the mother that his father repeatedly tries to prevent. Hence the father's insistence that women don't have "widdlers." The father (and Freud, for that matter) are rather desperately engaged in differentiating the geography of femininity from that of masculinity – anxious to indoctrinate Hans into the proper *spatial* relationship to his body. This is not a container for children; this is the bearer of the "widdler." In the production of "lumf" Hans is identifying with childbirth (being a penetrable body), and in that identification finds his sexual subjectivity sundered.

In the case history of Hans we learn how femininity is psychically linked to not only the founding of inside and outside (the pregnant body is the very image of the divergence of the inside from the outside) but also, through the child-product, a spatial relocation of the inside to the outside. It is perhaps because the woman gives birth that the feminine genitals come to be associated with (through an act of *displacement*) the primal (and founding) act of spatial disruption characterizing human subjectivity. This is the geography of the feminine as such – her spatially disrupted sexuality, at once concealed and revealed, the secret that is kept in a closet, as it were, even though everyone knows where the closet is located and the shape of its contents. Indeed, the closet door, the feminine body, is the scene of her sexuality that needs to be imaginatively reconceived as the mark of absence – as what is not presented. Hans worries that bodies are inside of everything, everywhere, dead bodies under the pavement, unborn children in his mother's body, in his father's, in his own. "But only women have children," Graf tells his son. Hans responds:

> "I'm going to have a little girl." "Where will you get her then?" "Why, from the stork. *He takes the little girl out*, and all at once the little girl lays an egg, and out of the egg there comes another Hanna – another Hanna. Out of Hanna there comes another Hanna."
> (Freud 1909: 125)

At one point he asks his father if there are "any more babies inside Mummy" (p. 130). His maleness finally comes to be determined as much from the fact that there are no (nor will there ever be) babies inside his body as it is determined through the "widdler" his father tells him only males possess. The not-having (babies) is transformed into the having (the penis).

Hans's parents disagree over whether the growing child should sleep in bed with them. His mother spoils the boy, complains Graf. Hans's desire to share a bed with his parents stages materially his felt status in between the body of the mother (where he has been) and the body of the father (what he will be) – and, in a further subjective disruption, the mother's body is where he will be again – as an adult man – the place to which he will return. As Irigaray puts it: "While remaining within the womb, the child changes place. While remaining within the woman, the man changes place. Both are smaller or greater in relation to the envelope which keeps them inside. Which is also them, in fact, as well as the relation between the two" (Irigaray 1993: 45). At the same time, his continuing need to sleep between them suggests his hesitancy to leave this liminal space where he can be both male and female. Hans's father holds his mother responsible for the phobia because of her "excessive display of affection" toward the boy (Freud 1909: 68). To the degree that Hans's spatial identification with *her* body interferes with his spatial identification with masculinity, his father is right; the mother *is* responsible for Hans's confusion that acts out as a phobia.

In order to "grow up" male, Hans's identifications – multiple, fluid – are being increasingly restricted. No wonder he does not want to leave the house. Out in the street where civilization happens, where time proceeds apace and the human race moves to national orderly heterosexualized time, Hans foresees what? the future? No, the cemetery: " 'I say, is there a man underneath? – some one buried?'" Smart boy – Hans recognizes from all the signs that when he leaves the mother's house, the door will slam and lock behind him.

In his repudiation of the centrality of the Oedipus, the analyst W.R.D. Fairbairn maintains that by claiming that desire for the parents gives rise to the child's libido, Freud is "putting the cart before the horse." Fairbairn claims that he, in contrast, foregrounds the developmental stage that *precedes* desire for the stage-appropriate object. Melanie Klein reports that at one point "Dick ran into the space between the doors and scratched on the doors a little with his nails, thus showing that he identified the space with the cart and both with the mother's body" (Klein 1987: 103). If, in the psychoanalytic economy, carts and spaces are representative of the maternal body while horses and trains are the phalluses that link sons to fathers, then it seems important that: (1) horses precede carts to confirm masculine precedence not only in reproduction but organ-value and that, (2) one can count on a certain sequence of events. The train arrives at the station. The cart follows the horse. It is not simply a matter of the timed and reliable developmental progress of the child; we see as well here a pervasive investment in a highly gendered account of progress. Indeed, the motive force for development is curiously always (for both sexes) masculinity – it is what is active, what moves – what draws femininity in its wake. Femininity, slothful, draped backwards in the cart, sunning, or casually loitering in the space between, must be propelled by masculinity in order even to become femininity. Masculinity, then, emerges in relation to origins and destinations.

Recall that it is the vision of the horse drawing a loaded cart that terrifies little Hans who understands all too well the nature of his journey and that there's no getting off once he boards that train. Fantasmatically, he can enter and exit the mother's box-womb, throw into disarray the developmental stages of his sister that reflect back to him horrifically his own very recently navigated path. He can pretend that the path is spatial and not temporal and that space (the box) is recuperable in the way time is not. But in grown-up linear, post-industrial, mechanized time (that is, train-time) Hans can only go forward to his heterosexual destination which is, of course, a predestination – suggesting, ironically, that linear time is merely a misrepresentation of a fundamentally circular route. He needs to leave the conjugal bed as well as the house itself because they are metonymies for the mother whom he must forfeit for the sake of his masculinity. When he arrives at the station, then, he will need to pretend, along with everyone else, that he has not simply returned home.

In their work on gender identity and gender roles, John Money and Patricia Tucker maintain that from the time we are conceived, we pass through various gender gates, both physiological and psychological. Before we enter the gate, the options remain open – but once the gate closes behind us, it is sealed shut. The gates stabilize gender for Money and Tucker. We all know who we are – Ladies or Gentlemen. The Oedipus, plotted to contain bodies in their separate but identical doors, remains intact – the imperturbable cartography of the body's ever-threatening polymorphous perversity. Money's work at the Johns Hopkins Sex Identity Clinic clarifies just how central it is to fix developmental stages – to parcel them out temporally and spatially, to harness time through space and vice versa, to lock gates, to schedule trains, to label clearly the doors, the entrance and exit routes.

Founder of "Circus Amok" traveling circus, Jennifer Miller, who has a full-length beard, describes her "fear of going to the bathroom in a public place. I always bring someone with me so people can hear my female voice and they won't look at the beard and say, "What are YOU doing in here!" Recently, in an airport, I used the men's bathroom for the first time. I just went in – it seemed less scary than the women's room. When I came out of the stall, a man was at the sink. We looked at each other in the mirror and he didn't gasp" (Miller 1995: 23). The women's room is a place Jennifer has trouble *being* in, inasmuch as her femininity does not conform to the understood expectations of urinary segregation. Always on the look-out for invaders (though never consciously), we police our public bathrooms – to sustain gender's territorial boundaries. Jennifer crossed over not as a transgressive act but rather because the men's room had become a safer place for her transgressive body. In a sense, her invasion of the sanctity of the men's room is remarkably terroristic in that her unrecognizability (the man at the sink didn't "see" her as an enemy) unsettles the very boundaries of the door, the overarching signifier of Ladies or Gentlemen, directional gender-segregated bodies.

On the train to Gmunden, accompanied by his sister in his mother's pregnant body, Hans enters gender as it though it were a stop along the way. Trains are motion, the possibility of traveling from one place to another. To be in the train is to be located in the movement itself. "There is always more place, more places," urges Irigaray, "unless they are immediately appropriated. The land cannot be laid waste if spatiality is produced by our bodies" (Irigaray 1992: 59). But the train's movement is harnessed in the service of the Oedipal trajectory – movement without a destination is dangerous. "Under the influence of genitality," writes Paul Schilder, "the separate space units are unified" (Schilder 1935: 295). That is, when you recognize your genital for what and *where* it is. Hans needs to learn to use the Gentlemen's room where, as Lee Edelman has pointed out, the revelation of the phallus is "an act of definitional display" (Edelman 1994: 272). Dick must choose a door – we can't permit him to wander forever in the space between.

When you arrive at Upper Sandusky, you may as well have arrived by a train that strands you in two-dimensional space that will forever after be the label of your gendered subjectivity.

> *The moment you leave the bathroom, some tentacled aliens transport you to the tiny planet Phobos, home of the Leather Goddesses. "Your head feels as if it's been run over by several locomotives, or at least one very large locomotive, and your clothes are now unrecognizable." If you are a woman, you are wearing a brass bikini; if you are a man, you are wearing a brass loincloth.*

(Mazursky 1986)

Go northeast or northwest.

Notes

1 Although Freud claims that this is not a true case of agoraphobia (Freud 1909: 151) we would now designate it as such.
2 Similarly, in *Leather Goddesses* you choose a door in order for this softcore pornography game to program opposite-sex partners for the various sexual encounters. In other words, the experience of gender is only important insofar as it designates an object.
3 I am referring here to Freud's account of the original "polymorphous perversity" of the infant who gains pleasure in a variety of ways, whose pleasures are not yet subordinated to the organizing effects of the genitals. See, for example, the essay on "Infantile Sexuality" in *Three Essays on the Theory of Sexuality*. What I am calling three-dimensional, then, means unrestricted to a particular "shape" (of, say, a bathroom door) that the culture both recognizes and validates *as* the proper place/shape for a gendered subject.
4 For example, Erik Erikson in the 1950s accounted for the personality style of the

Sioux Indians by pointing to delayed weaning. See *Childhood and Society* (Erikson 1963).

5 See Stephen Kern's *The Culture of Time and Space: 1880–1918* for an important account of how these changes in measuring (and theorizing) both time and space were experienced throughout Western culture.

6 In a culture increasingly concerned with the appropriate timing of various developmental achievements, it is important to note the often quite enormous delay between sexual maturity and marriage – a delay occurring precisely because of other social developmental pressures, e.g. the necessity for a man to be able to support a family before he married. At the same time, sexual initiation was expected to coincide with marriage. In a sense, the desiring trajectory was under strict instructions to wait for its regularly scheduled train!

7 In a footnote to the case history of Hans, Freud points out that a "station" is used for purposes of "*Verkehr* ["traffic," "intercourse," "sexual intercourse"]: this affords the psychological wrappings in many cases of railway phobia" (Freud 1963 [1909]: 123n).

8 In another child analysis, that of Rita in "The Oedipus Complex in the Light of Early Anxieties," Klein interprets the 3-year-old's insistence that "*she was not the doll's mother*" as her simultaneous refusal and desire to take away from her mother this surrogate baby brother. Klein needs to ignore what is the almost manifest content of the little girl's disavowal of the proffered position of femininity. I am not the doll's mother, no, I will not succeed my mother by becoming a mother myself. Being a woman doesn't look especially appealing to me. Rita's outright rejection of the imperatives of gender assignment is mangled by Klein into Rita's desire to be the mother of her baby brother. Klein forceably leads her into the proper Oedipal path. From mommy to daughter, the mommy–daughter relation repeated in the daughter–doll relation.

References

Casey, Edward S. (1993) *Getting Back into Place: Toward a Renewed Understanding of the Place-World*. Bloomington: Indina University Press.

Deleuze, Gilles and Felix Guattari (1987) *One Thousand Plateaus: Capitalism and Schizophrenia*, trans. Brian Massumi. Minneapolis: University of Minnesota Press.

Edelman, Lee (1994) *Homographesis: Essays in Gay Literary and Cultural Theory*. New York: Routledge.

Erikson, Erik (1963) *Childhood and Society*. New York: Norton.

Fairbairn, W. Ronald D. (1952) *Psychoanalytic Studies of the Personality*. London: Routledge and Kegan Paul.

Feldman, Sandor S. (1956) "On Homosexuality." In *Perversions: Psychodynamics and Therapy*, ed. Sandor Lorand and Michael Balint. New York: Random House, pp. 71–96.

Freud, Sigmund (1963 [1909]) *Analysis of a Phobia in a Five-Year-Old Boy*. In *The Sexual Enlightenment of Children*. Collier: New York, pp. 47–183.

Freud, Sigmund (1975 [1905]) *Three Essays on the Theory of Sexuality*, trans. and ed. James Strachey. (1993) New York: Basic Books.

Harvey, David (1993) "From Space to Place and Back Again." In *Mapping the Futures:*

Local Cultures, Global Change, eds Jon Bird, Barry Curtis, Tim Putnam, George Robertson and Lisa Tickner. London: Routledge, pp. 3–29.

Irigaray, Luce (1992) *Elemental Passions*, trans. Joanne Collie and Judith Still. New York: Routledge.

Irigaray, Luce (1993) *An Ethics of Sexual Difference*. Trans. Carolyn Burke and Gillian C. Gill. Ithaca: Cornell University Press.

Kern, Stephen (1983) *The Culture of Time and Space: 1880–1918*. Cambridge: Harvard University Press.

Klein, Melanie (1987) "The Importance of Symbol Formation in the Development of the Ego." *SMK*: 35–54.

Lacan, Jacques (1977) "Agency of the Letter in the Unconscious or Reason Since Freud." In *Ecrits: A Selection*, trans. Alan Sheridan. New York: Norton, pp. 146–78.

Lacan, Jacques (1994) *Le Séminaire, Livre IV. La relation d'objet*. Paris: Editions du seuil.

Lacan, Jacques (1988) *The Seminar of Jacques Lacan, Book I. Freud's Papers on Technique: 1953–1954*, trans. John Forrester, ed. Jacques-Alain Miller. New York: Norton.

Mazursky, Steve (1986) *The Leather Goddesses of Phobos*. Computer Software. Infocom.

Miller, Jennifer (1995) "Transgressive Hair: The Last Frontier." *Lilith* 20: 22–3.

Mitchell, Juliet (1987) Introd. *The Selected Melanie Klein*. New York: Free Press.

Money, John and Patricia Tucker (1975) *Sexual Signatures: On Being a Man or a Woman*. Boston: Little, Brown and Co.

Schilder, Paul (1935) "The Psycho-Analysis of Space." *International Journal of Psycho-Analysis* 16: 274–95.

Weiss, Edoardo (1935) "Agoraphobia and Relation to Hysterical Attacks and to Traumas." *International Journal of Psycho-Analysis* 16: 59–83.

4

ProjectingPlacesBodies

16

INSCRIBING DOMESTIC WORK ON FILIPINA BODIES

Geraldine Pratt, in collaboration with the Philippine Women Centre, Vancouver, Canada

'There's so much written about the body,' she groans, 'but . . . in so much of it, the body dissolves into language. The body that eats, that works, that dies, that is afraid – that body just isn't there. Can't you write something for my students that would put things into a larger perspective?' I said I would try.

Bynum 1995: 1

Consider the sketch that accompanied a recent "whimsical" piece on "The Working Women's nanny" in a national Canadian newspaper (Figure 16.1). Thick pouting lips, tattooed-arm, pierced navel, tight shorts and shirt that accentuate crotch and breasts, this is an image that suggests the embodiment of domestic work in Vancouver, Canada.

In pursuing this theme, I follow a contemporary academic fascination with "the body." While the reasons for this theoretical curiosity are no doubt multiple and complex,[1] Foucault's theorizing has been extremely influential in framing both the interest and terms of debate; it certainly underlies the analysis that follows. I am drawn to it because of the opportunities it allows to think through processes of subjectivation as both regulated and exceeding discursive regulation; I am also indebted to Foucault for his remarkably concrete demonstrations of the workings of power through the blurred material and discursive constructions of bodies. I start by briefly outlining my theoretical debts to Foucault, and then consider three concerns that have been raised by and against feminist scholars who have worked with his texts. I reiterate these concerns to both situate my analysis within feminist scholarship on the body, and to lay an agenda for the case study of domestic workers.

Figure 16.1 "The Working Woman's Nanny". A representation of a domestic
    worker in a national Canadian newspaper.
    Source: *The Globe and the Mail*, Tuesday, August 29, 1995.
    Drawn by Dominic Bucatto

With and against Foucault: bodies, race, work and space

From Foucault, I take the understanding that the workings of power shifted from the seventeenth century onward, from a spectacular, bloody, juridical power of sovereign ruler over subjects, to procedures of disciplinary power that work on and through the body, and processes of subjectivation.[2] Discourses of sexuality, in particular, became "dense transfer points" of power relations, through which subjects are formed, and bodies – individual and social ones – are regulated and shaped. Discourses of sexuality, that named and pathologized "deviant" sexualities, such as homosexuality and masturbation, arouse first within and were practiced upon the bourgeoisie. Foucault interprets these as a means of establishing class power; the bourgeoisie distinguished themselves from other classes by a moral, legitimate sexuality and healthful organisms: "for the aristocracy had also asserted the special character of its body, but this was in the form of *blood*, that is, in the form of the antiquity of its ancestry and the value of its alliances; the bourgeoisie on the contrary looked to its progeny and the health of its organism when it laid claim to a specific body" (1990: 124). Foucault is not claiming that the bourgeoisie self-consciously used this construction of itself to grasp class power but, rather, that diverse discourses (that he unravels historically) were tied up with the repositioning of the bourgeoisie and the consolidation of the nation-state. For my purposes, the important point is that the bourgeoisie constructed their class identity through the watchful regulation of their own bodies; eventually this self regulation was practiced on and by the working classes. Other identities practiced through the sexualized body emerged: hysterical woman, masturbating child, homosexual. Homosexuality, for example, which previously existed as a (relatively unremarkable and unremarked) sexual act, became a stigmatized, criminalized, and medicalized identity. Foucault thus dissolves the strict boundaries, and simultaneously explores the linkages, between discourses, individual bodies, identities, and social bodies (such as the nation-state). Disciplinary power "seeps into the very grain of individuals, reaches right into their bodies, permeates their gestures, their posture, what they say, how they learn to live and work with other people" (Foucault, quoted in Sheridan, 1980: 217).

Nevertheless, Foucault is frequently criticized for writing an account of European bourgeois discourses and nation-states as if this can be done in isolation from European colonial projects (e.g., Spivak 1988). Foucault's tendency to underplay the influence of non-European societies has been interpreted as part of and perpetuating colonial assumptions of European superiority and self-containment. Stoler rewrites what she sees as the colonial subtext underlying Foucault's analysis; that is that eighteenth and nineteenth century discourses of healthy, vigorous bourgeois bodies developed in the colonies first, and were shaped through exclusions of race and nation (citizenship), as much as class. Discussions of European self control were often shaped in relation to racist discourses about disorderly, sexually promiscuous, colonized bodies. As an example of this (from a later period) Anderson describes how Americans (as imperialists in the Philippines in the early twentieth century) distinguished (and legitimated) themselves in relation to Filipino bodies: "American bodily control legitimated and symbolized social and political control, while the 'promiscuous defecation' of Filipinos appeared to mock and to transgress the supposedly firm, closed, colonial boundaries" (1995: 643).[3]

Stoler speculates that Foucault is able to short-circuit this consideration of colonialism and race, in part, because he "absents" key actors from his account and possibly lets "the discourse of bourgeois sexuality stand in for the sociology of it" (1995: 110). She identifies servants living and working in colonial bourgeois homes as key actors who are vacated from Foucault's account, but who are nonetheless critical to the formation of bourgeois discourse. Stoler attempts to fill the gap in Foucault's history by looking at how colonial middle-class identity formation took shape in Dutch colonial Southeast Asia in relation to servants. This is not my concern here but, rather, to reverse the optic and to explore how domestic workers are themselves shaped through the experience of living in middle-class Canadian homes. This reorientation focuses a more general concern, which can be phrased as the question: "Whose identities and whose bodies are deemed significant for scholarly attention?"

This shades into a second criticism, leveled at both Foucault and feminists who draw on his work, which is nicely summarized by Stoler when she asks: "Is Baudrillard's snipe that Foucault merely replaces one fiction of *homo economicus* with another, that of *homo sexualis*, valid?" (1995: 170). There are two things being said here: (1) that Foucault emphasizes discourses of sexuality to the exclusion of others and is thus as vulnerable to the criticism of reductionism as the marxist theory that he criticizes, and (2) that labor and the economy are submerged in Foucauldian analyses. Ebert (1992–3) takes up the latter theme in her criticism of what she memorably (but perhaps unhelpfully) calls "ludic feminism:"

> From the site of materialist feminist critique, the most serious and damaging blind spot in ludic feminist theories of gender, sexuality, and the body is the occlusion of labor. It is largely a class blindness. The ludic notion of discursive subversion, semiotic activism, disruptive pleasures, and flexible notions of gender have a strong appeal to (upper) middle-class women. . . . Gender and sexuality are not simply the result of discursive and signifying practices performed on the body but also, and more importantly, they are the effect of labor performed by, on, and through bodies as historically determined by the division of labor and the unequal access to economic and social resources.
>
> (1992–3, 39–40)

While one could challenge Ebert's dichotomy between discourse and labor, I think that she is right in calling attention to the relative neglect of labor among feminists influenced by Foucault, and other poststructuralist thinkers. The request from Bynum's friend with which this chapter opens – for the body that "eats, that works, that dies, that is afraid" – signals the same issue; it is a request to which this chapter attempts to respond.

A third concern signals a focus that is less submerged in Foucault's own writing than in some subsequent feminist writing on the body. The body is a site where the material and discursive, the objective and the subjective, are intertwined; it has been reconceptualized as a site that defies the binaries that permeate Western philosophical and political thought (e.g., Grosz 1994). But bodies are also sited in ways that are critical to the formation of identities and bodies. Foucault's studies of prisons, hospitals, schools, and bourgeois homes displayed a keen awareness of this, and this has been an important point of contact between geographers and Foucauldian scholarship (e.g., Gregory 1994). Foucault explores, for example, how school design both reflected and concretized concerns about child masturbation. While this is certainly not true of all feminist treatments of the body (e.g., Grosz 1992, reprinted in this volume; McDowell 1995; Probyn 1996), I think it is fair to say that some feminists (e.g., Bartky 1988) have studied the body as a decontextualized site (of, for example, tattoos or feminine beauty regimes). A goal of this analysis is to foreground the geography of the inscription of domestic work through Filipina bodies, to convey the multitude of often painful ways that domestic work and racial identity are inscribed on Filipinas' bodies, mediated through the geographies of nation and home. What follows are stories of the discursive/material construction of domestic workers, taken from a series of focus groups with which I was involved through 1995–6.[4]

From R.N. to R.N.: national geographies

I was an R.N. in the Philippines and I'm an R.N. in Canada. Only in the Philippines I was a Registered Nurse and in Canada I'm a Registered Nanny.

April

Like thousands of others, the women I quote come from the Philippines to Canada, some via Singapore and Hong Kong, through the Live-in Caregiver programme. This is a visa programme, designed and implemented by the Canadian federal government, to create a supply of domestic workers at rates affordable to middle-class Canadian families. After two full years of live-in service, the domestic worker is eligible to apply for an open visa, and then landed immigrant status. The vast majority of domestic workers are women and, since 1984, over 60 per cent come from the Philippines. (For overviews of the history of domestic worker immigration programmes in Canada see Arat-Koc 1989; Bakan and Stasiulis 1994; Daenzer 1993; Mikita 1994.)

As participants in the Live-in Caregiver programme, domestic workers are both inside and outside the Canadian nation. They have only temporary and partial access to the rights of Canadian citizenship. Two practical limitations are, first, they are required to live in their employers' homes and are thus restricted in terms of mobility, and, second, they are prohibited from upgrading their educational qualifications while registered in the programme (and are thus denied certain rights to occupational mobility). Daenzer (1993) notes that race and country of origin are not incidental to the precarious and partial rights afforded to contemporary domestic workers. Tracing the history of domestic worker programmes in Canada from the 1940s, she argues that the rights of domestic workers have declined as the source countries for domestic workers have moved from Britain and Northern Europe, to Southern Europe, then the Caribbean, and most recently Southeast Asia.

As "racialized" women who come to Canada as temporary residents, domestic workers experience a radical homogenization of their skills and occupational experiences. Women who participated in the focus groups were trained in various occupations, including midwife, registered nurse, high school teacher, social worker, bookkeeper. Not surprisingly therefore (and despite stereotypical equations of Filipina and nanny), some had limited training as nannies, as the following conversation attests:

Mhay: And the baby. I don't know how to bathe the baby.

Susan: We're in the same boat Mhay.

Mhay: So, the first time I bathe the baby I called up April to ask for some pointers. There is no small bathtub or basin. I asked if they have a basin for the baby because I am worried that I might drop the baby while giving a bath. And besides I find the baby quite slippery. I told her my concerns, that I am worried for the baby. I was really afraid to give the baby the bath. My employer [later] showed me how to give the baby the bath.

Cecilia: So, they just assumed that you know everything . . .

Susan: On my part, I told them at the start that I don't know how to cook and I don't know how to take care of kids.

The point of recounting this snippet of conversation is not to suggest that these two women are unsuitable nannies. Rather, it demonstrates that two women trained as teachers have had to learn the skills of caring for small children on the job.

Domestic workers are also reminded in a more immediate, everyday sense of the naturalized link between Filipina and nanny in Canadian society, often read off the surface of their body.

Joergie: It's really difficult to integrate because . . . we're mixed into other cultures, so at the bus stop, because we're people of color, and this hasn't happened only once to me, they'll ask you: so you're a nanny, right? For me, they ask me, you're Chinese? (Laughs) I say no, I'm Filipino, and then they'll say, so you're a nanny, right? (Laughs) I find that racist, really.

In another segment of conversation, experiences of being marked as nannies are stated very succinctly:

Mhay: There's quite a lot of white women nannies. But they don't say that they are nannies. For us, Filipinos are always looked at as nannies.
Joergie: It seems that Filipinos are identified as nannies.
Susan: I don't know which dictionary, but it was written there that Filipinos are nannies.
Cecilia: The Filipina Woman.

Not only is their occupation read off their bodies in public spaces, filtered through the lens of visual racial classifications, some domestic workers experienced this radical reduction of identity to "nanny" in the workplace as well.

Cristy: If they introduce you to their visitors or their friends: "This is my nanny." They are very proud of it, that they have a live-in caregiver to look after their kids, because if they have a nanny, it's just like they're very rich.
Mhay: Of course, my employer, he doesn't call me a nanny. When he has a guest, he introduces me by my name. But my wards [the children], they introduce me to their friends, even to their adult friends, "This is my nanny." So, I can't do anything, since I really am their nanny. But it's somehow jarring, for them to call me by my identity as a nanny. So sometimes I tell them that I have a name, you can call me by my name. So they get gradually used to doing that.

In an extreme case, which took place in Hong Kong, a domestic worker's body and function as servant was further objectified by building an equivalence between her and the household decor. Cora's uniform was coordinated with her employer's dishes and she was asked, for example, to wear her black uniform when the black dishes were in use.

The racialization of diminished occupational opportunities,[5] the conflation of personal identity with occupational category, and the marginalization of domestic workers as outsiders within Canada – all of which occur within the tension of unequal North–South geopolitical relations – are also overlain with class considerations. In the following passage, Mhay speaks of the process of stigmatization within the Filipino community; Filipinos coming to Canada as immigrants with full citizenship rights[6] also stigmatize Filipino nannies. Landed Filipinos thus enact their own gestures of closure around citizenship and the nation-state, with subtle and not-too-subtle gestures instructing domestic workers that they lie outside the boundaries of respectability. What is particularly interesting about her description is that this is another case where Mhay's identity is read off her body.

> *Mhay:* It's really difficult being a nanny because every time they *look at me*, and see I'm Filipino, they think of me really as a nanny. . . . So sometimes, and I've a lot of experience of this, about "being just a nanny." But I've really thought it over, and I said to myself, why do I say I'm just a nanny? So I say: "I'm a nanny." So, I removed that "just" there. . . . One group will say, "Oh, there go the nannies, out on their day off together." It's mostly Filipinos like us who say that. And then there's that other issue about being, since you're a nanny, you're, you know, someone who steals husbands. That's why wives are angry with us. But I don't get [my identity] entangled with that issue, because I don't steal husbands. That's a dangerous situation. I'm not ashamed of myself, and think very low of myself because I'm a nanny, since I'm not doing anything wrong. I cook, clean, take care of children. I'm not just a nanny. I'm also a *nanay* [Tagalog for mother]. (Laughs.) (Emphasis added.)

* * *

> *Mhay:* I encountered someone once. My driving lessons were over, and we were in the park to eat, because we were hungry. There were many Filipinos in the park, and near the car were some Filipino men and women talking. My companion asked me: "Why are they smiling at you? Do you know them?" I said "No, and I didn't know why they were smiling." They must have heard me, so they said something . . . bad. They said, "Oh, those are nannies. And they're trying *to look like something else*." (Emphasis added.)

* * *

> *Ana:* Like this is how the Filipino people look at the domestic worker. They look at the nanny as just a sex object and a husband stealer.

* * *

290

Susan:　Outside the house, I haven't encountered whites who say, "Oh, a nanny" just because I'm Filipino. It's just that the difficulty is with Filipinos like us.

Home sweet disciplinary home

If the ways in which Filipina bodies are read and rewritten must be interpreted against the backdrop of the Canadian nation-state, which both draws domestic workers within the national territory for the purpose of gaining access to relatively cheap labor, and withholds key rights to them as workers, the geography of the live-in arrangement is key to how these labor relations are lived, and how Filipina women negotiate and resist both their labor conditions and the absorption of their identity within the category, "domestic worker."

The Canadian federal government stipulates that a domestic worker who comes to Canada through the Live-in Caregiver programme must live in her employer's home. Employers are required to provide a separate bedroom, with a lock, as well as access to a bathroom. The employer is eligible to subtract $300 from what they monthly pay to the domestic worker in compensation for the fact that they provide in-house room and board. The live-in requirement thus dampens the cost of employing a domestic worker, because the approximately $1000 a month that an employer pays to the domestic worker (approximately $700 net to the domestic worker after she deducts payments for taxes, Canada Pension Plan, and Unemployment Insurance) is clearly below subsistence in a city where monthly rents for a one-bedroom apartment currently average $640 (*Vancouver Courier* 1995: 6). Domestic worker advocacy groups have long recognized the potentials for abuse that result from the live-in requirement, including the tendency to stretch both the workday and work week.

Room . . .

Mhay tells of the test that she performs to determine whether her employer has entered her room, as well as her efforts to confront and resist this violation.

Mhay:　This is what I do to find out whether someone has gone into my room when I am away. Before I go, I would neatly arrange my bed and all my stuff and put them in a certain way, so that I would know if someone has come in and looked at my things or sat on my bed. If this happens, I don't confront them at once. I wait until they are in a good mood and tell them that someone has been in my room when I was away.

Susan:　Even when you close the door, when you return, the door is already open.

Cecilia:　So, you really hardly have any privacy, huh?[7]

291

As the conversation continues, the issue of privacy is developed, and the fact that some employers fail to provide a lock for the door and respect their room as a private space. The insecurity of their own room means that domestic workers cannot properly claim a place, and it communicates to them their lack of rights and the transitory nature of their "stay" in Canada. In the case of Susan, her employers grossly misunderstood the contractual arrangement and simply used her room as a guest bedroom in her absence:

> *Susan:* Then, when I'm on my holidays, and they had visitors, they let the visitors use and sleep in my room. One time when I was ready to return on Sunday evening, they asked me to just return in the morning because their friend is still around and staying in my room.

Others had experienced the same; the experience of violation was common.

> *Cristy:* Don't we have the right, when we go out, to lock our doors? Why should employers enter our rooms when they have given us those rooms to stay in, if not so that they can check things inside? Because I'm sure that when they go inside the room they check inside, right?
> *Joergie:* My employer says that that's because it's part of her house, and so she has the right to check what's in there, what you do in your room, whatever things you are hiding there. Yeah.

Some employers also maintain or attempt to maintain control over the domestic worker's space – and, one could argue, bodies and identities – through furnishings. The frequently-cited absence of a writing table is noteworthy, given the likelihood that a domestic worker will write letters to family and friends in the Philippines. The absent table speaks about employer's preconceptions, of domestic workers as bodies who work, but are without intellectual or personal pursuits.

> *Susan:* When I was still new, I didn't have a table in my room. The TV stands on the chest. I really needed a table because I love to write, draw, and do calligraphy. They did not provide me with a table, so I ended up buying one and putting it in my room.
> *Joergie:* For me, I don't have a table. If I write, I write on the floor, or on my bed.
> *Mhay:* I bought a picture with a frame and put it up on the wall. Prior to this, all four walls were bare. I did this without telling them because I thought that since I paid for this room, I should be allowed to do something about it. So I

arranged the room, put furniture and TV [the way I wanted them]. I would leave the door open so that they could see what's in my room, that it's not dull anymore.

Susan: I cannot make changes in my room. When I arrived there were already drawers, and there were paintings on the wall. I cannot change them anymore.

Cecilia: But you have no table, and drawers have no locks.

If their "private" space is not entirely their own, it should come as no surprise that a number of women spoke of their discomfort within the rest of the house when their employers were home, because of feelings that they must continue working and a sense of being surveilled.

Cecilia: Do you have the feeling that you are free in the house, that you can sit if you are not working?

Susan: Yes.

Mhay: Only when my employers are not around, yes. Once they're around, I prefer to stay in my room because I start seeing all of the mess. Sometimes I wonder why my employer cannot pick up her mess. So I prefer to stay in my room.

Joergie: They say that it is okay to stay in your room when they're around. But then you feel that they really don't want you to do this. They seem to want you around the house, except in your room.

Cecilia: Like sitting in the living room or in front of TV? Don't they tell you to feel at home, that you should treat this as your own house?

Susan: They tell me that. We have breakfast together. They don't leave at once. They read the papers. So my employer, my ward, and me, we all would be at the dining table together in the morning. They are quite good and nice to me. I could say the same thing about the child.

Joergie: My employer doesn't tell me that. But she watches me like a hawk.

One further instance of the ways that employers discipline domestic workers, through the control of domestic space and by designating whose raced, classed and gendered bodies have a place within the home, concerns domestic worker's rights to bring friends to what is in principle (and by right, given that they pay monthly rent) their home. Most domestic workers felt that they were unable to bring friends to their home and several employers had been explicit that male friends were disallowed at all times.

. . . *and Board*

Bodies not only need space, they need to be fed in order to survive. At the most basic level, hunger was a problem for some of the domestic workers who participated in the workshops. If starvation was not really the issue, their diminished status was nonetheless communicated and absorbed through food restrictions. A number of employers whom I interviewed in the summer of 1995 complained about the amounts of food consumed by nannies, always European or Canadian ones, in the manner of a humorous "She's eating us out of house and home" tale. One employer compared the cost of her Quebecois nanny's food consumption unfavorably to that of their previous Filipina nanny who "survived on toast." That the employer questioned the appetite of the Quebecois woman rather than that of the previous nanny is noteworthy, and may reflect "orientalizing" assumptions about the bodies and appetites of Asian women. It is instructive to consider this employer's remarks in relation to the following conversation:

Joergie: I have to finish the food. I don't ask for any more even if I still want to eat. I haven't tried to ask for extra food. Whatever she puts on my plate, that's all I get. . . . The food is quite good. The only thing is that you only get what she gives you. I've never attempted to ask for more. If it is not enough, that's all I ever get.

Mhay: That's true in my case. There are five of us and there would be five plates. And that's all that everyone gets. There is no second serving.

Joergie: But if they want more they can have a second serving. I am just too embarrassed to ask. And sometimes I will not eat with them because I still have to serve them. I will still be doing things in the kitchen, such as washing the plates. It's only when they have finished eating that I will start eating myself. It's like they are above there, while I am down here. . . . When I return after the weekend [away from her employer] I bring some food because I know that I will not have enough food in the house.

Mhay: It really is true that sometimes the food is not enough. And you don't have choice in the food. So if you want something special for yourself, you have to get it and prepare it yourself, and bring it to your room.

But beyond a shortage of food, some domestic workers felt a great awkwardness about introducing their own cultural habits in their household. Food, as a symbolic good that literally sustains the body, is a fascinating condensation of the symbolic and material and an important marker of cultural difference. But as such a dense "transfer point" of meaning and materiality it is perhaps difficult for the domestic workers themselves to articulate how they came to understand or assume their employers' distaste for their culture.

> *Susan:* I would be hesitant to cook when my employers are around. I would feel embarrassed. . . . In the morning, I can hardly eat their food even if I want to because they are still around. I would be able to taste it only after they have left the house. (Laughter.)
>
> *Mhay:* Once, I really wanted to cook my own food, which they do not eat. I opened the windows so they could not smell it in the house. Then I ate in the living room so that I could see them coming. Once I saw them, I would go to the kitchen and fix everything so that they would not find out that I cooked my own dinner or food. It's quite tough. (Laughs.)
>
> *Susan:* One time, after they had gone, I started eating, when all of a sudden they returned. So I had to throw into the garbage can the food that I was eating. I felt bad about it, especially that I did not have a chance to eat one of my favorite foods.
>
> *Cecilia:* You did not keep it in the refrigerator?
>
> *Susan:* I was not sure that they would not open the fridge.

These attempts to erase traces of their culture, both visual and olfactory, or to contain them within their rooms, stand in marked contrast to their experiences of literally ingesting another culture. Canadian culture is sometimes difficult to stomach.

> *Mhay:* My frequent complaint is about our food, which is pasta every dinner time. Within one week, pasta comes three times. Spaghetti. God, it's so boring!
>
> *Susan:* About food, we really differ from them in our fare. We are used to eating rice three times a day, while they eat vegetables and pasta. We look for rice, which they will not buy, but although we want rice, our salaries remain meager. So we're forced to go along with them, and eat what they eat. Even if sometimes we can't get it down (laughs) because it's all vegetables.

Not all employers are insensitive to dietary difference, but even in these circumstances the struggle of accommodating two dietary regimes in a single household is palpable.

> *Edrolyn:* I like an employer that gives and takes. Like my employer, she asked me what food I want to eat. Because some of the

employers don't want to ask you. She said, "It's your turn to do the list of the food this week, so that you can cook whatever you want." So I can go to Chinatown to go shopping. So this whole week I cook Chinese food for them. I know it's their first time to taste Chinese food. They are just trying to show me, "Edrolyn, I like this food." But I can see that they don't like it.

Ana: At least they are adjusting. Some are not.

Marlyn: Never.

Ana: You have to adjust.

Edrolyn: When we are at Whistler [a recreational property] , Monday to Thursday, I never eat rice because they always eat bread. Then I think that she observed that I feel strange, because I always eat rice. I love rice. Then she said, "Edrolyn, I have cooked rice. Here's the rice for your dinner." I said, "Oh, thank you." But I am shy because she did this for me. . . . My employer buys sacks of rice. The boy he loves to eat rice. The father said, "Oh Edrolyn, you teach him how to eat rice!" But his sisters don't want to eat rice.

Edrolyn's account of her employer's remark about transferring her food habits to his child is very interesting to consider in relation to Stoler's (1995) discussion of Dutch colonial concerns about such transfers. Dutch discussion turned around the transfer of milk from wet-nurse to child; there were fears that the wet-nurse would transfer blood (her race) along with her milk. The transfer of milk from wet-nurse to child seemed to heighten fears about the permeability of boundaries between white and non-white, and between bourgeois and colonial "others." While Edrolyn's employer seems to be phrasing the cultural transfer in positive terms, he is also noticing it (and communicating this to Edrolyn). Employer sensitivity about such transfers between nanny and child may account for the negative sentiment that nannies feel from their employers about their food.

We can also search for an explanation for domestic workers' sensitivity about employers' reactions to their food in the historical record. Anderson (1995) writes about Americans' portrayals of Filipino food and markets in the early twentieth century:

For many American colonialists, the Philippine marketplace provided a fascinating combination of sensuality and danger. . . . The market was readily represented as a locus of promiscuous contact and contamination. . . . Le Roy found that 'unless there be rigid and efficient supervision,' the markets were 'foci of infection.' Whenever he wandered through these places, Nicholas Roosevelt assumed that 'many varieties of intestinal germs and parasites may lurk in most foods.' For Daniel R. Williams, the markets were simply 'unwholesome and death-dealing plazas.'

While it is clearly speculative to track early twentieth century American medical discourses of filth and contamination to contemporary Vancouver homes, it is nonetheless possible that traces of historical discourses from one place are woven into contemporary ones in another. As Doty (1996) observes, although stereotypes are instantiated in particular contexts, they draw upon globalized representations, and histories of previous stereotypes are sedimented within contemporary ones. Whatever the explanation, it is clear that several domestic workers felt that employers hold their food (so basic to the nourishment of their bodies and selves, and the standard marker of culture) in distaste.

Resistance and reterritorialization

Domestic work is read off their bodies when they wait for buses and visit parks; they are objectified as their (often newly acquired) occupation; their temporary visa and non citizen status is daily enacted through their fragile occupancy of their rooms and houses and the long hours that they work within these private spaces; their cultural difference is daily negotiated by what they and the children they care for ingest through their bodies: these are all ways that domestic work is inscribed in, on, and through the body. This inscription process is also resisted as Filipina domestic workers exceed these closely written spaces of the body and identity.

In part Filipina domestic workers resist through the careful management of their bodies. Several women spoke of actively resisting two identities that are sometimes imposed on them as live-in domestic workers – those of sexually available object, and thief – through their bodies.

> *Edrolyn:* I don't comb my hair. Just wear a big t-shirt. Because I don't want my boss. My male employer talks to me when I am in the kitchen. Just talks to me, asking me everything. "What're you doing every weekend?" For me, that's fun. I just answer what he is asking me. But I don't want to be close to him. I always ask the wife if I have some question to ask. And the wife talks to him. Some nannies ask the man, and then the lady gets mad.
>
> *Yolly:* Especially the Chinese. They don't want their nannies to be close to the man.
>
> *Marlyn:* They are afraid that the nanny will steal their husband.
>
> *Ana:* Sometimes it's the nanny's fault too. Like if your boss is flirting to you. What you gonna do? You will not flirt to him

too. If he's flirting to you, back off! If you don't really like
him, just play hide and seek.

Vulnerability to accusations of stealing from employers was another
common concern. There were numerous stories of employers testing the
domestic worker's honesty by leaving jewelry and large sums of money in
clothing or in open view. Worse, several women recounted cases of domestic
workers who were "entrapped" by a disgruntled employer and outlined their
strategies to avoid a similar fate.

> *Edrolyn:* I don't touch their jewelry, even in Singapore. I am very
> conscious. I use a towel or gloves [when cleaning jewelry].
> Because some employers, they just set you up.
>
> *Marlyn:* When I quit my employer, I knew she would do that.
> Because she was so mad at me. So I said that I want someone
> to go with me [when I pack my belongings] and check
> everything I get. I said, "You have to see everything that I
> am packing." Because if I don't [do this] she might just put
> something in and will charge me later on.
>
> *Edrolyn:* That happened in Singapore. The nanny worked for so many
> years and she spent her money to buy jewelry. And she had
> lots of jewelry. So her employer was jealous. I guess because
> she was only a nanny and had lots of jewelry. You know what
> she did? [The employer] got her jewelry and transferred it
> to her nanny's bag. [The nanny] was leaving to go to the
> Philippines the following day, and [the employer] called the
> police and said that her nanny is stealing her jewelry. Then
> what the nanny did was say, "Okay, you trace whose finger
> prints are there." That's why I am always aware of that [and
> ensures that her fingerprints are not on her employer's
> jewelry]. . . . Even the attaché case I don't touch. I use a
> stick. You know why? They will take advantage. Because
> I've heard of this before.

Edrolyn thus controls her employers' access to her body by hiding its contours
beneath a big t-shirt, and taking measures (gloves and a stick) so as to prevent
the trace of her finger (body) prints on her employers' goods and spaces.

It is also resisted through discourse. In the most straightforward way it
is resisted through discourse by forcing employers to obey the law or
collectively by challenging provincial governments to extend basic labor
regulations to domestic workers[8] and lobbying the federal government to
lift the live-in requirement. Discursive resistance can also take more subtle,
individual forms that seem very close to Foucault. Consider, for example,
Mhay's decisive word play, removing the "just" from "just a nanny" and

replacing "nanny" with the Tagalog word, *nanay*, thereby reclaiming the meaning of respectable mother for herself as nanny. Other women told stories that seemed the physical equivalent of this productive re-use of discursive categories. So, for example, Edrolyn tells of complying utterly to her employers' instructions to clear the exterior of the rice cooker (a task that Edrolyn felt was unnecessary) by cleaning to excess, to the point where the paint was removed from the rice cooker.

They also resist and rework their subjectivity by coming to the Philippine Women Centre. The Philippine Women Centre is one of several organizations in Vancouver that they might turn to; others include The West Coast Domestic Workers' Association and Vancouver Committee for Domestic Workers and Caregivers' Rights. Each organization follows a different model of resistance.[9] The West Coast Domestic Workers' Association, for example, works within a legalistic mode and therefore (of necessity) casts domestic workers as individual plaintiffs standing before the law.

The geography of these different models of resistance is not insignificant. The West Coast Domestic Workers' Association has an office downtown, just off the business district. The Philippine Women Centre occupies a house, in South Vancouver, in the centre of the Filipino community in Vancouver. It is a social place where women come to eat,[10] can stay overnight or for extended periods, located in a neighborhood where they might begin to imagine and build concrete networks toward a life in Vancouver beyond "domestic worker." They can begin to reterritorialize identities that have been de-territorialized through the process of migration and by the radical tenta-tiveness with which they inhabit and are allowed to inhabit their (employers') homes. It is a place where they can refigure themselves as members of a group of exploited workers and citizens, rather than simply as responsible daughters, wives and mothers enacting an international but familial strategy for survival (which many are as well). It is a place where they can reclaim and nourish their bodies and learn new social, collective identities – identities as women that build bridges to other communities of women and feminists.

It is also a place where an academic feminist can learn new relations between bodies, identities and politics. During the focus groups, I was encouraged to see the physicality and collectivity of social existence in new ways. I had assumed, for example, that talking was the only way to tell stories; women at the Centre suggested sculpture, collage and role-playing as alternative ways of knowing.[11] I was surprised (and at first embarrassed – there is not much singing in the halls of academe!) by the songs that punctuated the workshops. The second workshop ended with a "performance" in which every woman was called upon to take her turn in the middle of the room to dance. Most dancers were extravagant, self-mocking, attempting to elicit laughter. Here bodies gyrated, gesticulated and claimed space in ways that seemed not to be tolerated, and which may not be offered, in their domestic homes. As I recall my self-conscious, stiff, contained "dance,"

I wonder that I am writing this paper on the body; my dance stands as a metaphor, rationale, and legitimation for the cramped space of authorship that I have chosen in this paper. It is also an expansive, shared space. I end with the song that began our first workshop:

Kung Alam Mo Lang Violy[12]

KORO:
Kung alam mo lang Violy (2 times)
Kung alam mo lang Violy ang totoo
Kung alam mo lang Violy (2 times)
Matagal ka na nilang niloloko

I   May midyum tirm development
Uutangin ang imbesmeynt
Panot na ang enbayrunmant
Tuwang tuwa ang gobernment

II   Pinay ay di komoditi
Por export sa ibang kantri
Wala namang sekyoriti
Lagi na lang gunugulpi

III Pangingilkil lang ang alam
Gobermit walang pakialam
Pinay pambayad ng utang
Sa IMF at World Bank

IV Hoy alam ko na Eddie (2 times)
Hoy alam ko na Eddie ang totoo
Hoy alam ko na Eddie (2 times)
Matagal mo na kaming niloloko

Acknowledgments

I would like to thank a number of people for their extreme generosity as critics of this paper: Suzanne Bausted, Trina Bester, Dan Hiebert, Jennifer Hyndman, Deirdre McKay, Heidi Nast and Steve Pile. I thank them for the various references, photocopied articles, newspaper clippings, and ideas that they passed along, as well as the serious intellectual and political challenges that they posed to me. Most, of course, I thank the women who participated in the project and offer this paper as a basis for further conversation. I would also like to acknowledge the generous support of SSHRC.

Notes

1 Emily Martin (1992) speculates that the preoccupation with the body results from the fact that in North America (and most likely elsewhere) we are experiencing a "dramatic transition in body percept and practice, from bodies suited for and conceived in terms of the era of Fordist mass production to bodies suited for and conceived in terms of the era of flexible accumulation" (121).

2 My account of Foucault's theorizing is extremely cryptic – for this I make no apologies. Numerous excellent extended accounts exist elsewhere (e.g., Butler 1993; Gregory 1994; Grosz 1994). I take Foucault's tool kit analogy to heart; in this paper I care less about finessing theoretical points than using some interesting social theory as a "tool" to open up a fuller understanding of the situations of Filipina domestic workers living and working in Vancouver in 1995.

3 It should be made absolutely clear that Anderson is not saying that Filipinos are thus but, rather, that this is the way they were portrayed by American colonial physicians.

4 As a white, Canadian-born academic, I am an awkward vehicle for writing such a cartography of Filipina domestic workers; over the last decade there has been considerable debate about the problems and possibilities that attend such a venture. (For a summary of these debates and a particular account in relation to sexuality, see England 1994.) What follows is a hybrid account, conceived and carried out in collaboration with fourteen domestic workers and organizers at the Philippine Women Centre in Vancouver through fall and early winter 1995–6. We met for six workshops, loosely organized around the project of telling stories about experiences as domestic workers. The topical agenda was set collaboratively at the first workshop; these themes were then discussed in focus groups of four or five women each in two workshops that followed, in very loose and unfocused ways. The first (six hour) workshop involved meeting with a core group of 12 women to establish the thematic agenda of the following focus groups. In the next (roughly 8 hour) workshop we broke into three small groups of 4 or 5 women each to begin talking through the established themes. In the next workshop a few more women were brought into the process, and we again moved into three smaller groups and worked through the same themes in more detail, with additional input from new women. Further workshops centered on analysis, coding and verification. All the women (except Cecilia, Pet and Jane) are presently domestic workers, in the Live-in Caregiver programme. The goal of the project was to encourage domestic workers' attempts to represent themselves through participatory focus groups.

I recognize the illusion of researcher "transparency" (the presumption that researchers can represent "authentic experience" by substituting others' voices through direct quotation.) The following conversational extract illustrates the negotiated, mediated nature of the accounts constructed in the focus groups in which I participated. This conversation took place in the third workshop, in a small focus group of five women.

Edrolyn: How about you, Gerry. Do you have a nanny?
Gerry: No.
Edrolyn: That's a personal question.

Marlyn: Do you want a nanny? Take me as your nanny.
Gerry: I send my kid to daycare.
Ana: She will be one of the good employers.
Edrolyn: You can see in her facial expression. My employer is good but a
          little bit Tupperware [plastic].

As we try to read each other's politics and sincerity through our bodies – our facial expressions – we also negotiate and mark a power dynamic: I am a white Canadian-born woman who could choose to hire a nanny. Other factors that divide us and complicate interpretation are those of language and translation. The discussions generally took place in Tagalog (except in groups in which I participated). They were transcribed and translated into English. Most of the quoted material has been translated, with the exception of instances when I am clearly present. I am also crafting this paper and it would be disingenuous to pretend otherwise. Nonetheless, and in collaboration with the research participants (who have read and commented upon this manuscript), I attempt to present conversations of domestic workers with a minimum of intervention.

5 Analyses of the 1991 census data indicate that Filipinas are the most occupationally ghettoized of all women (classified by ethnicity) in Vancouver. They are especially over represented in three occupations: housekeeper, child care worker (which are essentially the same occupation – domestic worker) and medical assistant (e.g., nurses' aid) (Hiebert 1994).

6 There are many routes into Canada as an immigrant. Relevant to this paper is the distinction between immigrant (which involves qualifying through a formalized point system) and coming to Canada via a special visa programme, such as the Live-in Caregiver programme. Immigration through the regular point system, in which points are accumulated for a whole range of specified criteria (e.g., having specified technical skills, which are often traditionally masculine ones, and educational qualifications from accredited institutions) is a classed, racialized and gender process.

7 Cecilia is the Chairperson of the Philippine Women Centre. Her comments are also interventions and illustrate the slippery terrain on which this "academic" project moved. It was meant to have educational and politicizing effects, for me as well.

8 In March 1995 a new B.C. Employment Standards Act was passed that includes domestic workers under provincial overtime provisions. Employers are now required to pay time-and-a-half for hours worked beyond a 40 hour week. That domestic workers were previously without this labor regulation, which most British Columbians take for granted, is itself indicative of their inside/outside status. Many domestic workers have had difficulty obtaining this right in practice. It was striking to me that a number of employers interviewed in the Summer of 1995 were adamant that these legislative changes had not been implemented.

9 I thank Suzanne Bausted for forcing my attention to this very important point. As she so rightly noted, to neglect this fact is to risk dominant representations of Filipina women. An important question concerns how individuals come to know about each organization, one that I cannot adequately answer. Certainly, members of the Philippine Women Centre recruit actively: domestic workers go

to shopping malls on weekends to recruit members and tell of organizing on the Sky Train (rapid transit). It is clear, then, that organizing also relies on visual inspection and assessment of national and cultural membership. Domestic workers also introduce friends that they might know through other contexts, for example, church.

10 Consider the following:

Mhay: I can only eat well here at the Centre during the weekend, and Friday evening.
Susan: It's only here that we can really replenish ourselves.

11 Since writing this, the centre has moved to a downtown location: 451 Powell Street, Vancouver, B.C. V6A 1G7. Ultimately we mostly told stories, although one group engaged in role-playing. There is a tension between what is "useful" as scholarly documentation and what works for community organizing. This tension deserves to be thought through more carefully than I am able to within this text.

12   If you only knew, Violy

If you only knew, Violy (2 times)
If you only knew, Violy, the truth
If you only knew, Violy
You've been fooled for a long time.

1 There is a medium term development plan
Borrowing money for investment
And exploiting our resources
The government is rejoicing.

2 A Filipina is not a commodity
For export to other countries
Without security
and always abused.

3 Corruption is all that they know
The government doesn't care
Filipinas are used for debt payment
To the IMF and World Bank

4 I already know Eddie (2 times)
I already know Eddie, the truth
I already know Eddie (2 times)
That you have been fooling us for a long time.

As testimony to the materiality of the body, the next workshop began with the song "Hoky Poky."

References

Anderson, W. 1995. Excremental Colonialism: Public Health and the Poetics of Pollution. *Critical Inquiry* 21, 640–69.

Arat-Koc, Sedef. 1989. In the Privacy of Our Own Home: Foreign Domestic Workers as Solution to the Crisis in the Domestic Sphere in Canada. *Studies in Political Economy* 28, 33–58.

Bakan, Abigail B. and Daiva Stasiulis. 1994. Foreign Domestic Worker Policy in Canada and the Social Boundaries of Modern Citizenship. *Science and Society* 58, 7–33.

Bartky, Sandra. 1988. Foucault, Femininity, and the Modernization of Patriarchal Power. In Irene Diamond and Lee Quinby (eds) *Feminism and Foucault*. Northeastern University Press: Boston, pp. 61–86.

Butler, Judith. 1993. *Bodies That Matter*. Routledge: New York.

Bynum, Caroline. 1995. Why All the Fuss about the Body? A Medievalist's Perspective. *Critical Inquiry* 22, 1–33.

Daenzer, Patricia. 1993. *Regulating Class Privilege: Immigrant Servants in Canada, 1940s–1990s*. Canadian Scholars: Toronto Press.

Doty, R. L. 1996. *Imperial Encounters*. University of Minnesota Press: Minneapolis.

Ebert, Teresa L. 1992–3. Ludic Feminism, the Body, Performance, and Labor: Bringing Materialism Back into Feminist Cultural Studies. *Cultural Critique* Winter, 5–50.

England, K. 1994. Getting Personal: Reflexivity, Positionality, and Feminist Research. *The Professional Geographer* 46, 80–9.

Foucault, M. 1990. *The History of Sexuality. Volume 1: An Introduction*. R. Hurley (trans.). Vintage Books: New York.

Gregory, Derek. 1994. *Geographical Imaginations*. Blackwell: Oxford, UK and Cambridge, Mass.

Grosz, L. 1992. Bodies-Cities. In Beatriz Colomina (ed.) *Sexuality and Space*. Princeton University Press: Princeton, pp. 241–54.

Grosz, E. 1994. *Volatile Bodies*. Indiana University Press: Bloomington.

Hiebert, D. 1994. Labor-market Segmentation in Three Canadian Cities. Paper presented at the annual meetings of the Association of American Geographers, San Francisco, March 1994.

Martin, Emily. 1992 The End of the Body? *American Ethnologist* 19, 121–40.

Linda McDowell. 1995. Body Work: Heterosexual Gender Performances in City Workplaces. In David Bell and Gill Valentine (eds) *Mapping Desire*. Routledge: London, pp. 75–95.

Mikita, Jeannie.1994. The Influence of the Canadian State on the Migration of Foreign Domestic Workers to Canada: A Case Study of the Migration of Filipina Nannies to Vancouver, British Columbia. Unpublished MA thesis, Department of Geography, Simon Fraser University.

Probyn, Elspeth. 1996. *Outside Belongings*. Routledge: New York and London.

Sheridan, Alan. 1980. *Michel Foucault: The Will to Truth*. Tavistock: London.

Spivak, Gayatri C. 1988. Can the Subaltern Speak? In Cary Nelson and Lawrence Grossberg (eds) *Marxism and the Interpretation of Culture*. University of Illinois: Urbana.

Stoler, Ann Laura. 1995. *Race and the Education of Desire*. Duke University Press: Durham and London.

Vancouver Courier: 1995. West End Top Choice. 3 December, 6.

17

MAPPED BODIES AND DISEMBODIED MAPS

(Dis)placing cartographic struggle in colonial Canada

Matthew Sparke

Place-map-body

In a recent and valuable introduction to the critical study of geography and empire, Anne Godlewska and Neil Smith conclude their remarks by noting how necessary but difficult it is to re-construct the so-called "rival geographies" of those who have struggled against imperialism.[1] This paper explores the substantive difficulties facing such projects of reconstruction using an example of a rival geography that has thus far been interred in the archives of colonial Canada. Following Edward Said's original formulation of rival geographies my effort at reconstruction has been informed by the post-colonial notion that in the struggle to "to reclaim, rename and reinhabit the land," the impulse is "cartographic."[2] However, in contradistinction to Said and many of the post-colonial theorists who have "remapped" his often polarized picture of colonial struggle with a discourse of "cartographies of struggle," "interstitial spaces," "Third space" and the like, this chapter is focused on an actual case of cartographic struggle in which maps were involved. This said, I am centrally concerned with at least two sets of questions raised by so-called post-colonial criticism: first, about the *epistemic violence* implicit in metropolitan representations of "pure" native space; and second, about the contested *translations* between space and political (including national) identity. In the case examined here cartography is the translating link, but in order to understand the power relations implicated in this link I focus most especially on how bodies are variously mapped and left out of maps, how they are represented and dissimulated with political effect by different cartographies of colonialism.

In interrogating the Canadian cartographic archive and in questioning the disembodying translation effect that is the archive's colonial inheritance,

I learn from Gayatri Chakravorty Spivak's essay on reading the archives.[3] In this exemplary essay Spivak displaces Fredric Jameson's discussion of "worlding" in Heidegger by pulling it, if only for a moment, into the effaced arena of colonial cartography. Here, she says,

> what emerges out of the violence of the rift [between earth and world] . . . is the multifarious thingliness of a represented world on a map, not merely "the materiality of oil paint affirmed and foregrounded in its own right" as in some masterwork of European art.[4]

Reworking Heidegger Spivak argues that one of the ways through which colonialism "worlds" or constructs the colonized world is through "the necessary yet contradictory assumption of an uninscribed earth."[5] It is such a disembodying, earth-evacuating assumption which in turn, she says, makes "the 'native' see himself as 'other.' "[6] The fact that such lands were previously far from uninscribed not only reveals the mistakenness of the colonial assumption, it also indicates – to use Spivak's gratingly displaced translation of Heidegger – how "the multifarious thingliness of a represented world on a map" might also open up the possibility of other, native maps mapping and inscribing the land otherwise. In a more recent article discussing the fiction of Mahasweti Devi, Spivak finds a powerful example of such "native" inscription in the bodily form of what she calls "the socially invested cartography of bonded labor."[7] The bonded labor in question is that of the prostitute Douloti who, at the close of Devi's story and Spivak's essay, lies down and dies vomitting blood on a map of India drawn in the clay courtyard of a schoolhut. "Douloti," writes Devi describing the dead body, "is all over India."[8] In this violently embodied cartographic reinscription of the postcolonial map of India, Spivak suggests that Devi is providing her readers with a reminder about the sexually, ethnically and economically marginalized subjects that have been left behind by the move from colony to supposedly independent nation-state. In short, "[t]he space displaced from the empire-nation negotiation now comes to inhabit and appropriate the national map, and makes the agenda of nationalism impossible."[9]

The displacing "space of difference" evoked by Douloti's coporeal cartographic reinscription of the national map shares much with the example from colonial Canada I wish to elaborate. This example also ends with the death of a subaltern woman, a blood vomitting end (in this case due to tuberculosis) that is marked on the Canadian landscape by a monument in St. John's harbor Newfoundland. Her name was Shawnadithit, and, in the words of the monument, she was "very probably the last of the beothics who died on June 6th 1829".[10] The Beothuk[11] (otherwise known to the early colonialists as "Red Indians" because of their practice of painting their bodies with red ochre) were the native population of the island before it became Newfoundland. In fact, their mass deaths from disease, starvation and murder can be interpreted as a form of embodied inverse of the island's colonial reinscription as new found land. This, in other words, was a colonial "worlding" that brought "othering" with a vengeance to the native inhabitants. Yet even here, in the form of a series of story maps drawn by Shawnadithit, a native cartographic reinscription of the land charts a story that is eccentric to and undermining of the sweeping narrative running from native land to colonial newfoundland to modern nation.

To be sure, Shawnadithit's non-fictional history also departs in certain ways from the example of Douloti examined by Spivak. Devi's story takes place in decolonized terrain, albeit, as Spivak concludes, terrain that is simultaneously subject to the extraction of wealth that constitutes contemporary neo-colonialism. It is a story that displaces the post-colonial national dreams of a colonized people not, as with Shawnadithit's legacy, the post-colonial national imaginary of a white settler-colony. And it is a story where the body of Douloti serves as a vehicle in the metaphorization of cartographic inscription. As Spivak has elsewhere noted (in relation to the metaphorical fate of another subaltern figure from Devi's work), the strict relationship between vehicle and tenor in metaphor means that such metaphorical moves are directly related to the simultaneous underplaying of substantive connections between the vehicle, in this case the woman's dead body, and that which the body as vehicle is employed to evoke, here, the situation of similar subaltern subjects "all over" India (the "all over" here evoked by the further instrumentalization of the map as another metaphorical vehicle).[12] By contrast, Shawnadithit's cartography introduces immediate and politically significant links between bodies and maps. In fact the contrast is a double one because in addition to being actively produced by Shawnadithit's own hand, which is to say the "native" body as observer and inscriber, her map work is also distinguished from all dominant Western genres of cartography precisely insofar as it features representations of bodies inhabiting and moving through space. The mapped bodies discussed in this chapter, then, are not just dead or metaphorical vehicles, although in one case they are both, nor are they the passive objects of the masculine and imperial gaze, although the bulk of the discussion is concerned with how the body of Shawnadithit has previously been positioned in precisely this way. Instead, they are bodies depicted cartographically *by* a native woman, bodies which, representing the encounters between the colonized and colonizers, inhabit and thereby represent a space of embodied beween-ness.

The disjunctive quality of the space of embodied between-ness can be better understood if it is compared with that against which I have opposed it in the title of this chapter. By "disembodied maps" I am referring first of all to the abstract colonial mapping of Newfoundland as such. In this sense, the "New" as it is commemorated in the contemporary toponymy reminds us that as the island began to emerge on the horizon of the European geographical imagination Newfoundland was something radically discontinuous and disembodied from the place as it had been understood, inhabited and experienced by those who had lived there previously. From its outline on the detailed Desceliers map of 1550, and onwards, as Fabian O'Dea has documented, into and through the seventeenth century, the emerging European depiction of Newfoundland was as a basically empty space, a space seen from sea, delineated by coastline, and as a peopleless void within.[13] Even coves and bays were not fully explored, and, as O'Dea notes in studiously eurocentric terms, it was not until the 1760s that Captain James Cook came "to establish a scientific basis for an understanding of the true shape of Newfoundland".[14] Whatever the merits of Cook's theodolite and telescopic quadrant, this "true shape" was still even then largely just a matter of outline (Figure 17.1). Joseph Banks, who was traveling with the famed surveyor, admitted that "we Know nothing at all of the Interior Parts of the Island."[15] Yet even as such acknowledgements were made, the abstracting and disembodying effect of the Cartesian cartography simultaneously presented the interior as known as empty as uninhabited. Less than a century later, the anticipatory aspect of this colonial enframing effect became disembodied reality. By 1830 there were no living native bodies left.

Shawnadithit's cartographic work survives as a rival geography that directly contests the disembodying abstraction of the colonial maps. However, in order to be read this way it also needs to be critically reinterpreted as the work of an embodied agent of knowledge. This means coming to terms with how "the last of the Red Indians" has been dug up and moved discursively in much the same way as Shawnadithit's actual grave was dug up to make way for a new road. Disinterred and disembodied, she has been transported into the pantheon of modern Canada's famous figures, a pantheon that otherwise includes explorers, fur traders, politicians and railway-men whose more general discursive duty in death remains as heroes and heroines in the romanticization of the very processes that caused the genocidal demise of native people like the Beothuk in the first place. One of the specific side-effects of this exercise has been to have turned the few statements and drawings Shawnadithit made for her custodians in St. John's into anthropologically nationalized relics.[16] Treated as disembodied relics of a national tragedy, they tend to be entombed in books and articles as just so many mawkish momentoes, the last fading gasps of a native informant whose people have gone forever. By contrast, I want to question this process of disembodying entombment, problematizing the convention by carefully examining just one of her maps in order to draw attention to the agency as opposed to the tragedy it embodies.

In attempting to reconstruct the cartography of Shawnadithit as an embodiment of agency I do not want to pretend that I am simply presenting a pure native perspective on spaces past. As Spivak has famously argued and as Jonathan Crush has recently reminded geographers, we cannot speak unproblematically about the subaltern speaking.[17] Late-in-the-day posturing

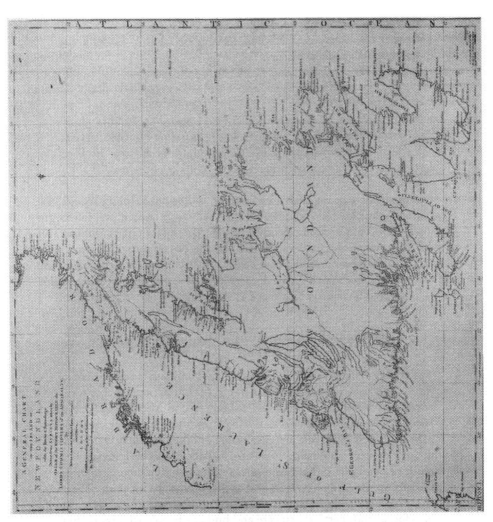

Figure 17.1 Cook's general chart of the island of Newfoundland.

by Western academics to either "allow the natives to speak for themselves" or to "give the natives voice" fails to come to terms with how in cases such as Shawnadithit's the gendered and racialized subaltern subject is always being spoken about by others: she is always being represented and these representations demand a persistent vigilance as to their own politics. In short, to considerably modify Spivak's discussion of the *Srimadbhagavadgita*,

> [t]here is no historically avaliable authentic [Red] Indian point of view that can now step forth and reclaim its rightful place in the narrative of [Canadian] history.[18]

Instead,

> we might teach . . . the way to informed *figurations* of that "lost" perspective. The geopolitical postcolonial situation can then serve as something like a paradigm for the thought of history itself as figuration, figuring something out with "chunks of the real."[19]

In the colonial situation discussed here it is Shawnadithit's cartographic legacy that provides the "chunks of the real" from which we can begin the work of reconfiguration. Examining her representations of the space of embodied between-ness it does indeed become possible to imagine a "lost perspective." However, in order to do this it is first necessary to problematize the work of the discursive grave-diggers that have previously disembodied and reconfigured Shawnadithit's maps as the work of a native informant.

Artifacts and appropriation: the native body and disembodied discourse

The story of the Beothuk Indians of Newfoundland is indescribably tragic – and on several levels. It is tragic, of course, because they were wiped out by white furriers and fishermen and Micmac Indians – hunted down like wild geese during a two-hundred-year cycle of haphazard genocide. But it is also tragic in other, subtle ways. There is, for instance, convincing evidence that the blood bath began as a result of an accident; that had it not been for a chance encounter these Indians might easily have survived as partners in the fur trade. There is also the tragedy of good but failed intentions. There were, in Newfoundland, perfectly sincere and humane people who tried after their fashion to save the Beothuks from extinction; they simply had no sensible idea of how to do it. Finally, there is the tragedy of a culture lost. Our knowledge of the Beothuks is abysmal because scarcely anybody bothered to find out anything about them until they were all gone. There is something inexpressably forlorn in the

final picture of the lovely Shawnadithit, the last surviving member
of her race (her features already sallowed by the ravages of consumption), trying as best she could, through a series of story-maps, to
explain how her people lived, hunted and worshipped and how, one
by one, they died.

Pierre Berton[20]

So begins the eloquent lament to what he calls "The Last of the Red Indians"
by one of Canada's more popular and populist historians. Pierre Berton's
humanistic account of the genocide in Newfoundland – boldly inserted in *My
Country* between colorful stories about the cultic Brother XII of British
Columbia and the Catholic turned Protestant Charles Chiniquy of Quebec
– manages with all the skill of a best-selling national raconteur to piece
together a story-book strength narrative from rather patchy evidence and, in
driving that tragic narrative home, repeat all the basic gestures of benevolent
but Eurocentric appropriation. Just as with the earlier painful accounts of
Harold Horwood, Ernest Kelly and Keith Winter, there can be no doubting
the sincerity or anguish evident in Berton's criticisms of what he describes as
"the slaughter" of the Beothuk.[21] Like Horwood who in a 1959 *Macleans*
"Flashback" made an angry appeal to Canadians to remember "The people
who were murdered for fun,"[22] Berton describes for another generation how
fishermen "set out on shooting parties, as they would for deer or wolves, to
bag themselves a few head."[23] Like Winter who romanticized the Beothuk
as stereotypical noble savages – "They had a mellifluous language, loved to
sing and dance, and made a habit of welcoming all strangers with feasts
and friendship"[24] – Berton too portrays them in terms that say more about
the narcissism of the Euro-Christian imagination: "They were tall by earlier
European standards, handsome, fairer than most Indians, with large expre-
ssive eyes . . . ; a shy, unwarlike race, who treated their women with respect,
believed in the importance of marriage and were said to reject both adultery
and polygamy." Finally, as such unsavage-like savages, they made for what
Berton, like Kelly, goes on to describe[25] as "perfect victims." They were, he
says, "ripe for killing."[26] However, as he accomplishes all this, Berton, like the
other three authors, seems oblivious to the way in which he himself is in turn
appropriating and manipulating the story of the Beothuk as "perfect victims"
in the staging of a national tragedy.[27]

Clearly the general re-examination of the brutality of colonialism in
Newfoundland has not been without value. It has sensitized some of the
scholarly anthropological and historical writing about the island to the
politics of colonial knowledge,[28] and it has also led to some more detailed
research – most especially the work of Ralph Pastore – aimed less to dismiss
than to complexify the monocausal, intentionalist accounts of extermination
offered by the popular historians.[29] Moreover, it has usefully upset the facile
justifications proffered by apologists for empire who would, following

311

Diamond Jenness, the most imperial of Canada's anthropologists,[30] prefer to blame the bloodshed on the Beothuk themselves: "But very soon trouble arose for the Indians stole . . . [and t]he fishermen retaliated by shooting every native that dared to show his face."[31] This said, even after Berton's book – indeed, it seems, as a partial response to what are dismissed as its "intemperate allegations" – Frederick Rowe repeated this same tired alibi, claiming like Jenness that "if there was any chance of a permanent friendly relationship . . . the Beothuks themselves probably destroyed it through their persistent habit of stealing."[32] It is not any individual critic's fault that Rowe, a senator from Newfoundland, still saw fit in 1977 to defend the island's settlers by rehashing the rhetoric of empire and blaming the Beothuk. However, because Berton's own account takes on so many of the characteristics of a morality play, he is complicit in setting a scene for an all too tokenistic argument between himself and the likes of Rowe. The tokens in such arguments are the bodies of the Beothuk themselves. Most problematically, the stage is set in such a way that their agency counts for nothing but drama. Instead, a singularized script is written for them, their corporeal experience becomes downplayed, and their place in the process of colonization is diminished to that of either a cute or criminal bit-part in the drama of their own destruction. In the more romanticized dramas of Berton and his three predecessors the script so written at least makes for a picture of the Beothuk with which the modern Canadian reader can partially sympathize. Yet, it is this very same appeal to the contemporary citizen that holds within it the problematic gesture of disembodying misappropriation. Presenting Shawnadithit as an attractive tragic figure and focusing on her gendered and racialized body as an object, as "[h]andsome, with beautiful teeth and a swarthy complexion,"[33] the historians simultaneously dig up and disembody her cartography as a hallowed artifact, relocating it to the national pantheon.

Key to the appeal that Berton's story affords his contemporary Canadians is the gesture of epistemic colonization that sets up the Beothuk as a special form of what Spivak, glossing Edward Said, has dubbed the West's "self-consolidating Other".[34] As Barbara Godard usefully argues in specific relation to the Canadian context, "it is through this encounter with the Other who is Native to this land, that a 'totem transfer' occurs and the stranger in North America 'goes native' to possess the land, to be Native."[35] In this role as Other, the Beothuk do indeed make for "perfect victims." They are, after all, tall and "fairer than most Indians," the almost white but not quite natives who have nuclear family values and who, more fortunately still, are not around any more to be asking troubling questions of the Canadian state about returning their ancient lands. Moreover, the terrible documented horrors comprising the historic genocide create such an overpowering picture of violence that they seem almost to give an alibi to other, supposedly more "civilized," forms of colonialism in the rest of Canada.[36] It is as if by repeating the litany of tragedy and by loudly and angrily berating Newfoundland's fishermen that the modern historians can somehow find a more general form of redemption. In the moment of passionately narrating the tragedy, they can, as Godard put it, go native.

At the heart, perhaps one might say the bleeding heart, of this whole process remains Shawnadithit herself, and, most especially, the tension between her position as a native informant and what Berton describes as the fourth and final part of the tragedy, "the tragedy of a culture lost." There is indeed something inexpressibly forlorn here, but it is not what must have been Shawnadithit's very real bodily pain. Instead, it is the tragic lengths to which the modern historians will go to try and recapture and retell her story for the modern Canadian audience. Berton criticizes what he calls the "unctuous meddling" of the so-called "Boeothuck Institution" founded in 1827 by settlers who, having failed in their imperial mission of finding and civilizing the "Red Indians," began instead to try and educate Shawnadithit for the purposes of transforming her into a native informant.[37] However, it is equally possible to criticize the way in which Berton himself describes the "one valuable by-product" of the meddling, namely, the "remarkable series of drawings."[38] Compare his romantic turned ethnological description, of "the lovely Shawnadithit," "trying as best she could through a series of story maps, to explain how her people lived, hunted and worshipped . . . ," with the account of William Epps Cormack, Shawnadithit's custodian in her last few months in St. John's. Exchanging notes with Bishop John of Nova Scotia, Cormack commented:

> Shawnandithit is now becoming very interesting as she improves in the English language, and gains confidence in people around. I keep her pretty busily employed in drawing historical representations of everything that suggests itself relating to her tribe, which I find is the best and readiest way of gathering information from her.[39]

In both cases the appropriative assumptions of what the anthropological critics George Marcus and Michael Fischer call "salvage culture" are in full operative force.[40] The notion that Shawnadithit is "interesting" insofar as she begins to learn English, highlights the more general imperial angst played out in nearly all the historical treatments of her people since. It is this angst, one that is ultimately preoccupied more with the lack of ethnological information than with the lack of people, that repeatedly transforms Shawnadithit's maps into disembodied artifacts. Their connections to a lived process of cartographic inscription and struggle are thus neglected, and instead they are turned into a quaintly visual source of data for narratives detailing the ill-fated journeys of Captain David Buchan (who was sent out by the Governor to communicate with the Beothuk) and John Peyton (who became Shawnadithit's master). Like the Chipewyan woman "Thanadelthur" whose position in the nation-building deployment of the historical archive has been critiqued by Julia Emberly, Shawnadithit thus becomes transformed into another Canadian native informant, another slave woman in a national narrative for which she serves simultaneously as "heroic proxy and sacrificial victim."[41] It is, I think, the responsibility of the contemporary critic to refuse such nationalizing artifactualization by returning to study the cartography as the work of an embodied agent of knowledge.

Clever with a pencil: Shawnadithit's new-found-land

All savage Nations, whose language is necessarily defective, are accustomed to symbols; ingenious in the use of them and quick in ascertaining their meaning. . . . Any that more particularly belong to the Boeothuk may probably be painted out and explained with Mr. Peyton's help by Shawnawdithit. She may also assist in depicting her own tribe and their dress and habits as she is clever with a pencil.

Bishop John of Nova Scotia[42]

Who would have thought death was warm
plump with meat and men who smile too much,
who ask questions with pencils, wanting you to draw the canoes, the
tents, the chasms
dug for winter houses. They ask you to speak your language so they
can study its sound.
How full of holes it is, subterranean tunnels
echo around your failing lungs.
Can they hear?

Joan Crate[43]

Joan Crate's evocative poem, itself – like this chapter – imposed on the past, nevertheless asks the critical question: "Can they hear?" Bishop John's comments, along with the endeavors of the information hunters described above, would suggest a chronic and interested deafness. In what follows, I want to challenge this failing, and contest in particular how the assumption put forward by the Bishop – of a "necessarily defective" language – has also, in various ways, been extended to treatments of Shawnadithit's cartography. Introducing here her map depicting the River Exploits and the journeys of the Beothuk and Captain Buchan's party, I will proceed to outline four ways in which the map's treatment as an artifact has suppressed attention to its lively, embodied and contested context. Such suppression, I will argue, effectively silences Shawnadithit's surveyor's voice. Although this attempt to reconsider a native informant (a subject that can only be read) as an agentic surveyor (someone capable of producing definitive descriptions) demands a certain "(im)possible perspective,"[44] it is, I think, vital in order to counter the more

314

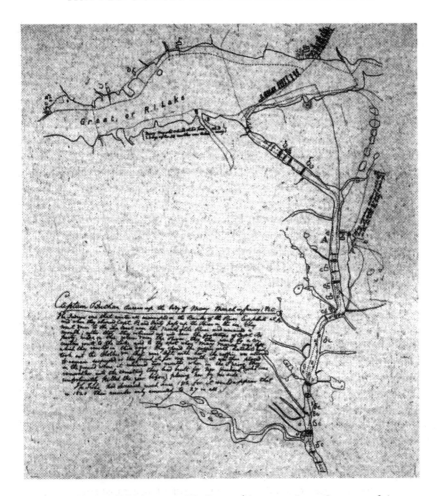

Figure 17.2 Shawnadithit's map with Cormack's annotations. Courtesy of the
Newfoundland Museum. The original map's black and red markings
are lost in this reproduction; yet another indication of the "(im)possible
perspective" of representing Shawnadithit's agency.

conventional sublation of the subaltern subject as body as object into the
lower strata of evolutionary historical narratives. Following the pattern of
Bishop John's self-contradictions, these are the same narratives that adduce
the very skills of those like Shawnadithit to be the final proof of skillessness,
savagery and general historical backwardness.[45] At one level, therefore, they
self-deconstruct. Yet, at the same time, they also dominate and contaminate
the historical record as a whole. They prescribe a disciplinary structure
with which critics inevitably have to negotiate if they are to disclose some of
the embodied processes of cartographic inscription. In figuring out these

processes that were situated *in* but not totally controlled *by* colonialism, Shawnadithit's cartography itself maps a route in-between (Figure 17.2).

The body drawing the map

Shawnadithit drew the map (along with at least three others) in the winter of 1829.[46] Afterwards, it was only a matter of months before she died on 6 June, her burial being recorded in the register of the Anglican Cathedral. By this time, her presence in St. John's and her work had attracted a great deal of curiosity, so much so that *The Times* in faraway London posted an obituary on 14 September.

> Died. – At St. John's Newfoundland on the 6th of June in the 29th year of her age, Shawnadithit, supposed to be the last of the Red Indians or Beothuks. This interesting female lived six years a captive amongst the English, and when taken notice of latterly exhibited extraordinary mental talents.[47]

The primary way in which her cartography was first received and reconstituted as a disembodied artifact was as testimony to such "extraordinary mental talents" reported by *The Times*. *The Times* did not explicitly say "extraordinary, for a savage, a 'Red Indian,'" that was an understanding carried by the colonial context. It was also the guiding (mis)understanding of the Englishmen in charge in Newfoundland. Captured Beothuk women were often referred to in objectifying ethnological terms as "interesting females,"[48] and the Rev. Wilson, much like Bishop John, had already testified to Shawnadithit's "surprisingly" clever skills with a pencil.

> She made a few marks on the paper apparently to try the pencil; then in one flourish she drew a deer perfectly, and what is most surprising, she began at the tip of the tail.[49]

Cormack, the University of Edinburgh trained anthropologist whose written annotations cover Shawnadithit's cartography, thought in turn that he was getting an "interesting" information source when he took charge of her as president and founder of the salvage oriented "Boeothuk Institution." Even contemporary writers still speak of his good fortune with an eager anthropological gusto: "Luckily he arranged to have Shawnadithit sent to him in St. John's where he questioned her extensively about her people."[50] However, what gets lost in all this astonished talk of luck, extraordinary talents and surprising interest are the interests of Shawnadithit herself. Like the ethnocentric fascination with Australian Aboriginal relics described by Paul Carter, such "activities divorce the project of study from the context of [its] production, that living space in which places have histories and implements are put to use."[51] I do not think it is useful, indeed it is just as arrogant, to diagnose Shawnadithit's interests long distance in the sensationalist style of Winter's *Shananditti*.[52] Nevertheless, it is possible to at least take more careful cognizance of the circumstances, the living embodied spaces, in which she put pencil to paper.

Up to the time that Cormack "sent for her," Shawnadithit had lived almost six years as an unpaid servant in the house of John Peyton Jr., the magistrate of the fishing community of Twillingate. This was the same house from which John Peyton Sr. had set out on many expeditions to kill the Beothuk, and also from which Peyton Jr. had began his own infamously brutal assault in which the apparent Beothuk chief, Nonosabasut, was shot and bayonetted to death while pleading with Peyton not to take his partner, Demasduit, as a captive (a story mapped out in another of Shawnadithit's maps). Her baby left behind to die as well, Demasduit was nonetheless captured and duly taken to the house where she was named Mary March. Shawnadithit's own arrival at this house was in turn no less forced. By this time, Demasduit had died of consumption, her dead body having been carried back by Captain Buchan to the site of her capture. Shawnadithit, as her map shows, and as I shall discuss in the next section, observed this pathetic funeral cortege move up and down the river in 1820. After that, though, it was only three more years before she, along with her mother and sister, were taken captive, presented as reward material to Peyton, and transported to the courthouse in St. John's. A failed attempt to let them go back to the interior only led to the more speedy deaths of her mother and sister from consumption. Shawnadithit, alone, returned to Twillingate to work for nothing but her keep from one of the families centrally involved in the final demise of her people. Very little is recorded in the archives about her experiences in this house. There is no mention, to introduce one plausible scenario, that she was raped or brutalized. Neither, however, is there any specific mention that she was not. It is impossible to know what living day to day in this family home of known Beothuk killers might have been like for her. However, since this was the place from which Beothuk hunts had started out and in which Demsaduit had contracted tuberculosis, and since the pre-colonial Beothuk moved "camps" on a seasonal basis, being trapped in this one house for five years may well have felt like bodily incarceration – even if there was no cell as such.

As if this situation was not already grossly traumatic, she was then, five years later, taken to St. John's a second time to be made the object of further governmental, missionary and philanthropic turned anthropological interest. This, then, was the context in which she drew the map. Clearly these were very different circumstances to those enjoyed by a cartographer like Cook, and, for that matter, different from the conditions experienced by native mapmakers like the Inuit described by Robert Rundstrom.[53] Shawnadithit's cartography was instead drawn under duress, when she was already ill, her world and family destroyed. Considering this context at least makes it possible to move beyond treating the map as a curio. The situation was alive with colonial power relations. Urging her on was Cormack eager to salvage information; reporting how through his "persevering attention" and constant tending of "paper and pencils of various colours," Shawnadithit "was enabled to communicate what would otherwise have been lost."[54] Far from enabled, however, Shawnadithit may very well have felt disabled by such circumstances, surrounded by people who while soliciting and gathering information treated her language as "gibberish."[55] From this perspective, other topics than simply camping sites, numbers of bodies and colonial expeditions can be read in her map. There is a record of pain and misery, perhaps; a will to mark the truth about the interactions of bodies and their spaces of travel; and, most of all, a need to communicate through and of an embodied Beothuk representation of space.

The embodied spaces of colonized and colonizer

Shawnadithit's cartography's expression of a Beothuk geographical imagination has also been hidden because of how the maps have been treated as artifacts of History. I use the capitalized form of the word here, following Robert Young, in order to represent the sort of Western history that disavows its geographical provenance and claims the space of universal truth as the whole world's History.[56] Converted into an artifact in the service of such History, Shawnadithit's work has been treated as a source of raw information to be drained of data about the various expeditions of Peyton and Buchan into the interior. Howley set the pattern for instrumentalizing the maps in this way, and others like Berton, Winter, Such and Rowe have followed in his stead. The History of these ill-fated missions is, as a result, often recounted in great detail by the historians. The directions taken by the different settler sorties are listed with febrile exactitude. From Shawnadithit's map of the River Exploits, for example, the information gatherers follow her depiction of the British party led by Buchan – the group she depicts crossing the lake. They retrieve data about the numbers in the group, the number of stops that were made, the camp sites, the place where Buchan left the coffin of Demasduit, the vain search for the Beothuk around the edge of the frozen lake, and the trip back. Occasionally, as in Berton's account, it is noted that all of this adventuring seems pathetic and ill-considered. But in this too there is a disembodying gesture. The irony of Buchan's mission, a mission to return a dead body of a Beothuk woman who had been captured to be used as a go-between, becomes appropriated and disembodied as a tragic metaphor of the wider tragedy of miscommunication.

Only very occasionally, and then only briefly, are the actual positions of the Beothuk mentioned. It seems that because the historians find it impossible to corroborate these map depicted positions and journeys with any written records their acceptance as History is denied. Because, by contrast, Buchan's journey is also recorded in conventional archives, Shawnadithit's mapping of it is used as a supplement. In this way, the treatment of her cartography parallels the more general way in which British imperialists have looked upon indigenous geographical knowledge elsewhere. As Matthew Edney notes in the case of India:

> The British had a very low opinion of Indian knowledge as a whole, and of their geography in particular, thinking it too indefinite, too imprecise, or completely false. They had no qualms about using geographical knowledge from native sources, but they discarded it as soon as even the sparsest survey had been completed.[57]

The same seems to be true in Newfoundland. There is evidence that explorers used maps drawn for them by the Beothuk and/or Micmac for moving around.[58] However, when it comes to the writing of History, such maps can only ever serve, it seems, as artifacts supplementing the true fount of knowledge, the colonial archive. Nonetheless, as Derrida has famously pointed out, there is a disruptive potential lying within such moments of supplementation,[59] and, in the case of Shawnadithit's cartography, such disruption comes in the shape of the embodied geography that History hides.

Clearly marked on the map are the bodies, the travels and the camping sites of the Beothuk, a historical geography inscribed in red. Shawnadithit's drawing shows where they camped; the places from where they observed Buchan's party; and the tracks along which they followed him. It documents the overland routes the Beothuk made to move back and forth from the lake; the way in which they arrived to take down Demasduit's coffin; and, the place they took her body to bury it next to that of Nonosabasut. Moreover, apart from just these movements, Shawnadithit's cartography also renders thematic a series of critical epistemic dynamics neglected by the information hunters. Most obviously, it documents the fact that the Beothuk were also observers and agents of geographic interpretation. Moreover, it reminds the contemporary critic that the Beothuk not only interpreted the land but the impact of colonialism on that land. Between the mythical purity of pre-colonial space and the emptiness of post-colonialism Newfoundland-style, her map lies on the threshold as a representation of colonial contact from a native point of view.

Despite the numerous references made to her maps, no effort is ever made by commentators to draw attention to the processes of observation and territorial inscription that they reflect. There is a point in his book at which Howley notes in passing how:

> One tall birch tree on the summit of [Canoe Hill] was pointed out by Shawnadithit as the lookout from whence the Indians observed Peyton's movements.[60]

However, he makes little of this as an indication of Beothuk knowledge making, and makes no connection at all with the more systematic observation and inscription of the land represented in Shawnadithit's work. It seems to me, however, that we can see in her map a whole geographical imagination, a coherent assemblage of conceptions of space and time that relate intimately to bodies, lives, deaths and travel.

Again, I do not think it is useful to read too much into the map and anthropologize these conceptions. Rather, the important point is to simply acknowledge their existence. From this perspective we can at least begin to take direct account of how the island whose New "true shape" Cook inscribed with such diligence from the sea, was already far from uninscribed by the Beothuk on the land. Shawnadithit's cartography documents how their inscriptions had another shape, one that was constituted and understood in the terms of another, less sea-bound geographic discourse. Her mapping of Buchan's party's route also documents how the arrival and movements of the colonialists presented new spatial developments that were nonetheless interpretable within these older and more landed geographic terms. Clearly, the Beothuk observed the men in Buchan's party – men who were to become the new and official observers of Newfoundland's interior – and they saw them not just, as Cormack notes, from point "A" on the river, but also, as the map itself underscores, from a specifically Beothuk point of view. The map does not tell us much about how the party and its route were interpreted in such terms as fear or friendship. However, it is possible to note that, despite the dreadfully ominous scene of the coffin-carrying cortege, Buchan's group is still depicted by Shawnadithit as made up of people, albeit people with smaller bodies, and that the geography of their journey, their new track across the land, is marked with great care. In the context of a cartography that charts so close a connection between knowledge of the land and travel through it, these new colonial tracks surveyed by Shawnadithit constitute new found land: traveling that records the space of between-ness in the new colonial geography. Such new found land as seen from the perspective of those who were colonized is so rarely documented that historians entranced by the "New" of History seem to have forgotten that it can exist.

The context of cartographic recognition

It could be argued that Cormack himself, unlike the latterday historians, was more respectful of the autonomy of the native knowledge presented to him. Certainly he took more careful note of Shawnadithit's observations, annotating the work to mark observations she must have mentioned. He also seemed to take her knowledge as a form of reliable knowledge about the Beothuk from which he could estimate such exact records as the numbers of those surviving in 1820. However, this same attention to exactness also betrays how Cormack in turn misappropriated the spatial specificity of

Shawnadithit's maps by treating them as anthropological artifacts. The care, the concern, the attention to detail was all directed towards the higher Historical purpose of anthropopological salvage. It is interesting in this respect to see how his careful annotation of Shawnadithit's map with its close attention to the fate of Beothuk bodies, concludes by describing Beothuk deaths in the passive voice – "her Husband (who was unfortunately killed the year before)" and "The Tribe had decreased much since 1816." No mention is made in this disembodied voice of the colonial clash of bodies and of who might therefore have been responsible for the deaths. It, History, just happened, and the map simply facilitates the accouting of this History's consequences for the number of living Beothuk bodies. In general, then, it seems that Cormack's regard for Shawnadithit's drawing became his way of satisfying a need that he had outlined in *The Edinburgh New Philosophical Journal*: a way for him to procure what he called "an authentic history of th[is] unhappy race of people, in order that their language, customs and pursuits might be contrasted with those of other Indians."[61] Comparative anthropology was in this sense another formula for sublating Shawnadithit's geography into History.

More recent treatments have continued this transfiguration of the cartography as anthropological artifact. In such treatments there generally appear to be two basic rhetorical gestures, in fact, the two same basic gestures that were part of Bishop John's backhanded compliment about Shawnadithit's cleverness with a pencil. One of these elevates the maps, praising their topographical accuracy, their artistic skill and their generally high quality as specimens of the supposed genre. The other gesture radically diminishes the work, pouring scorn on its "Indian" character, intimating its apparent sameness *vis-à-vis* some homogenized "Indian" norm, and firmly marking its difference as inferior *vis-à-vis* the strict scale and grid-oriented rigor of properly drawn Cartesian cartography. Howley, who was himself a celebrated cartographer of Newfoundland's geology,[62] begins and ends his own description of Shawnadithit's maps with the second of these two gestures.

> Although rude and truly Indian in character, they nevertheless display no small amount of artistic skill, and there is an extraordinary minuteness of topographical detail in those having reference to the Exploits river and adjacent country. These latter bear a striking resemblance to Micmac sketches of a similar character.[63]

By contrast, Winter's description puts the negative comparison with Western cartography in the middle.

> The drawings are accurate in topographical details, but they lack regular scale: rivers and lakes appear larger than they really are. Nevertheless, the details of shoreline, islands, bends in the river, falls,

rapids, and junctions of rivers are accurate, and the relation of each of these to the other is correct.[64]

It seems unfortunate that Winter's avowedly pro-Beothuk account should end up sounding more like the patronizing praise of a paternalistic school report. Indeed, perhaps Howley's racist view of the maps' "rude and truly Indian" character is more honest, it makes the epistemic violence clear. Winter does not say the details are "correct" by Western standards, but the implication is there all the same.

Another more scholastic turn to Shawnadithit's maps by the cartographic historian G. Malcolm Lewis also suffers the same schizophrenic ethnocentrism.[65] Lewis includes the map depicting the Beothuk observations of Buchan's party as one among many other examples of mapping by what he collectively homogenizes as "North American Indians."[66] He too praises the "remarkable and readily interpretable detail"[67] in the work. However, he also begins his essay with a sweeping assessment of the deficiencies in "Indian" mapping:

> Indian representations of actual networks (whether drainage, routes or boundaries) were not drawn to scale and were characterised by gross distortions of direction.[68]

Overall the general approach of such commentaries shares the same imperial will-to-knowledge evident in Cormack's search for an "authentic history." It also operates like the White Australian accounts of Aboriginal spatial history critiqued by Carter. In these accounts, argues Carter, "[p]ools, pastures and tracks were taken out of context and used like quotations."[69] The comparisons that treat Shawnadithit's work as a decontextualized, disembodied anthropological artifact are no less cooptively appropriative. Theirs is an approach that delights in the artifact's quaint and quotable accuracies, all the while placing it low down in a supposedly larger comparative hierarchy.[70] Such a hierarchy inevitably elevates the "true" cartography of those like Cook to the top, setting it up implicitly or explicitly as a standard against which all below is judged deficient. Lewis's own more recent multicultural epiphany may at first blush seem to afford a way of avoiding such imperial hierarchization.[71] But his overarching desire to compare, contrast, and most of all, codify maps as artifacts representing "different cultures . . . at different stages of development" remains the same.[72] If this is the end point – a reconfirmation of an epistemic empire that places European maps at the end point of some developmental time-line – there seems little point in even beginning such a multicultural project. Nevertheless, an attention to the same differences noted by Lewis may lead somewhere else once the project is freed from the temptations of timing and taxonomy.

I find a more useful way forward in Brody's account of what he entitles

Maps and Dreams.[73] His work relates to a late 1970s land rights struggle in British Columbia in which the Beaver people represented their lands cartographically. Although he was there as a researcher and assistant in the production of contemporary maps concerning the people's land use patterns, Brody reports Atsin – a Beaver hunter – telling him the following.

> Oh yes, Indians made maps. You would not take any notice of them. You might say such maps are crazy. But maybe the Indians would say that is what your maps are: the same thing. Different maps from different people – different *ways*.[74]

It is by maintaining Atsin's unhierarchical attention to different maps and different ways that Brody goes somewhere different from the anthropologizing and antiquarian treatments of Shawnadithit's maps I have cited above. In this way, comparative questions like those broached by Lewis might still be pursued, but without the invidious desire to establish a developmental chronology.

In relation to Shawnadithit's work, Brody's attention to what he underlines are the different "dreams" behind different maps opens a whole series of questions about the guiding assumptions of recent cartographic and cultural history. Some of this scholarship has examined and critiqued the connections between the imperial gaze, cartography and what Timothy Mitchell calls the "enframing" of territory.[75] Mitchell's work itself functions primarily as a critique of modernity. However, as Sami Zubaida has noted, there are some unfortunate assumptions about "pre-modern" culture – including the historicist notion of pre-modernity itself – implied by the critical argument.[76] One of these is the suggestion that "pre-modern" cultures do *not* have a capacity, however differently organized, to picture people and places from a distance. What Brody might label the geographical dream of Shawnadithit's map clearly problematizes any such simple generalization: it does picture and yet it does not empty or disembody the colonial landscape. It therefore demands of critics a more nuanced reading of the differences distinguishing the disciplinary picturings of Europeans from the picturings of the people they colonized. In this vein, we can go back to Cook's map and compare its exact outline and empty center with Shawnadithit's detailed depiction of pathways through the interior.

Cook's map can clearly be argued to represent an enframing moment in which we see, in Mitchell's terms, a "conjuring up [of] a neutral surface or volume called 'space.'"[77] Such colonial conjuring of abstract space provided a seemingly blank frame of reference through which important parts (ports and other information vital for fishing and navigation) could be distinguished from the unimportant (the empty interior). Moreover, the emptiness so conjured can also be read as a representation of the lack of British interest in the interior, a gap in governmental knowledge, which combined with the related lack of disciplinary power in this arena, ultimately made possible the violent processes of its actual emptying.[78] Shawnadithit's map, by contrast, pictures a peopled interior, a space of bodies, movement, life and death. Certainly, this is not the modern "world as exhibition" approach to picturing examined by Mitchell – although, as I have shown, this is precisely what the historians have attempted to make it. However, it does nonetheless embody a particular world view that divides the important from the irrelevant. It is in this vein that I think we can return to the assessments of the taxonomists about the lack of scale. Rather than a deficiency, the uneven scale in Shawnadithit's map can be read as a rigorous and reliable picturing of the uneven possibilities for travel by foot across so uneven a landscape. Rather than set up an imaginary abstract grid of space on which all else falls into synchronic place, the complex time-space the map depicts traces the spatiality of bodies and movement. It is still a representation of space, still a picturing, and still an embodiment of a geographical imagination. Yet it is also a spatiality rooted specifically in the corporeal pathways of Beothuk journeying and knowing.

Bodies in between colony and nation

I have already outlined the ways in which national historians have transformed Shawnadithit's maps into national relics. In this last section on the context of her cartography, I want to continue this process of problematization by examining the more substantive connections between mapping and national identity. As Thongchai Winichakul and Richard Helgerson have illustrated with book length discussions, there is a long and globally varied history connecting mapping and nation building.[79] Clearly the connections go way beyond the practical business of charting the length and breadth of national territory. They also extend to the complex power relations underpinning the imagination and organization of the nation as a spatially coherent community. In Elizabethan England, for example, Helgerson argues that cartography was key to the developing break with the regime of dynastic loyalty. "Maps," he says, "opened a conceptual gap between the land and its ruler, a gap that would eventually span battlefields."[80] Even today, national maps and atlases continue to serve as a major vehicle for teaching citizens the spatial reach of their nationality, allowing them – as Benedict Anderson emphasizes in the revised edition of his famous book – to dream the secular, national dream of continuous, horizontal community.[81] Beneath such unbroken horizons many other dreams of community may, of course, exist, and yet the flatness of the map, what Anderson calls its "logoization of political space," would seem to hide them away. Returning to the differential spatiality of a map like Shawnadithit's allows such dreams of alienation (and, indeed, of alien nations) to break the unbroken surface. Comparing her cartography with attempts to map Canada as a complex and regionalized yet mappable whole, illustrates a form of negotiation with nationalizing discipline that Brody evokes in his account of the Beaver people's struggle. "Dreams collide: new kinds of maps are made."[82]

In Canada in particular, as the chair of the Bank of Montreal noted in Volume Three of the *Historical Atlas of Canada*, "[m]apmaking and nation-building are inextricably linked."[83] As the banker proceeded to suggest, and as numerous reviewers have agreed, the *Historical Atlas* itself has ably continued this tradition.[84] At the same time, its pursuit of national cartographic representation (and, through this, integration) in the contemporary period of first nations' struggle, has taken an uneven but sometimes new approach to the question of first nations within the nation of Canada. Unlike most previous national atlases of Canada, the *Historical Atlas* departs from a geographically exclusionary understanding of the country as just a white settler-colony and, for the first time, goes a considerable way towards putting native peoples on the map.[85] Indeed, despite the biblical boast of mapping Canada "From the Beginning to 1800" – the subtitle given to the English edition of Volume One by the publishers – there is a sense in which the thorough attempt made to document a native presence on the land in the first volume fundamentally displaces the white settler colony thesis, and, with it, the very notion that the land had always been Canadian.[86] Unhappily, the teleology of the three volume series as a whole concludes in Volume Three by addressing the twentieth century with only one obvious attempt – a Gitksan and Wet'suwet'en map – to mark the continuation of the same native presence into the places and politics of the present. But this, I think, creates all the more reason to re-examine the effort in Volume One to put native people, and amongst them the Beothuk, on the map. The embodied question-marks of Shawnadithit's cartography, I think, ask us to consider how even this effort itself is not without dangers.

Putting first nations on maps that chart a story of national development may well be inclusionary, but it may also be incorporative, cooptive and controlling too. Put another way, it may historically nationalize people and places who now, seeking decolonization, refuse this very nationalizing principle. In the miserable context of traplines, reserves and rejected land claims, being on a Canadian (as opposed to a first nations map) has most often coincided for native people with the violent disciplinary force of colonialism. It was with this same disciplinary structure of violence that the scholars working on the *Historical Atlas* inevitably had to negotiate themselves. Committed, as they were, to only using the most reliable and academically respected data, they turned chiefly to the disciplined information sources of Canadian archeology and anthropology rather than to the oral histories of first nations. According to Conrad Heidenreich, one of the *Historical Atlas's* most conscientious researchers of native movements, "memory ethnography . . . was ruled out as being virtually worthless."[87] In its stead then the *Historical Atlas* sought to put native people on the map by turning first to archeologists'

SPARKE

assessments of so-called "Prehistory" and, second, to colonial records. Such assessments and records are of course interested, and as Cole Harris, the editor, warns in his preface to Volume One: "[m]ore than good will is required to penetrate an Indian realm glimpsed through white eyes." Certainly there was plenty of good will in Volume One – "we have tried to accord full place to native peoples" notes Harris – but in negotiating how "to penetrate an Indian realm" the scholars negotiating with the penetrative power/knowledge apparatus of the colonial archive still continued to gaze through white eyes – their own and those of the colonial record keepers before them. Thus, albeit marginally and supplementally, Shawnadithit's gaze provides a different and less penetrative perspective on a small part of the terrain mapped by the *Historical Atlas* as Newfoundland.

Apart from the inclusion of a Beothuk pendant and an archeological generalization of Beothuk space in the "Prehistory" section,[88] the first major treatment of the Beothuk in the *Historical Atlas* comes in the second section on "The Atlantic Realm."[89] The most detailed part of the plate in question illustrates the findings of an incredibly comprehensive attempt by Ralph Pastore to map Beothuk habitation and burial sites (Figure 17.3). This represents a notable effort "to accord full place to native peoples," and placed as it is between a plate showing the routes of various European explorers[90] and a plate describing the migratory fisheries of Newfoundland[91] it has a certain displacing force on the *Historical Atlas's* narrative of nation. Nevertheless, this displacement is brought back into Historical line by the scene settings that enframe Pastore's map. The scene is seen or at least glimpsed, in the editor's words, through "white eyes." In the introduction to the so-called "Atlantic Realm" it is noted, for example, that the "advantages and disadvantages of settlement in Newfoundland were argued throughout the seventeenth century in both England and France." Nothing is said here about the possibility of a considerable discourse amongst the Beothuk about these same "advantages and disadvantages" of European settlement. The way that Shawnadithit's surveys represent a continuation of such a Beothuk discourse on into the nineteenth century is neglected, and this despite the fact that as maps they employ the same cartographic medium as the *Historical Atlas* itself. The narrative of national development in the Atlantic Realm instead concludes that "ravaged by malnutrition and disease and lacking guns, their summer fishing stations pre-empted by fishermen and their winter villages subject to the depredations of furriers, the Beothuk of Newfoundland became extinct." To be sure, the words above Pastore's map also reemphasize the Beothuk suffering: "Harassed by trappers and fishermen, weakened by European disease, and excluded from coastal resources, they died out early in the nineteenth century." In other words, the colonial violence is definitely noted. But at the same time, the suggestion that the colonialists might be stealing the land – a suggestion evoked by the painful ironies surrounding the coffin in Shawnadithit's map – is never brought up, all the while the atlas

326

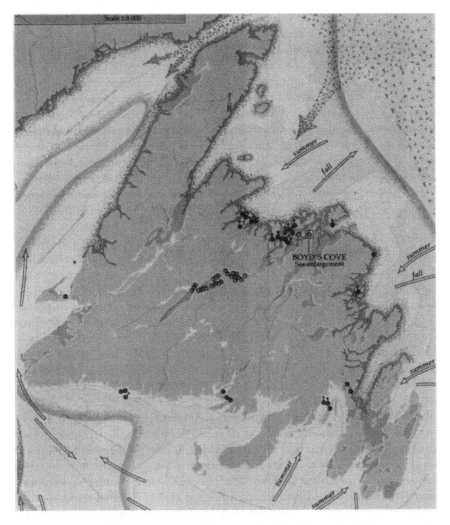

Figure 17.3 A black and white copy of the Newfoundland portion of Plate 20 in the *Historical Atlas of Canada, Volume One*. Plain circles represent habitation sites. Circles with protruding additions represent burial sites. The arrows in the sea represent the directions of seasonal cod migrations

twice mentions that the Beothuk themselves stole.[92] For the reviewer Paul Robinson, this stress on Beothuk stealing came two times too many:

> Somehow I found this explanation of the Beothuk's eradication too simple, too convenient, as if it were expedient to skirt nimbly a topic that most of us choose to ignore. The memory of Shawnadithit, her mother and sisters – the last of a race of people – is too strong in my

mind. They are deserving of better treatment, particularly in a book as majestic in its approach to our complete heritage as is the Historical Atlas of Canada.[93]

Robinson's criticisms – although they ignore the simultaneous attention to harrassment in the proferred explanations – are well taken. However, I would argue that it is precisely the question of the *Historical Atlas*'s "majestic" approach to cartographic completeness that is at issue. The nationalizing desire to produce a complete cartography that accords full place to native peoples is brought into crisis by the very disciplined lengths the research and narrative go to secure completeness.[94] This epistemological tension becomes clearer in the second major moment in which a Beothuk presence is acknowledged, namely the final map of the volume entitled "Native Canada, ca 1850."[95]

In many ways the map "Native Canada, ca 1820." is an incredibly disruptive way to conclude an atlas subtitled "From the Beginning to 1800." Instead of a singularized colonial genesis story that might conclude by depicting the originary capitalist movement of the fur trade and fisheries in 1800, the map evokes a vast multiplicity of native movements. Instead of homogenizing native experiences, it charts diversity. And instead of mapping points of colonial control – which might be readily (mis-)used in court as testimony to the vigor of colonialism and the extinction of native rights – it highlights the complex dynamics of survival in the face of advancing colonial settlement. Here, then, perhaps more than in any other place in the *Historical Atlas*, we find a real radicalization of the plural "Des Origines" in the doubled-up sensitivity in the subtitle of the French edition of Volume One.[96] Yet the epistemological tension noted above also works here too. In the end, the diversity and disruption is all recalled within a coherent map of the whole of Canada, a map that no native cartographers would ever have drawn let alone imagined as their national community. The various historical native movements – including those of the Beothuk – are charted so comprehensively and pictured so coherently in one map that the final result is their recollection on the single unbroken horizon of the modern Canadian nation. They become not first nations as such, but "Native Canada, ca 1820," a synchronic moment in Canada's progressivist march from archaic past into modern nationhood. Colliding with this coherency, the embodied space of in-betweeness that comprised the new-found-land of Shawnadithit's map introduces "a newness that is not part of the progressivist division between past and present, or the archaic and the modern."[97] Instead, this newness is, to quote Bhabha again, "the "foreign" element that reveals the interstitial."[98]

Most immediately we can note that the mapping of encounters in 1820 made by Shawnadithit casts some doubt on the accuracy of the Newfoundland portion of "Native Canada, ca 1820." In this portion the likely seasonal routes taken by the Beothuk from the interior to the sea are shown by small arrows while larger arrows depict the advance of colonial settlement (Figure 17.4). In the context of the events portrayed by Shawnadithit the Beothuk's seasonal routes, and particularly the route towards the south-west, seem quite unlikely. Moreover, as Cormack's annotations to Shawnadithit's map suggest, the number of Beothuk around this time was probably about 27. Such reduced numbers – based in large part on Shawnadithit's testimony – would suggest that by the end of the 1810s the whole traditional pattern of life and movement was equally reduced. By the 1820s then, survival in the interior, and most immediately in the winter of 1820, surviving the foray of Captain Buchan's party up-river, were probably the more likely historical and geographical concerns. The fact that Shawnadithit mapped these events and not (as far as we know) some seasonal hunting and gathering patterns also suggests their priority from the perspective of the colonized.

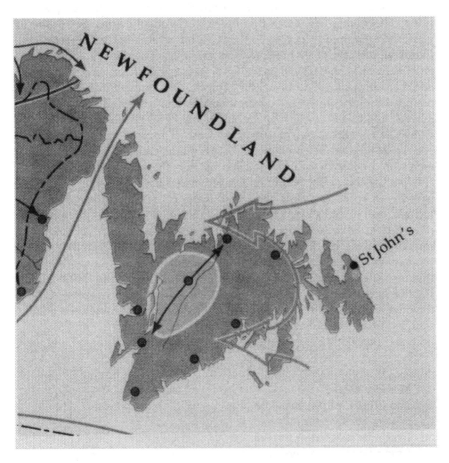

Figure 17.4 A black and white copy of the Newfoundland portion of Plate 69, "Native Canada, ca 1820", in the *Historical Atlas of Canada, Volume One*.

More significant than these questions of seasonal movements is the issue of how the whole colonial scene is represented. Shawnadithit's map records the traveling bodies of the colonizers and the colonized. At the level of content, it portrays a middle ground, an interstitial space shared but traveled through and experienced differently by the two groups. At the level of how this embodied space is represented, Shawnadithit's depiction of the scene is itself also something of an interstitial work. It employs the paper and pencil, "outline ordered" mapping format of the Europeans with whom she was communicating. Yet, as I have argued in the preceding three sections, it also reflects a series of Beothuk perspectives that problematize abstract European notions of space, scale, and time. The *Historical Atlas* perspective on "Native Canada, ca 1820" – as with the earlier map of burial and habitation sites (Figure 17.3, p. 327) – does *not* throw these same notions into question. It directly counters the myth that by 1800 Canadian history was primarily a matter of European struggle on the land, but it does not abandon the pan-Canadian perspective of the European explorers. Its representation, unlike Shawnadithit's non-national picturing of people, is abstract, disembodied and detached. It is mapped out on the template of the modern nation. It still usefully represents the grand dynamics of contact – the small arrows of native movement outsized by the large incoming arrows of colonial settlement – but it does so completely within the framework of modern cartography. There is no space here for the embodied in-between moments of contact and conflict represented by Shawnadithit. Indeed, because the whole national scene of "Native Canada" must be brought together as such on a single sheet there is not even space, let alone the (possibilities of) non-synchronous time, to include the complexities of exchange, timing and native interpretation presented in Shawnadithit's story map. Instead, with a flurry of colors and images befitting the European war-room, "Native Canada, ca 1820" allows the modern citizen reader to take in with a single glance a last synchronous slice of Canada's non-modern past before the modernizing, nationalizing impulse of the next two volumes obscures the genealogy of that presence from the present. Shawnadithit's map of the embodied space of inbetween-ness, by contrast, urges us to remember otherwise.

A Conclusion

Memory speaks:
You cannot live on me alone
you cannot live without me . . .
I can't be restored or framed
I can't be still I'm here . . .
in your mirror pressed leg to leg beside you
intrusive inappropriate bitter flashing
Adrienne Rich

330

When memory speaks here, there is no less limitless diversity of the self; there is a self that occupies a space of ambivalence, a space of agonism:

Homi Bhabha

Not all migrants are powerless, the still standing edifices whisper. They impose their needs on the new earth, bringing their own coherence to the new-foundland, imagining it afresh.

Salman Rushdie[99]

Since Shanawdithit's mapping embodies memory, refuses colonial enframing, speaks from a space of ambivalence, and brings its own coherence to new found land, I have found the above quotations an uncannily relevant summary of her colonial, as opposed to post-colonial, struggle. Lifting them as a single piece from one of Homi Bhabha's more recent essays, I do not want to enter into the poetics of Bhabha's own post-colonial self-fashioning, but rather concentrate on the implications of the few words he inserts in between those of Rushdie and Rich. "When memory speaks here, there is no less limitless diversity of the self; there is a self that occupies a space of ambivalence, a space of agonism," he says. This is exactly how I interpret Shawnadithit's cartographic testament to the space of embodied between-ness. There is no limitless diversity here, and her story is not, as I have shown, so easily accommodated into a multicultural remapping of the Canadian past. Her mapping of the embodied colonial encounter instead reimagines the new found land of colonialism as a space of native observation and inscription. Unframed, it obliges the reader to adopt a different perspective from that required to understand the enframing and disembodying cartography of Cook. As a native cartographic narrative, though, it is not a story of migrant power. Nevertheless, it is a reminder of the possibility of mapping migrancy otherwise, of mapping new found land through the body. Memory in this sense does touch us in the present, refusing to go away, an embodied absent presence, a painful record of a lost space of between-ness, intrusive inappropriate bitter flashing.

Notes

1 Anne Godlewska and Neil Smith, "Introduction: Critical Histories of Geography," in Anne Godlewska and Neil Smith, eds, *Geography and Empire*, New York: Blackwell, 1994, pp. 1–8, at p. 8.

2 Edward Said, *Culture and Imperialism*, New York: Alfred A. Knopf, 1993, p. 266.

3 Gayatri Chakravorty Spivak, "The Rani of Sirmur: An Essay in Reading the Archives," *History and Theory*, 24 (3), 1985, 247–72.

4 Ibid. p. 253.

5 Ibid. p. 254.

6 Ibid. p. 254.
7 Gayatri Chakravorty Spivak, 'Woman in Difference: Mahasweta Devi's "Douloti the Bountiful', in Andrew Parker, Mary Russo, Doris Sommer, and Patricia Yaeger, eds, *Nationalisms and Sexualities*, New York: Routledge, 1992, pp. 96–117, at p. 98.
8 Quoted in ibid. p. 113.
9 Ibid. p. 113.
10 A photograph of the monument is included in Peter Such, *Vanished Peoples: The Archaic, Dorset and Beothuk People of Newfoundland*, Toronto: N C Press Limited, 1978, p. 86.
11 William Kirwin describes three historical phases in the English rendering of the name "Beothuk": the first (1828–85) as "Beothick"; the second (1908–52) as "Beothuck"; and the third as "Beothic": see "A Note on Beothuk Names in Newfoundland," *Onomastica Canadiana*, 74 (1), 1992, 39–45.
12 See Gayatri Chakravorty Spivak, *In Other Worlds: Essays in Cultural Politics*, New York: Routledge, 1988, p. 244. I have elsewhere discussed how the instrumentalization of space in spatial metaphors can lead to an anemic geography that systematically underplays the substantive links between space and those things (the economy, the body, politics etc.) which it might be used to evoke. See Matthew Sparke, "White mythologies and anemic geographies," *Environment and Planning D: Society and Space*, 12, 1994, 105–23.
13 Fabian O"Dea, "The 17th Century Cartography of Newfoundland," *Cartographica* Monograph No. 1, 1971, Toronto: University of Toronto Press.
14 Ibid. p. v.
15 Quoted in Ralph Pastore, "The Collapse of the Beothuk World," *Acadiensis* xix, 1, 1989, 52–71, at p. 55.
16 Her maps and other drawings were first published as part of the mini-archive produced as a book by the geological cartographer of Newfoundland, James P. Howley. See J. P. Howley, *The Beothuks or Red Indians: The Aboriginal and Inhabitants of Newfoundland*. Cambridge: Cambridge University Press, 1915.
17 Gayatri Chakravorty Spivak, "Can the Subaltern Speak?" in Cary Nelson and Larry Grossberg, eds, *Marxism and the Interpretation of Culture*, Chicago: University of Illinois Press, pp. 271–313; and Jonathan Crush, "Postcolonialism, De-colonization, and Geography," in Godlewska and Smith, op. cit., pp. 333–50.
18 Gayatri Chakravorty Spivak, "Time and Timing: Law and History," in John Bender and David E. Wellbery, eds, *Chronotypes: The Construction of Time*, Stanford: Stanford University Press, 1991 p. 116.
19 Ibid. p. 116.
20 Pierre Berton, "The Last of the Red Indians," chapter seven of his book, *My Country*, Toronto: Penguin, 1976, p. 165.
21 Ibid. p. 168.
22 Harold Horwood, "The people Who Were Murdered for Fun," *Macleans Magazine*, October 10, 1959, 26–44.
23 Berton, op. cit., p. 168.
24 Keith Winter, *Shananditti: The Last of the Beothuks*, Vancouver: J.J. Douglas Ltd, 1975, p. 4.
25 Ernest Kelly, *Murder for Fun: The Rape and Slaughter of the Beothuck Indians of*

Newfoundland, Cobalt, Ontario: Highway Book Shop, 1974. "The tall, handsome and meek Beothucks," he said, "were an easy target for the roudy, vicious and brutal fishermen," p. 6.

26 Berton, op. cit., p. 166.

27 As Scott Watson notes in relation to the "Group of Seven" artists, lamenting the "dead Indians" that "inhabit the wilderness as ghosts" is a classic gesture of Canadian nationalism. In "Race Wilderness, Territory and the Origins of Modern Canadian Landscape Painting," *Semiotext(e)*, 17, 1994, 93–104, at p. 98.

28 See, in particular, Leslie F. Upton, "The Extermination of the Beothucks of Newfoundland," *Canadian Historical Review*, lviii (2), 1977, 133–53; and Rançoy Raynauld, "Les pêcheurs et les colons Anglais n'ont pas exterminé les Béothuks de Terre-Neuve," *Recherches Amérindiennes au Québec*, xiv (1), printemps, 1984.

29 See Ralph Pastore, "The Collapse of the Beothuk World," *Acadiensis*, xix (1), 1989, 52–71; and, Ralph Pastore, "Native History in the Atlantic Region during the Colonial Period," *Acadiensis*, xx (1), 1990. Even Pastore, though, never elaborates in these pieces on the cultural politics of why the question of extermination is of such contention.

30 For a valuable contemporary critique of Jenness's construction of "native barbarism" see Peter Kulchyski, "Primitive Subversions: Totalization and Resistance in Native Canadian Politics," *Cultural Critique*, 21, 1992, 171–95.

31 Diamond Jenness, "The Vanished Red Indians of Newfoundland," *Canadian Geographical Journal*, 8 (1), 1934, 27–32, at p. 27.

32 Frederick W. Rowe, *Extinction: The Beothuks of Newfoundland*, Toronto: McGraw Hill Ryerson, 1977, pp. 2 and 25 respectively.

33 Berton, op. cit., p. 182.

34 Spivak uses this phrase repeatedly. See in particular her essay "Can the Subaltern Speak?" op. cit.

35 Barbara Godard, "The Politics of Representation: Some Native Canadian Women Writers," *Canadian Literature*, n. 124–5, 1990, 183–225, at p. 190.

36 This was certainly Howley's prefatorial contention in 1915. He describes the "blotting out" of the Beothuk, concluding with a comparison: "It is a dark part in the history of British colonization in America, and contrasts very unfavourably with that of the French nation in Canada and the Acadian provinces, where the equally barbarous savages were treated with so much consideration, that they are still met with in no inconsiderable numbers, and in very appreciable condition of civilization and advancement." Howley, op. cit., p. xx.

37 Berton, op. cit., p. 184.

38 Ibid. p. 184.

39 Quoted in Howley, op. cit., p. 210.

40 See George Marcus and Michael Fischer, *Anthropology as Cultural Critique*, Chicago: University of Chicago Press, 1986. How far these authors themselves avoid salvaging anthropology as cultural critique on this same basis is open to question.

41 Julia V. Emberly, *Thresholds of Difference: Feminist Critique, Native Women's Writings, Postcolonial Theory*, Toronto: University of Toronto Press, 1993, p. 101.

42 Bishop John of Nova Scotia to William Cormack, quoted in Howley, op. cit., p. 207.

43 Joan Crate, "Shawnandithit (Last of the Beothuks)," *Canadian Literature*, 124–5, 1990, at p. 17.

44 Gayatri Chakravorty Spivak, "Time and Timing: Law and History" in John Bender and David E. Wellbery, eds, *Chronotypes: The Construction of Time*, Stanford: Stanford University Press, 1991, pp. 99–117, at p. 105.

45 Ibid. pp. 105–6.

46 The originals are in the Newfoundland museum in St. John's: reference numbers NF 3304, NF 3308, NF 3307, and NF 3300.

47 Quoted in Howley, op. cit., p. 231.

48 The touted care and consideration of Captain Buchan was, for instance, couched in these terms: "I am much pleased to find that these interesting females are under the care of Mr. Peyton." Quoted in Howley, op. cit., p. 169.

49 Ibid. p. 171.

50 Such, 1978, op. cit., p. 83.

51 Paul Carter, *The Road to Botany Bay: An Exploration of Landscape and History*, New York: Knopf, 1988, p. 345.

52 Such personalist story-telling seems only to lead to assumptions, and, with it, in Winter's case, sexist generalization. "Like women the world over," he avers, "Shananditti was interested in clothes." Winter, op. cit., p. 100.

53 Robert Rundstrom, "A Cultural Interpretation of Inuit Map Accuracy," *Geographical Review* 80 (20), 1990, 155–68.

54 Quoted by John Hewson, *Beothuk Vocabularies, Technical Papers of the Newfoundland Museum*, 2, 1978, p. 7.

55 Ibid. p. 7.

56 Robert Young, *White Mythologies: Writing History and the West*, New York: Routledge, 1990. I discuss the relations between history and geography in *Negotiating Nation-States: North American Geographies of Culture and Capitalism* Minneapolis: University of Minesota Press, 1998.

57 Matthe Edney, "The Patronage of Science and the Creation of Imperial Space," *Cartographia*, 30/1, 1993, 61–7, at p. 63. As Barbara Belyea also notes: "Native maps were useful until a survey could be made which would anchor their geographical features to a spatial grid." In "Inland Journeys, Native Maps: Hudson's Bay Company Exploration 1754–1802," unpublished manuscript, p. 17.

58 There are four copies of anonymous, undated map parts kept in the Canadian cartographic archives in Ottawa. The catalogue entry reads: "thought to be of the Exploits River, . . . drawn on birch bark by a Newfoundland Indian." A further annotation suggests that, according to Edward Tompkins of the Newfoundland Provincial archives, this "Indian" might have been the guide to Cormack, whose name might have been "Sylvester Joe." Ottatwa: H3/110/ R. Exploits/ n.d.

59 See for example, Jacques Derrida, "The Supplement of Copula: Philosophy Before Linguistics," in Alan Bass, trans., *Margins of Philosophy*, Chicago: University of Chicago Press, 1982, 175–205.

60 Howley, op. cit., p. 96.

61 Ibid. p. 189.

62 See Edward Tompkins, "The Geological Survey of Newfoundland," *Museum Notes: Information Sheets from the Newfoundland Museum*, 11.

63 Howley, op. cit., p. 238.

64 Winter, op. cit., pp. 128–9.

65 For more on what she calls Lewis's "schizophrenic approach," along with its tendency to operate with a "working sense of 'true' and 'false'" founded in a "European cartographic convention," see Barbara Belyea, "Amerindian Maps: the Explorer as Translator," *Journal of Historical Geography*, 18 (3), 1992, 267–77, especially pp. 269–70.

66 G. Malcolm Lewis, "The Indigenous Maps and Mapping of North American Indians," *The Map Collector*, 9, 1979, 25–32.

67 Ibid. p. 28.

68 Ibid. p. 25.

69 Carter, op. cit., p. 344.

70 In Belyea's words such approaches "still adhere, unconsciously if not deliberately, to a notion of progress towards increasing cartographic accuracy, and they persist in measuring that accuracy in European terms." Belyea, op. cit., p. 270.

71 G. Malcolm Lewis, "Metrics, Geometries, Signs, and Language: Sources of Cartographic Miscommunication Between Native and Euro-American Cultures in North America," *Cartographica*, 30 (1), 1993, 98–106.

72 Ibid. p. 98.

73 Hugh Brody, *Maps and Dreams: Indians and the British Columbia Frontier*, Vancouver: Douglas and McIntyre, 1988, 2nd edn.

74 Ibid., pp. 45–6.

75 See Timothy Mitchell, *Colonising Egypt*, Cambridge: Cambridge University Press, 1988. For an example of such scholarship in geography, see Derek Gregory, "Between the book and the lamp: imaginative geographies of Egypt, 1849 – 50," *Transactions of the Institute of British Geographers*, in press; and, Edney, op. cit.

76 Sami Zubaida, "Exhibitions of power," *Economy and Society*, 19 (3), 1990, 359–75.

77 Mitchell, op. cit., p. 44.

78 The chief justice of Newfoundland, Mr. John Reeves, reported this non-space of disciplinary order to the Parliamentary Committee in Westminster. "This is a lawless part of the island, where there are no Magistrates resident for many miles, nor any control, as in other parts, from the short visits of a man-of-war during a few days in the summer; so that people do as they like, and there is hardly any time of account for their actions." Quoted in Howley, op. cit., p. 55.

79 J. B. Harley, "Deconstructing the Map" in *Writing World: Discourse, Text and Metaphor in the Representation of Landscape*, Trevor Barnes and James Duncan (eds), New York: Routledge, 1992; Benjamin Orlove, "Mapping Reeds and Reading Maps: The Politics of Representation in Lake Titicaca," *American Ethnologist*, 18, 1991, 3–38. Richard Helgerson, "Nation or Estate? Ideologist Conflict in the Early Modern Mapping of England," *Cartographica*, 30(1) 1993, 10–20; Thongchai Winichakul, *Siam Mapped: A History of the Geo-Body of a Nation*, Honolulu: University of Hawaii Press, 1994.

80 Richard Helgerson, "Forms of Nationhood: The Elizabethan Writing of England," Chicago: The University of Chicago Press, 1992, 114.

81 Benedict Anderson, *Imagined Communities: Reflections on the Origin and Spread of Nationalism*, London: Verso, 1991, revised edition.

82 Brody, op. cit., p. xx. I think this notion of negotiation with a disciplinary structure is missed by Huggan in an otherwise interesting discussion of the

book's "manipulation of time–space metaphors" (p. 58). See Graham Huggan, "Maps, Dreams and the Presentation of Ethnographic Narrative: Hugh Brody's 'Maps and Dreams' and Bruce Chatwin's 'The Songlines'," *Ariel*, 22 (1), 1991, 57–69.

83 Donald Kerr and Deryck W. Holdsworth, eds, *Historical Atlas of Canada. Volume Three: Addressing the Twentieth Century 1891–1961*, Toronto: University of Toronto Press, 1990. Donor's page.

84 For a valuable institutional analysis of the Atlas as a federally funded national project, see Anne B. Piternick, "The Historical Atlas of Canada/The Project Behind the Product," *Cartographica*, 30 (4), 1993, 21–31.

85 On the limited geographical imagination of the white settler-colony thesis, see Fances Abele and Daiva Stasiulis, "Canada as a 'White Settler Colony': What about Natives and Immigrants?" in Wallace Clement and Glen Williams, eds, *The New Canadian Political Economy*, Montreal: McGill-Queen's University Press, 1989, 240–77.

86 Cole Harris, ed., *Historical Atlas of Canada: Volume One: From the Beginning to 1800*, Toronto: University of Toronto Press, 1987.

87 Conrad E. Heidenreich, "Mapping the Location of Native Groups 1600–1760," *Mapping History/L'Histoire Par Les Cartes* 2, 1981, 6–13, at p. 6.

88 *Historical Atlas of Canada: Volume One*, Plate 9.

89 Ibid. Plate 20.

90 Ibid. Plate 19.

91 Ibid. Plate 21.

92 Ibid. Plate 20, and p. 48.

93 Paul Robinson, "Mapping Canada's Early Years," *Atlantic Provinces Book Review* 14 (4), 1987, 5.

94 Alan Green put the problem like this: "Native peoples are not slighted . . . , [y]et the national scope of the atlas and *a fortiori*, Cole Harris's nationalistic and Inisian preface, fly in the face of the non-national realities of native history. The pre-history maps all tend to point to the artificality of any conception of "Canada" before the advent of the French. . . . [T]here is some contradiction between Cole Harris's desire to do right by the Indians and the patriotic themes that he stresses in summing up Volume 1." In his review published in *Labour/Le Travail* 22, 1988, 273–4, at p. 274.

95 *Historical Atlas of Canada: Volume One*, Plate 69.

96 *Atlas Historique Du Canada, 1: Des origines à 1800*, R. C. Harris (Directeur), L. Dechêne (Direction), M. Paré (Traduction) et G. J. Matthews (Cartographe), Montréal: Les Presses De L'Université De Montréal, 1987.

97 Homi Bhabha, "How Newness Enters the World," in his book, *The Location of Culture*, New York: Routledge, 1994, pp. 212–35, at p. 227.

98 Ibid. p. 227.

99 All three quotations are from Homi K. Bhabha, "Unpacking My Library . . . Again," in Iain Chambers and Lidia Curti, *The Post-Colonial Question: Common Skies, Divided Horizons*, New York: Routledge, 1996, pp. 199–211, at p. 205.

18

EMBODYING THE URBAN
MAORI WARRIOR

Gregory A. Waller

Soon after its theatrical release in 1993, *Once Were Warriors* had become the most commercially successful film ever produced in New Zealand. Based on a novel by Alan Duff that won the 1990 PEN Best First Book award and helped to make Duff a highly visible neo-conservative Maori polemicist, *Once Were Warriors* was adapted for the screen by playwright Riwia Brown and directed by Lee Tamahori, who had been responsible for several award-winning television commercials. The project received substantial financial backing from the government-funded New Zealand Film Commission, which touted the production as its fourth "Maori feature film."[1] Although previous Maori films like Barry Barclay's *Ngati* (1987) had gained accolades and some measure of prestige on the international film festival circuit, *Once Were Warriors* was the first such feature to turn a profit and garner a North American theatrical, video, and cable television release, no doubt in part because only Tamahori's film approximates the high production values associated with mainstream Hollywood movies, even as it gains a marketable measure of difference by focusing on an endangered "indigenous" family in violent, contemporary, urban New Zealand.[2]

Once Were Warriors begins, however, not with a glimpse into life on the mean streets, but with another view entirely, an opening image of New Zealand as a pristine, unpopulated, undeveloped landscape, where mountains, water, and sky balance in sublime harmony. This is immediately recognizable as the type of image long-favored in travel brochures, coffee-table photography books, postage stamps, and documentaries like the TVNZ/BBC production, *Land of the Kiwi* (1987), which was telecast as part of American public television's "Nature" series. Literally within seconds, Tamahori's camera begins to move back and then down from this landscape to reveal that the New Zealand we have seen is in fact a colorful billboard for ENZPOWER overlooking a busy freeway. Below, where the camera heads, is visually and aurally dense urban terrain: traffic whizzing by, wired-in walkways flanking the freeway, and graffiti artists at work on decrepit

337

buildings that seem to no longer have any productive function, while the soundtrack buzzes with guitar feedback à la Jimi Hendrix. This type of representation is also familiar – though perhaps not specifically in relation to New Zealand – in, for example, journalistic exposes of how the urban other half lives.

The camera movement that begins *Once Were Warriors* is not about the substitution of image for image, of one New Zealand for another. Instead, it presents itself as something of an optical illusion and ironic revelation that enacts several related binary oppositions: billboard tableau versus moving cinematic image; fabricated primeval nature versus real urban present; picturesque, corporately sponsored lie versus less inviting (unsponsored?) truth. To descend into the city is only the necessary first step. The film then sets out to chart urban space, mark its zones and pathways, test its boundaries, and explore whether it can be reconnected to a repressed and seemingly more authentic version of nature and the past. In so doing, *Once Were Warriors* dramatizes how Maori people traverse and pass time in this postcolonial space, how they – collectively and individually – find an occupation or calling or identity and come to inhabit a city which at first sight seems so oppressively restrictive, so lacking in resources and opportunity.

If we assume with Grosz that, first, "each environment or context contains its own powers, perils, dangers, and advantages" (1992: 250) and that, second, "the body must be regarded as a site of social, political, cultural, and geographical inscriptions, production, or constitution" (1994: 23), then what happens to the Maori body under the specific circumstances defined by *Once Were Warriors*? What possibilities exist for agency – in mastering, gendering, performing, constructing the body – in this contemporary city beneath the billboard? If once *were* warriors, can and should there again *be* warriors? And how would modern warrioress be embodied?

Forming and informing these questions and the answers that *Once Were Warriors* provides about warriors and the city is the recent cultural, social, and political discourse in New Zealand concerning things Maori. The most relevant aspect of this discourse for my purposes are those understandings of *Maoritanga* (Maori identity) and *taha Maori* (the Maori "side" of New Zealand culture) that accord privileged status to tribalism, native land, and the issue of "land alienation" (Awatere 1984; Mead 1984; Sullivan 1995; Walker 1990; Yoon 1986) For example, Awatere in 1984 began her influential, highly controversial call for Maori "sovereignty" with the claim that, "in essence, Maori sovereignty seeks nothing less than the acknowledgement that New Zealand is Maori land, and further seeks the return of that land" (1984: 10). The rights of *tangata whenua* (original inhabitants of the land) and the legal status of Maori land claims and other formal grievances against the government figured prominently through the ensuing decade, which saw public discourse give much attention to the current ramifications and abiding "principles" of the 1840 Treaty of Waitangi (McHugh

1991; Oddie and Perrett 1992; Sharp 1991; Walker 1991). Furthermore, as Greenland notes, the "Maori cultural revival" that began in the late 1960s was and continues to be often overtly essentialist in emphasizing the "cosmological, mythological and spiritual nexus that joined *tangata* (people) and *whenua* (land)" (1991: 93).

More recently, Sullivan refined the by-now familiar argument that "land is the very essence of Maori identity" by focusing specifically on tribalism, rather than some broadly collective notion of "Maori:" "regardless of colonization and all deprivations caused by the introduction of capitalism, each tribe has its own regions which ancestrally . . . they still feel is their *turangawaewae* – that is, land with which they strongly identify" (1995: 44). Sullivan narrates Maori history as a saga of heroic, counter-hegemonic, anti-colonialist resistance. Even though by the early twentieth century Maori had lost much of their land and moved in great numbers (particularly after the Second World War) away from *turangawaewae* to cities like Auckland, "tribalism did not disappear nor was the tribal nature of Maori society destroyed" (1995: 48). "In contemporary times," Sullivan insists, "the strength of the Maori people still lies in the concept of tribal affiliation" – though she associates this present-day "tribalism" much less with specific tribal land than with a set of "traditional" values (1995: 57).

What would this "tribal affiliation" mean in terms of the cinematic representation of Maori and New Zealand space? What type of tribalism does the postmodern city encourage or necessitate or allow for? How would this affiliation mark and be marked by Maori bodies? To what extent does the postcolonial narrative of tribalism – so essential for *Maoritanga* – involve the reclamation of rural, ancestral land and the return to the village as authentic, even maternal site? *Once Were Warriors* takes up these questions by focusing on the members of one fragmenting, self-destructing, poor, nuclear family: wife and husband (Beth and Jake Heke, both in their late thirties), three teenagers (Nig, Boogie, and Grace), and two younger children. Following the film's lead, I will center my reading on the precise way these characters – who, across lines of age and gender, each embody *taha Maori* – represent and enact the interplay between body and place.

Grace: words and body of the community

Thirteen-year-old Grace Heke, around whose corpse the family and the larger community will eventually gather for the first time, is often presented in or immediately about the family's state-provided home, where she reads to her younger siblings, helps prepare and serve them meals, and puts them to bed. Most important, Grace comforts the younger children, as well as her mother and her brother Boogie, who is close to her own age. While she is said not to be "grown up," she in fact carries out maternal responsibilities in the domestic sphere more effectively than does her mother, in part, it seems, because Grace has no appetite for food, beer, or sex. And unlike her brothers, she exhibits no desire to bare her body or to train, decorate, or display it.

As Tamahori's camera insists, Grace does indeed have a body, which means that she has both the capacity for mobility and the likelihood of becoming a victim. In three significant sequences Grace serves as a guide to and negotiator of urban space, traversing a terrain that defies any clear cognitive mapping on the viewer's part. Twice she visits her more materially dispossessed friend, Toot, who lives in an abandoned car beneath an overpass – an underside to the freeway that is not visible in the film's opening image. In these treks Grace knows precisely where she is going, while the editing here is at once continuous (in following Grace's motion, maintaining clear screen direction, and relying on a constant musical theme to bridge distinct images) and discontinuous (in the temporal and spatial ellipses between shots of different urban sites). Her walk links her home and busy neighborhood with this isolated and ramshackle haven for the homeless, where Toot's small campfire flickers like a feeble light in the immense darkness of the bush.

Taking her seat in a car that will never move, Grace's first words to Toot are: "Think we'll ever get out of here?" What is "here" for Grace? What kind of an outside does she dream of, since, initially, at least, she seems at home in the city? How can an outside be imagined in a postcolonial context? She gains a destination later when she gazes from afar at her mother's ancestral village, which looks at first sight strangely depopulated even though it is within easy driving distance of the city. The only way for Grace and the film to actually enter (and repopulate) the village is with her suicide after she has been raped. And this violation occurs not somewhere out on those city streets but in her own home by "Uncle" Bully, one of her father's closest drinking mates. After the rape and Grace's futile attempts to wash clean her body and bloody clothes, she embarks on her third – now more aimless and much more troubling – trek through the city. On this nighttime harrowing of hell Grace becomes viewer of skid-row grotesqueries and object of propositioning gazes. From "here" the path leads back home for the last time. Then to hanging herself from the leafless tree in her family's dusty backyard and so to Grace's journey to her mother's village, which the film takes to be somehow a *return* to her real home. Only with this movement – the restoration and recognition of Grace – can warriorhood for the family and the community likewise return. The abused and then ritualistically cleansed body of a pure girl serves as the necessary sacred object that for the first time brings all the disparate elements of the larger Maori community together, gathered for her funeral, a traditional *tangi* out where no freeways roar. All of the community, that is, except for her father and his drinking mates.

Grace does not aspire to becoming a warrior, does not speak of it as a goal or a life-or-death ultimatum, though her life and death seem to be preconditions for the (re)awakening and tangible manifestation of warriorness by her brothers and her mother. Grace knows her way around the somewhat

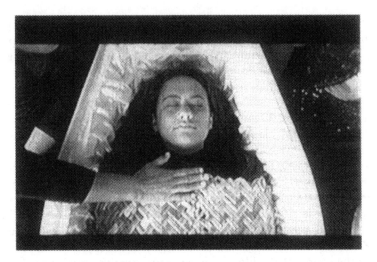

Figure 18.1 Grace's corpse at the *tangi*, around which the community gathers

generic cityscape and knows, too, about globally circulated images. Posters for Hollywood bi-racial urban comedy/action-adventures like *Lethal Weapon 2* and *White Guys Can't Jump* hang in her room. But, much more significantly, Grace enters the film already imbued – "naturally" – with *Maoritanga* and with the *mana* that comes from being a sort of tribal storyteller.[3] Her medium of expression is her journal in which she writes mythological fables to read to Toot and her younger siblings. Judged in terms of effect, however, the most significant part of her journal is her last, more confessional entry, which is not read by her mother until after the funeral. In this way Grace – now disembodied completely, out of "here" except in word and spirit – is made to communicate from beyond the grave, naming her rapist and precipitating the complex act of *utu*, or vengeance, that ends the film, repaying Uncle Bully and damning her father as well.

Jake: father and mongrel

The father's damnation in *Once Were Warriors* looks like this: "Fuck all that warrior shit!" Jake screams in drunken rage in the film's final image, as he is cut off from family and community both, diminutized by an extreme long-shot that literally renders him small, then drowns out even his defiant words with the sound of a siren. Such is the well-deserved, self-generated dead end of Jake the Muss, performer of muscular masculinity, failed father, and – with his mates – stand-in for a lost generation of urbanized, enslaved Maori whose city has shrunk beyond reclamation.

Jake's fate is hardly prefigured in his first appearance in the film. Standing with supreme self-confidence in the bright sunlight of an outdoor market, ready to swallow an oyster, Jake with a sneering smile stares down a group of stylish young punks who strut to the beat of street-corner hip-hop. The lively, "ethnic," urban neighborhood feeds his appetite, offers a stage for the enactment of his public identity. He has no need or desire to train or redecorate his body, which is already bulked up and adorned with tattoos: a scorpion on his neck, links of barbed wire on his upper arm, a snake-skin design on his forearm. These pointedly non-traditional tattoos, like his black sleeveless t-shirt and black leather pants, Jake shares with what his wife calls his "mongrel mates." The Mongrel Mob, one of the major Maori gangs to emerge in the late 1960s and early 1970s, was another manifestation of resurgent tribalism, in the guise of aggressive defiance and outlaw ethnicity (Walker 1990). *Once Were Warriors* thus could be said to ask what has twenty more years in the city – two decades marked by a much-proclaimed Maori renaissance and a much-debated policy shift toward official biculturalism in New Zealand – brought to Jake and this urban tribe? (Pearson 1991)

Over the course of the film, Jake's appetite remains insatiable – for beer, sex, food, attention – while his little corner of contemporary New Zealand contracts until it contains only his state-owned home and the Royal, the warehouse-style beer bar he haunts. And, as a corollary of or compensation for this spatial contraction, Jake repeatedly loses control of his performance as Jake the Muss, particularly when he impulsively, violently terrorizes his wife and daughter and then beats Grace's rapist to a bloody pulp. Initially, Jake's power and prestige (his *mana*, we might say) seemed to arise from the fact that he was streetwise enough to know when to throw his punches. By the end of the film there is no doubt that Beth is correct in condemning Jake for being a mere "slave" to his "fists" and hence no warrior at all.

Figure 18.2 Jake and his mates at the Royal.

Jake's home-away-from-home, the Royal, looks more like some oversized Wild West saloon or biker hangout than a working class pub. Disconnected spatially from anything resembling a neighborhood, this bar functions as a primarily homosocial, exclusively Maori arena for camaraderie, serious beer drinking, and performance in the form of tale telling, karaoke singing, and exhibitionist violence. Here men are separated from "useless pricks." Someone is clearly profiting – economically, hegemonically – from providing a space for this self-destructive, hyper-masculine urban tribalism. The Royal is without question the most thriving business enterprise in this world, a piece of private property which Jake naively assumes to be his turf to patrol and monitor as he sees fit. "Get the fuck out of here!" he orders Beth, as if he owns the place.

With one notable exception, Jake's movement through contemporary New Zealand only loops back on itself, from home to Royal and Royal to home. In fact, he is never even shown making the actual transit from one location to the other. If Jake has his way the two places will become perfectly continuous. *Once Were Warriors* rejects this destructive and self-deluding way of mastering and conflating space, offering in its stead Grace's instinctively felt longing for a separate, inviolable domestic sphere and for a return to ancestral Maori land.

The exception to Jake's repetitive motion is when he and Beth rent a car so that they can take the family to visit Boogie, who is in a juvenile detention center. The day is fine, the roads clear, and Jake smiles with pleasure at being behind the wheel of a "flash" car. He even proposes a leisurely drive which takes them beyond the city limits to a quiet vantage point near the water from which they can see Beth's postcard-like village in the distance. This sight – not some glimpse of *pakeha* ("white" New Zealanders) wealth and privilege – prompts Jake's insecurities and his defiant affirmation of his power as Jake the Muss. Seeing the village and hearing Beth's story of their courtship, Jake feels trapped by his supposedly inherited, essential identity as the product of "a long line of slaves, fucking slaves." When he gets back behind the wheel, the road delivers them to the Royal – always accessible, inviting, present, looming large against the blue sky in a wide-angle image. Soon, again, Jake will bring the party home and Grace will be raped by his close mate and the next night she will hang herself during one more mongrel get-together.

Nig and Boogie: brothers in arms

Once Were Warriors contrasts – often quite schematically through cross-cutting – Nig and Boogie's stories so as to pose two distinct ways that a young Maori male could elect to take a place in the world, by-passing, almost leapfrogging past, the "slavish" way of the father. (That such by-passing is possible, that young, urbanized Maori males have access to individual agency, and that there

Figure 18.3 The first image of Boogie, model of hip-hop cool

is indeed a place for them to take are basic assumptions of the film.) Boogie is smaller, younger, less muscular, less defiant than Nig; at first they almost seem to be a full generation apart. In fact, the brothers do not literally share the same screen space until late in the film when they carry Grace's casket at her *tangi*.

In his introduction, Boogie looks to be a model of hip-hop cool, his hair a stylish mix of thin braids falling over shaved linear patterns, his arm draped over an attentive teenage girl who holds a cigarette to his lips. He confidently, casually stands (or, better put, poses) near a classy stolen car – without even the need, it seems, to drive off. In a flash, cops arrive, the panicked teens have nowhere to run, and Boogie is thrown on the hood of a patrol car and commanded to "settle down!" And so he will, but where?

Boogie is returned home by the (*pakeha*) police, then removed the next day by a (*pakeha*) juvenile court judge who decrees that the boy should be placed under "social welfare custody." Incarceration and forced removal from the family home are absolutely the best things that could happen to him; a change in space is here the necessary precondition for all other "interior" changes. Under the tutelage of Mr. Bennett (social worker cum youth advocate cum teacher and embodiment of *taha Maori*) in the state's institutional, segregated space, Boogie chooses to become a warrior. And Bennett's first lesson is that breaking windows out of anger with a *taiaha* (traditional spear) has nothing to do with being a warrior: "you think your fist is your weapon?" Bennett declares. "When I have taught you, your mind will be, you'll carry your *taiaha* inside you." This transformation – Maori-ization – of the "inside," it turns out, can only be achieved by the disciplining of the body.

Bennett instructs Boogie and a group of other shirtless teenagers in the ways of the *haka*, a highly formalized, physically enacted, chanted exhortation in which the male participants invoke and literally reach for ancestral inspiration. The *haka* aggressively challenges the spectator, who takes the position of the enemy or opponent, in part because its gestures – protruding tongues, chest slapping, stamping feet – so completely emphasize the collective and individual body of the Maori warrior. What also marks this particular *haka* and sets it in opposition to once-familiar touristic tableaus is that the boys involved are quite diverse in age, physique, size, weight, and "blackness" – significant marks of difference which are not effaced in the disciplined, collective performance.[4]

Perhaps surprisingly, the reformatory seems perfectly suited for practicing the *haka*. Quite the opposite of both the Royal and the Heke's cramped public housing, the large, sparse, naturally-lit, wood-floored state building is a space that looks to be unbreachable and unadulterated by consumer culture and urban noise. Nowhere is this more evident than during Boogie's solo performance in the empty practice hall, after he has been let down by his parents yet again. His *haka* is a personal gesture of self-mastery and self-assertion, of sublimation and aggression, enacting and creating an embodied self that is at once individual and tribal. In Boogie's case, at least, the preconditions for such embodiment are state intervention, paternalistic tutelage, and quasi-monastic space.[5]

In contrast to Boogie, Nig's path to warriorhood does not lead through the roomy corridors of the state social welfare system. At times it seems as though Nig is fated to inhabit Jake's severely constricted city, much as he is initially marked as the "pretty" son of Jake the Muss. For example, unlike Boogie, Nig

Figure 18.4 Boogie's solo *haka*, performed in a space provided by the state.

twice visits the Royal. However, in the final scene he is able to walk away – once and for all, the film suggests – from his father's haunt. And, until the end of the film, every time Nig is shown at home he lingers only long enough to make an emphatic exit from what he calls a "fucking hole."

Like Jake (and Beth as well) Nig is first shown during the opening credit sequence as part of the active panorama of syncretic, diverse, urban Maori street culture. Whereas Jake enacts his identity as the Muss, Nig – in a black vest and no shirt, eyes covered by dark shades – is at work with two mates, pumping iron in a tin-roof, open-air, wire-fenced, makeshift gym where graffiti covers the wall and an engine block serves as a counter-weight. Under his mother's admiring gaze, Nig is building his body using the discarded material at hand, in a site that much more clearly calls to mind a prison yard rather than a fitness club.

What Nig wants, however, is not a muscularity or an individual reputation to rival Jake the Muss. He wants the "patch," the gang insignia or "colors" that will signify his membership in Toa Aotearoa, a Maori gang. ("Toa" is Maori for gang; "Aotearoa" is now in common use as the preferred Maori name for the land that came to be called New Zealand.) This male tribal group is marked by their long, pulled-back, braided hair, black sunglasses, vests, leather pants and, most spectacularly, by their elaborate tattoos (or, simply, "tats"), spreading in intricate patterns over arms and chest, neck and face. Toa Aotearoa clearly borrows much from the look of *Road Warrior*-styled post-apocalyptic punk. Unlike the tattooed facsimiles of scorpion and barbed wire that Jake wears, the gang's tats resemble the *moko* (facial tattoos) that are so prominent in nineteenth-century representations of Maori warriors and in the carved figures on the doorways, gateways, side posts, and interior panels of traditional Maori *wharenui* (meeting houses) (Bell 1992; Mead 1984; Thomas 1995). While Alan Duff can only ironically contrast the *moko* of ancient – and thus for him, true – warriors with the debased, self-aggrandizing, unearned, formally imitative tattoos on the faces of modern-day gang-bangers (1990: 172–7), the film sees the designs that cover half of Nig's face as proof of the desire for tribalism and perhaps of his capacity for incorporation into what it will define as the Maori community. His tats do not mark him as a warrior, but they also do not exclude him from the ranks of warriors.

Given the available options for urban Maori masculinity, it is not surprising that Nig is willing to pay the bodily price required to gain a patch and a redecorated face. In a world of abandoned vehicles and busy freeways, Toa Aotearoa unquestionably have the capacity for auto-mobility – witness their first appearance in *Once Were Warriors* whizzing around a police car and down an empty street. What is noteworthy (*pace* American representations of bikers, urban youth gangs, and post-apocalyptic outlaws) is that this gang is not really interested at all in speeding through space or battling for turf. Toa Aotearoa has staked out its own *pa* (fortified village), an enclave removed from the Maori neighborhood but still somewhere in the city. Jake and his

Figure 18.5 Nig cruises with his Toa Aotearoa brothers-to-be.

mongrel mates belly up to someone else's bar as if they owned the place, which they clearly don't. Toa Aotearoa has taken over what seems to be a deserted junkyard or factory site, installed their own security system, rolled out the barbed-wire-topped gates. They have reconstructed industrial space, turning what was abandoned, scarred, and debris-filled into a tribal compound, an urban *pa*.

It is significant that the film does not return more often to this compound, preferring instead to transport Toa Aotearoa to Beth's still pristinely rural village and then to return Nig, along with Boogie, to the domestic space of the Heke household, now headed by his mother. But the compound is the site for one crucial scene, Nig's ritualized initiation into the gang. Cross-cutting here parallels Boogie in the hands of the seemingly well-intended judge with Nig in the hands of his all-male, shirtless "new family," who surround and beat him until he collapses, then embrace him as a brother. In other words, before his "pretty boy" face can be marked with Toa Aotearoa tats that will forever attest to his new-tribal affiliation, Nig must prove his worthiness and willingness by putting his body on the line.

Removed from urban space, Boogie is tutored by Bennett in *Maoritanga*, learning his lessons and internalizing discipline so well that he is able to perform the *haka* by himself. Sequestered within the city, Nig submits to and endures Toa Aotearoa's ritualized punishment, a new tradition that the gang has constructed, borrowing from some aspects of *Maoritanga*.[6] In Duff's novel, Boogie can be remade as *"different,"* "proud and ramrod-straight" under Bennett's tutelage, while unsalvageable Nig dies in a senseless gang rumble (1990: 126). The film accords both brothers the status of warriors, regardless of how diverse the paths are that take them from the street to the village. In this way the sons are not destined to become the father. Warriorness, it seems, exists in potential in all Maori, who thus must choose to be what this text deems to be Maori. Whether or not Jake's ancestors were slaves is irrelevant. The point is that he refuses a ride to the *tangi*.

Beth: home and/in the city

Specifically, in the way Tamahori cross-cuts between Boogie in court and Nig being initiated into the gang, and more generally, in the way the film constructs a city by following the mobility and the settling down of Jake, Grace, Nig, Boogie, and Beth, *Once Were Warriors* corresponds to what Massey calls "the existence in the lived world of a simultaneous multiplicity of spaces: cross-cutting, intersecting, aligning with one another, or existing in relations of paradox or antagonism" (1994: 3). However, the revelatory camera movement that opens the film charts a different geography as it descends from the ENZPOWER billboard to the "real" urban terrain. Rather than affirming "simultaneous multiplicity," this shot establishes hierarchical opposition: lie versus truth, the picturesque natural versus the forbidding urban, and so on. Then, to complicate matters further, on down the line Beth's ancestral village – the desirable alternative to the city – is imaged and understood as the sort of "place" that Massey critiques in *Space, Place, and Gender*, that is, as "a site of an authenticity, as singular, fixed and unproblem- atic in its identity" (1994: 5). Meaning, in this instance, that the village is offered as veritable well-spring and embodiment of *Maoritanga*. But this "fixed" place does undergo transformation during the film. At first it is to be seen only in the distance, unpopulated and unapproachable. But with Grace's death the village suddenly becomes inhabitable, teeming with life. How is this transformation accomplished?

As noted earlier, we get within the scopic vicinity of the village when Jake drives the rental car to a sort of promontory from which the ancestral site can be seen. Ironically echoing the camera movement that opens *Once Were Warriors* (from billboard to street life), the camera again cranes down from a natural tableau, only this time there is revealed no signifier of corporate authorship and no assaulting urban "reality." Instead, sunlight glimmers on the trees and still water, the family relaxes on a rare plot of green grass, and Grace asks: "what's that place over there?" Beth responds by beginning to narrate the village into existence, telling the story of her "piece of dirt." Then follows a point-of-view extreme-long-shot focalized through the entire family, though it has been Grace who directs this gaze by her question: like a mirage or a visualized memory, the *wharenui* (main carved house) and two other buildings stand framed against the blue sky. The distant village is visible but not visitable, unpeopled and unchanging, a static place that disappears from view after only four seconds. Yet the talk about the village continues. By describing the interior of the *wharenui* and the elders

Figure 18.6 Fleeting image of Beth's ancestral village, seen from afar.

who guided her life there, Beth associates the village, as *home*, with personal reminiscence and attests to its affecting, if residual, presence. Then Jake adds his more bitter, cynical coda: "this place gives me creeps," which we might say – borrowing Massey's phrase – opens the identity of this place to "contestation" (1994: 169), even though the film does not seem inclined to allow much credence to Jake's critique.

Given this first view of Beth's ancestral village – cut off by Jake, unseen by either Nig or Boogie – it becomes something of a triumph when urban Maori can actually move across the water to revivify and reinhabit this site, allowing Beth to announce, "we're home." Grace's death provides the occasion; around her body the new community gathers. But it is via Beth that the key spatial transition occurs. During the film's opening credits Beth had moved freely, at least around the immediate vicinity of the family home. But until the abortive visit to see Boogie, Beth never again leaves the house. This does not mean that she somehow claims as her own this domestic space or that the private home functions as a refuge. Far from it. Here she is beaten and, it is implied, raped; here Grace is raped; here Jake invites his mates over whenever he so desires. Under some protest, Beth puts up with what she calls a "woman's lot" until the morning after she has held Grace's corpse in her arms. Then Beth stands up to Jake's curses and threats of violence, takes it as her prerogative to define what truly is "home," and insists that she will "take Grace back home, to the *marae*" (the open assembly area of the village or, more generally, the village as communal site). A slow dissolve bridges the spatio-temporal distance and foregrounds an almost exact graphic match between Beth's defiant expression and the face of a carved figure on the gable of the *wharenui*. The *moko* on this carved figure become, while the dissolve

Figure 18.7 Dissolve links Beth's face with the *moko* of a carved figure in the village.

lingers, tattoos on Beth's face, which bears no scars from Jake's beatings. (In the novel, Beth carries the "scars from a hundred hidings" (1990: 87).) This dissolve rewards her for her heroism and underscores her status as warrior. It also beautifully underscores the complex interplay between body and place: Beth simultaneously guides us to and becomes "back home;" her place both marks and emerges from her face. This dissolve also makes literal what Massey calls "the association between place and a culturally constructed version of 'Woman' " (1994: 10). Beth is figuratively tattooed because of her new, selfless commitment to motherhood, home, and cultural tradition, thereby becoming a Woman, which *Once Were Warriors* would have us take to be the bedrock of her identity, reached at last.

"A sense of home," Featherstone notes, "is sustained by collective memory, which itself depends upon ritual performances, bodily practices and commemorative ceremonies" (1993: 177). Grace's *tangi* is ritualistic in that it includes a *haka*, chants, and formal oratory. These traditional performances – coupled with a chorus of plaintive cries and apologetic confessions – give life to the village. In its quite detailed representation of the *tangi*, *Once Were Warriors* closely corresponds to Maori funeral rituals as described by Sinclair, who argues that "of all Maori ceremonials, it [the *tangi*] has made the fewest concessions to modern pressures," and therefore it "provides an arena in which Maori may proclaim their autonomy and unity" (1990: 231–2). The commemorative ceremony begins with a procession moving through the arched gateway, led by Nig and Boogie carrying the coffin. This is the first time that the two brothers have appeared on-screen together, underscoring the inclusive sense of community enacted at the *tangi*, which also includes Toot, Mr Bennett, Beth's friend Mavis (who had sung in the Royal earlier), other members of Toa Aotearoa, older folks dressed in suits and black funeral attire, and Boogie's companions from the state institution. The utopian aspirations of *Once Were Warriors* could hardly be more self-evident: on the *marae* age groups are no longer segregated, women become as prominent as men, neo-traditionalists stand next to gang members, mutual respect rules the day, all voices are worth hearing.

Inclusion, as always, is by definition exclusion, even if diversity is what is being celebrated. What is excluded from the Maori home in *Once Were Warriors*? The village's Other could be taken to be *pakeha* New Zealand, which is categorically absent. The rest of the film, in fact, includes no more than a handful of *pakeha* who appear on-screen. But perhaps a more telling point is that this privileging of a non-urban, traditional, yet still inclusive Maori space necessitates the exclusion of an Other that is also Maori. During the *tangi*, cross-cutting repeatedly juxtaposes the village with the Royal, where Jake drinks himself ever-deeper into self-pity and futile violence. Unlike the absent *pakeha*, Jake chooses exclusion, since he refuses the offer of a ride to the *tangi*. But it is exclusion nonetheless. *Once Were Warriors* therefore seems to perpetuate what Massey describes as the problematic understanding of "places as bounded enclosed spaces defined through counterposition against the Other who is outside" (1994: 168).

And the exclusion that marks this village extends further, beyond Jake the profligate mongrel and failed yet unrepentant father. On the one hand, the three-piece suits, leather jackets, hip hairdos, and gang colors in evidence at the *tangi* both underscore diversity among the Maori present and also, in Massey's phrase, link the "place to places beyond" (Massey 1994: 156). On the other hand, the diegetic music so prominently featured elsewhere in the film is absent from the village sequence. Because this music – performed and/or listened to by Maori – incorporates and adapts a range of internationally available styles, including hip-hop, reggae, doo-wop, electric blues, and techno-pop, it provides a complex, important link between present-day urban Maori New Zealand and places (and peoples, cultures) beyond, as well as a link to this culture's own syncretic past. Perhaps the absence of this music on the *marae* implies that the indigenous home necessarily requires some substantial measure of cleansing, purification, or policing if it is to become inhabitable by *tangata whenua*. To return home would then mean to escape not only from urban New Zealand but also from the contemporary global marketplace. This escape, the film suggests, is possible only during the *tangi*, with its ritualized performance. For after the funeral, the Hekes move cityward again, back to the state-provided housing and back, even, to the Royal.

Considering the price paid and the narrative energy expended in getting to the village, this return is somewhat surprising. Why return at all? Does the experience of visiting ancestral land somehow inscribe itself on or purify the body? In the novel, Beth is born again as a *Maoritanga*-believing, grass-roots

351

activist, pulling her community up by its bootstraps without demanding anything from the *pakeha* powers-that-be. Tamahori and Brown temper Alan Duff's socially conservative fantasy, giving Beth only the responsibility of being a mother who reclaims her house in the city (ritualistically cleansed of "ghosts" by her elderly uncle from the village) and gathers together her children – now including Toot as well – under one roof for a wholesome, bountiful meal, the first such meal in a film filled with food. Excessive appetite, desire, thirst, emotionality, physicality seem to have departed the house when Jake stormed out. This is as true for the mother as for her warrior sons, who – tats and muscularity notwithstanding – have become upright and obedient, happy to be members of the fatherless family circle. Then Beth reads Grace's journal entry describing the rape, which demands that justice be meted out both to Uncle Bully and to Jake. This necessitates one last trip to the Royal, where Beth controls Nig's anger and proves more than capable, again, of standing up to Jake, affirming in her words and pose her own "pride," "spirit," "*mana*." She has found her identity and her calling. The family, with Nig at the wheel, drives off; "we're going home," Beth announces.

If at the end Beth is word and "spirit" in the service of family and home, then Jake is more than ever merely muscle, the masculine Maori body become uncontrollable, self-destructively raging, killing its own. The film's only solution is to render Jake small, distance him as the camera moves up and away until the Muss is virtually disembodied and lost in space. Yet the Royal – Jake's arena and refuge – still stands. Whatever effects the pilgrimage to that place Grace calls "out of here" has on Beth and Nig (making them both, for instance, more willing to shoulder familial responsibility), it surely does not alter the distribution of wealth, property, land, and power in New Zealand, on-screen or off-screen.

That the final image of *Once Were Warriors* leaves us with the sound of approaching sirens and the sight of the Royal rather than the communal ritual of the *tangi* tempers what I have called the film's utopian image of an ancestral, restorative, communal, unproblematically "Maori" place apart from the city and the postmodern present. Another way of putting this is to say that as much as Beth is willing to risk her life to take Grace's corpse home, as much as Beth's identity derives from the village, she also elects to return with her children to the city, which could almost be taken as a narrative re-staging of the opening shot of the film, when the camera descends into the urban present. Of course, the film's resolution does not negate what I have called its utopianism, which speaks particularly to the anxiety of Maori as *tangata whenua* in a postmodern, "free market" New Zealand that underwent dramatic social and economic restructuring during the 1980s (Cocklin and Furuseth 1994). Nor by the same token, does the privileging of the village somehow erase the film's often celebratory images of a heterogeneous, vital street culture. In this respect *Once Were Warriors* offers a cinematic version of what Massey describes as "a simultaneous multiplicity of spaces" – spaces this film structures and prioritizes both narratively and thematically.

In quite literal ways, the spaces of Aotearoa/New Zealand in *Once Were Warriors* constitute and are in turn constituted by the Maori body: the village is only a chimera until Grace's corpse is returned home; the monastic reformatory provides the site for the Maori-ization of Boogie's body; the Royal is the arena for Jake's muscularity; Toa Aotearoa reconstitutes urban detritus and reinscribes traditional *moko*; across generations and gender, the bodies of Mongrels, street punks, and elderly traditionalists are, to return to Grosz's formulation, "site[s] of social, political, cultural, and geographical inscriptions" (1994: 23). That said, it is likewise clear that *Once Were Warriors* – in spite of and because of the conditions of postmodernity and post-colonialism – privileges one (utopian?) version of this complex process: the belief that inscription is a matter of individual agency, that constitution is chosen, that production is self-production. And for *Once Were Warriors* the Maori body in question is or should be the warrior body – disciplined, assertive, unbending, inspired by spirit and tradition, unpolluted by appetite, unbeholden to the fist. Then there's Jake the Muss.

Notes

1 Dennis and Bieranga (1992), offer the most interesting history of New Zealand film. Blythe (1994), surveys the representation of Maori in New Zealand film and television up to the late 1980s. Waller (1996), examines the industry in New Zealand in terms of the activities and public discourse concerning the New Zealand Film Commission.

2 Previous "Maori films" have had very little, limited release outside (and sometimes inside) New Zealand, unlike certain literary texts written by Maori authors. Fuchs notes that a common complaint voiced by critics about Maori fiction like Patricia Grace's *Potiki* is that is "too candid, not sufficiently complex, or overly senti-mental" (1995: 208). Rejecting this view, Fuchs attempts to reveal the "traces of 'Maoritanga' that nudge at the surface story" (1995: 215); in particular, the way *Potiki* "literally performs Maori ideas of life and death and the passing of time" (1995: 219). Such a critical maneuver might be appropriate for *Ngati* or *Mauri*, but it seems of little relevance to *Once Were Warriors*, which, it seems to me, takes up "Maori ideas" without attempting to "perform" them formally in terms of image and narrative structure.

3 Screenwriter Riwia Brown's influence is obvious here. In Duff's novel, Grace is insecure and utterly powerless, simply overwhelmed by the distance between white, middle-class domesticity and the horrible conditions of her own life (1990: 78–83).

4 See, for instance, the obligatory *haka* photograph that represents Maori masculinity in Moore (1936: 176).

5 Perhaps by turning Boogie over to Bennett and providing this site and occasion for Maoritanga, the social welfare system in *Once Were Warriors* could be seen as being analogous to the New Zealand Film Commission, with its support of "Maori" film

projects. Duff, in contrast, time and again endorses a "free-market" ideology of self-help: "Work! We work our way out. Same as we lazed ourselves into this mess," says the principal figure of Maori authority in the novel (1990: 185).

6 Toa Aotearoa and the quite different gang in Duff's novel both could also be analyzed in terms of recent debate over the "invention" of Maori tradition by Dominy(1990), Hanson (1989), and Sissons (1993).

References

Awatere, D. 1984. *Maori Sovereignty*. Auckland: Broadstreet.

Bell, L. 1992. *Colonial Constructs: European Images of Maori, 1840–1914*. Auckland: Auckland University Press.

Blythe, M. 1994. *Naming the Other: Images of the Maori in New Zealand Film and Television*. Metuchen, New Jersey: Scarecrow.

Cocklin, C. and Furuseth, O. 1994. "Geographical Dimensions of Environmental Restructuring in New Zealand," *Professional Geographer*, Vol. 46, pp. 459–67.

Dennis, J. and Bieringa, J., eds. 1992. *Film in Aotearoa New Zealand*. Wellington, New Zealand: Victoria University Press.

Dominy, M. 1990. "Maori Sovereignty: A Feminist Invention of Tradition," in *Cultural Identity and Ethnicity in the Pacific*, J. Linnekin and L. Poyer, eds, pp. 237–57. Honolulu: University of Hawaii Press.

Duff, A. 1995 (originally 1990). *Once Were Warriors*. New York: Vintage.

Featherstone, M. 1993. "Global and local cultures," in *Mapping the Futures: Local Cultures, Global Change*, J. Bird, B. Curtis, T. Putnam, G. Robertson, and L. Tickner, eds, pp. 170–87. New York: Routledge.

Fuchs, M. 1995. "Reading toward the Indigenous Pacific: Patricia Grace's *Potiki*," in *Asia/Pacific as Space of Cultural Production*, R. Wilson and A. Dirlik, eds, pp. 206–25. Durham, North Carolina: Duke University Press.

Greenland, H. 1991. "Maori Ethnicity as Ideology," in *Nga Take: Ethnic Relations and Racism in Aotearoa/New Zealand*, P. Spoonley, D. Pearson, and C. Macpherson, eds, pp. 90–107. Palmerston North, New Zealand: Dunmore Press.

Grosz, E. 1992. "Bodies and Space," in *Sexuality and Space*, B. Colomina, ed., pp. 241–53. New York: Princeton Architectural Press.

Grosz, E. 1994. *Volatile Bodies: Toward a Corporeal Feminism*. Bloomington: Indiana University Press.

Hanson, A. 1989. "The Making of the Maori: Cultural Invention and its Logic," *American Anthropologist*, Vol. 91, pp. 890–902.

McHugh, P. 1991. *The Maori Magna Carta: New Zealand Law and the Treaty of Waitangi*. Auckland: Oxford University Press.

Massey, D. 1994. *Space, Place, and Gender*. Minneapolis: University of Minnesota Press.

Mead, S., ed. 1984. *Te Maori: Maori Art from New Zealand Collections*. New York: Abrams.

Moore, W. 1936. "New Zealand 'Down Under,' " *National Geographic*, Vol. 69, pp. 165–218.

Oddie, G. and Perrett, R. eds. 1992. *Justice, Ethics and New Zealand Society*. Auckland: Oxford University Press.

Pearson, D. 1991. "Biculturalism and Multiculturalism in Comparative Perspective," in *Nga Take: Ethic Relations and Racism in Aotearoa/New Zealand*, P. Spoonley, D. Pearson, and C. Macpherson, eds, pp. 194–214. Palmerston North, New Zealand: Dunmore Press.

Ritchie, J. 1992. *Becoming Bicultural*. Wellington, New Zealand: Huia Publishers.

Sharp, A. 1991. "The Treaty of Waitangi: Reasoning and Social Justice in New Zealand?" in *Nga Take: Ethnic Relations and Racism in Aotearoa/New Zealand*, P. Spoonley, D. Pearson and C. Macpherson, eds, pp. 131–47. Palmerston North, New Zealand: Dunmore Press.

Sinclair, K. (1990). *"Tangi*: Funeral Rituals and the Construction of Maori Identity," in *Cultural Identity and Ethnicity in the Pacific*, J. Linnekin and L. Poyer, eds, pp. 219–36.

Sissons, J. 1993. "The Systematisation of Tradition: Maori Culture as a Strategic Resource," *Oceania*, Vol. 64, pp. 97–116.

Sullivan, A. 1995. "The Practice of Tribalism in Postcolonial New Zealand," *Cultural Studies*, Vol. 9, no. 1, pp. 43–60.

Thomas, N. 1995. "Kiss the Baby Goodbye: *Kowhaiwhai* and Aesthetics in Aotearoa New Zealand," *Critical Inquiry*, Vol. 22, pp. 90–121.

Walker, R. 1990. *Ka Whawhai Tonu Matou: Struggle Without End*. Auckland: Penguin.

Walker, R. 1991. "Maori People Since 1950," in *Oxford History of New Zealand*, 2nd edn, G. Rice, ed., pp. 498–519. Auckland: Oxford University Press.

Waller, G. 1996. "The New Zealand Film Commission: Promoting an Industry, Forging a National Identity," *Historical Journal of Film, Radio, and Television*, Vol. 16, pp. 243–62.

Yoon, H. 1986. *Maori Mind, Maori Land*. New York: Peter Lang.

19

SEX, VIOLENCE AND THE WEATHER

Male hysteria, scale and the fractal geographies of patriarchy

Christopher Lukinbeal and Stuart C. Aitken

weather (girl) . . . possible suspect . . . blonde temptress . . .
I love . . . my wife would kill me . . . **she'd kill you** . . . she
planned to take me to Hollywood . . . she thought about me all
the time . . . James, Don't you want to see it? . . . **sordid drugs** . . .
lost my virginity . . . type of sexual act . . . **gruesome**
scene . . . **sexual act** . . . Suzanne is all smiles with Walter,
her dog . . .

my motive

the young man

seduced him

lingerie

did

"it"

. . . act . . . virginity . . . Sex, Violence and the
Weather

Here's what I found out; that all of life is a learning experience. Everything is part of a big master plan, but sometimes it's, well it's hard to read, I mean it's like, if you get too close to the screen all you can see is a bunch of little dots. You don't see the big picture until you stand back. But when you do everything comes into focus. Hi, my name is Suzanne Maretto.

Gus Van Sant's *To Die For* (1995) begins with the camera panning over various pieces of news script framed at odd angles and at different scales of resolution. This visual imagery, combined with a quirky musical score, gives the impression of a Hitchcock-like thriller. As the opening credits fade, the dots of news script change to video lines and a close-up image of Suzanne Maretto's eye which then resolves into her face through zoom-out. At this point, the Hitchcock intrigue contracts into a relatively simple narrative about a woman who will stop at nothing to make a name for herself in news television. Suzanne begins her career by reading the weather on a cable station in the small town of Little Hope, New Hampshire. Various attempts to enhance her standing in the male dominated profession of news media include sleeping with a television executive while on her honeymoon and planning her husband's murder when he suggests that she should focus on having children rather than a career. Suzanne's main gambit at news stardom revolves around "Teens Speak Out," a chaotically amateurish documentary video that she makes with the help of three local high-school students. The main product of this liaison is Suzanne's seduction of the teenagers and their subsequent murder of the husband who might limit her career.

Several themes emerge from *To Die For* that may be intertextually connected to Van Sant's earlier and perhaps more serious works: *Drugstore Cowboy* (1989) and *My Own Private Idaho* (1991). Van Sant uses male hysteria, the voyeuristic gaze, and transgressive space to highlight the infinite repetition and obsession of what we suggest are fractal geographies of patriarchy.[1] In this context, fractals relate to social patterns within patriarchy that obsessively repeat themselves and defy the constraints of scale. Patriarchy brands fractal geographies upon bodies, and moving outward through the body, it stifles agency at larger scales by infinitely deferring, hiding and naturalizing the constraints and psychology embedded in geographic scale. Consequently, geographies of patriarchy control bodies and agency at multiple scales by reifying the same fractal representations at all scales.

Van Sant's reference to Hitchcock in *To Die For* provides a springboard to the theoretical arguments we want to make about the gaze and fractal geographies. We argue that hidden within fractal geographies are hysterical males whose sexual and political identities are unbound and unstable. Male hysteria is theorized by Lacanian and post-Lacanian feminists in two ways. First, institutional male hysteria, which focuses on castration anxiety and misogamy, empowers particular patriarchal orders by allowing some men to dominate others through the process of recoding them as effeminate. For example, many games, wars, and corporate battles are forms of institution- alized male hysteria where certain types of masculinity are acceptable, and others are not. To be unable to play these games or to refuse to participate is to be labeled effeminate. We argue that institutionalized male hysteria not only permeates and reinforces patriarchy at all scales (through bodies, families, communities and nations to the global) but also reifies the micro to macro hierarchical structure of patriarchy. In contrast, effeminate male hysteria, which focuses on the loss of oneness with the mother, works to uncode structures created by hegemonic masculinity and institutionalized male hysteria.[2] As a challenge to patriarchal structures, we argue that effeminate male hysteria dissolves the fractals of political and sexual identity by creating transgressive spaces. For the purposes of this paper, transgressive spaces are defined as sites where patriarchy and scale are deformed, where men are uncoded as men, and where the gaze is genderless. While both forms of male hysteria revolve around desire and anxiety over the mother, we contend that institutionalized male hysteria constructs patriarchal scale whereas effeminate male hysteria works to break apart these fractals. Our central thesis, then, is that male hysteria paradoxically defies the construction of the patriarchal scales that it supports.

This paradox embedded within the two forms of male hysteria helps create fractal geographies that simultaneously reify and transgress patriarchal structures. Parts of *My Own Private Idaho* and *Drugstore Cowboy* are eloquent presentations of men's resistance to community, city, and citizenship and so we use these films to unpack some of the ways patriarchies construct themselves through reification at all scales. To do so, we assume that multiple forms of patriarchy are omnipresent at the scale of bodies, families, com- munities, cities, and societies, but their crippling effects are hidden, in part, by the fractal imagery that is simultaneously independent of scale and responsible for disempowerment through scale. Van Sant's work weaves a course through, between, and around patriarchal scales insofar that the male sexualities he presents simultaneously resist and appropriate fractal patterns of identity. By so doing, Van Sant disrupts the seemingly natural relationship between hegemonic masculinities and scale. In what follows, we first chart the way hysteria is constructed from interplay between bodies and sexuality. We then examine how male hysteria produces and transgresses patriarchal scale through focusing on four themes in Van Sant's movies: misogamy and the gaze, transgression and intimacy, jumping scale, and territorial control.

The body hysteric

Many social theorists and historians have pointed out that hysteria is the primary vehicle from which all psychoanalytic theory arose (Bjørnerud 1996; Micale 1995; Smith-Rosenburg 1972; Kahane 1995). The etymology of the word hysteria comes from the Greek *hystera*, meaning the uterus. Egyptians and Greeks similarly theorized hysteria as the wandering womb, an autonomous uterus which roamed around the body plaguing a women's health. The meaning of hysteria has shifted over the centuries relating to sexuality in the Greek and Roman times, to demonological origins during the Middle Ages in Europe. Mark Micale (1995) suggests that Freud's studies on hysteria are the second sexualization of the term. He also suggests that the rise and fall of hysteria studies mirrors society's anxieties over gender instability. In the late 1960s and early 1970s feminist researchers began re-examining the latent misogamy, social control, and gender issues surrounding hysteria (see Fischer-Homberger 1969; Smith-Rosenburg 1972; Wood 1973). One of the most prominent theories arising from these studies was by Carroll Smith-Rosenburg (1972) who maintains that hysteria is an intolerable ambivalence to gender roles created by society. At the same time, she suggests that hysteria provides an unconscious mechanism through which the body can resist gender roles. More central to our arguments, Luce Irigaray (1985a, 1985b) evinced that Freud never adequately constructed a theory of female sexuality and this lack allows women to be labeled as hysteric. Consequently, she suggests that female hysteria refers to a woman who mimes her sexuality in masculine form, where she is recoded into the oedipal masculine. Claire Kahane (Bernheimer and Kahane 1985, Kahane 1995) combines Irigaray's theorization and other feminist theories to posit that hysteria is a bodily representation of repressed bisexual conflict, an "unconscious refusal to accept a single and defined subject position in the oedipal structuration of desire and identity" (Kahane 1995: x–xi). She theorizes that the hysteric does not give up anything, but simply defies and contests the symbolic boundaries and laws that define the scale of the body.

Male hysteria is used to theorize Freud's relation to Dora (Bernheimer and Kahane 1985), white fear and anxiety surrounding the black male (Saint-Aubin 1994), the Gulf War (McBride 1995; Showalter 1997), the invasion of Panama (Weber 1994), and men's relation to death and cyborg life (Fuch 1993). Following Irigaray, male hysteria may be constructed as the uncoding of men as men (emasculation) rather than the effemination of men (Kirby 1988; Weber 1994). Emasculation can be both a liberating and a political event: it removes the fractals through which masculinity is reified. Once a man is uncoded as a man he may be re-coded as feminine. Within the context of patriarchal logic, effeminization is a key process through which hierarchy and difference is established. Male hysteria is not only about the domination of men over women but, also, it is a means of empowering specific types of masculinity. By signifying a man as a woman, he is castrated through language, a discursive construction powerful enough to produce scales within masculinity and patriarchy. The effeminate other is not strictly defined as female or "woman" because males may also be marked and signified as effeminate (McBride 1995).

At one level, male hysteria is an oedipalization because it comprises rivalry with the father (hegemonic masculinity and patriarchy) and fear of castration (emasculation and removal from social interactions that constitute and construct masculinity). While Freud traces hysteria to the oedipal fear of the father, we argue that the origins actually lie with *the* mother (Irigaray 1985a). *The* mother reminds *the* son of his own castration, the separation between the part (himself) and the whole (the mother). Anxiety is inscribed in a desire that can never be fulfilled: a return to oneness with the phallic mother. As the boy passes through the Lacanian mirror stage, desire is transposed to the father and later to patriarchal society. Desire for the phallic mother never ceases, but upon threat of castration (finding out that the mother/woman is not the bearer of the phallus) he must defer desire to the father to have the phallus.[3] The motivation for identifying with the father cannot be fulfilled because it is based on the assumption that the father possesses the phallus of the phallic mother (Bjørnerud 1996). However, the boy desires the intervention of the father to "delimit the seemingly boundless power of the mother" (Caput 1994: 40). Consequently, the desire to possess the mother's phallus is never satisfied. Irigaray (1985a, 1985b) argues that castration fear is in actuality displaced separation anxiety and desire for the mother. As we will argue in a moment, the illusion of the father's psychological prominence (and his possession of the phallic mother's phallus) is reinforced through institutionalized male hysteria.

Desire and anxiety are bound in a psychological struggle to maintain balance between "manhood" and masculine subjectivities on the one hand and, on the other hand, power and dominance over the effeminate (m)other. The key forces of desire and anxiety construct a fractal geography of patriarchy within which male hysteria fitfully reposes. But male hysteria is also about the disruption of the defenses set up to protect the boundaries between the spaces of the self and spaces of the other and so it parodies desire and anxiety for the oneness of the pre-oedipal where there is no separation between subject and object. With male hysteria there is a collapse of the scalar fix in subjectivity; the subject is in crisis (according to the norms of hegemonic masculinity) and able to float outward and inward without bodily resistance. This is not a moral issue, but rather a permutation of identity. Not only is the hegemonic subject in crisis but the reality produced by the subject is also in crisis. When this occurs, the hegemonic subject will attempt to reaffirm the primacy of the masculine phallus thereby insuring and maintaining the "natural" patriarchal order. However, the individual's subjectivity is also in crisis which allows them the possibility to redefine their relationship with body, world and patriarchy. Thus, male hysteria simultaneously undermines the hierarchies of patriarchal space, penetrates the fixidity of patriarchal scales, and transgresses social and identity politics.

From an etiological perspective, male hysteria, to be hysteria at all, had to be modeled after female hysteria. The diagnosis of hysteria in men is a sign of weakness; it labels a man as a non-man, in short, it is a castration in a word (Showalter 1993). Effeminate male hysteria does not fit within the hegemonic order of patriarchy and masculinity. This form of hysteria undermines the construction and re-production of hegemonic masculinity and patriarchal scale by showing how masculinity may be restructured and nurtured outside the bounds of the oedipal conflict of masculine power struggle and female subjugation. Effeminate hysteria expresses itself most forcefully at the scale of the body and it consequently unhinges the process of propriety and the patriarchal enforcement of "knowing one's place."[4] Effeminate hysteria exposes the unnaturalness of the production of patriarchal scales.

Institutionalized male hysteria is that form of the psychosis that is condoned by most patriarchal structures. It often appears in the form of territorial games that are ideologically laden with the suppression and subjugation of the effeminate other. The marking and production of meaning within a defined territory naturalizes patriarchal claims to power. In this sense, hysteria is a means to justify control over spatial and behavioral practices. Latent homoeroticism and sexual phobias are hidden in the more obvious territorial conflicts that permeate baseball and soccer fields, in the violence of street battles between police and gangs, in the tension between corporate and labor power mongers, and in geopolitical conflicts (see Showalter 1997; Weber 1994; McBride 1995). Similarly, Heidi Nast (forthcoming; Blum and Nast 1996) suggests that oedipal metaphors, latent with heterosexuality and racism, serve "varying but strategic geopolitical ends" (Nast, forthcoming). America's war on drugs and much of the homophobia that surrounds the AIDS crisis also suggest institutionalized male hysteria (see Micale 1995). Male hysteria, in this form, is condoned by patriarchal society and is hidden by institutional rhetoric. A refusal to participate in these struggles may mark behavior as effeminate or deviant.

Institutionalized male hysteria originates from "unresolved desire for and aggression against the mother" (McBride 1995: 179). Since *the* mother signifies unification and the oneness from which life sprung, she is also feared for that oneness, and she is the object of aggression for having destroyed the oneness the male child at one time experienced (cf. Aitken and Herman 1997). *Woman* is castrated in the place of man so that his anxiety may be allayed. The phallic mother must be controlled, dominated and marked with the signifiers of patriarchy. Without the masculine phallus as the primary signifier, the male becomes hysterical, disassociated from body, language and the scales of patriarchy. Consequently, the roots of territorial games lie in castration and a displaced separation anxiety. Continuity is the fleeting prize where masculinity is briefly aligned with the illusion of obtaining the true phallus. However, the rhetoric underlying territorial games is castration, and a struggle for possession of the phallus. This gendered power struggle has a darker, more formidable structuration:

Men can both fantasize what it is like to be the mother, to be a woman, and resist that temptation. For losing the phallus, losing the game, losing the war, they lose themselves as men. And for this frightful thought they punish their enemy, desire revenge against the mother, and seek to destroy her image as woman ... patriarchal culture is established on the victimization of women *qua* woman.

(McBride 1995: 156)

If hegemonic masculinity embodies an ideal patriarchal order and reinforces the status quo then, within this configuration, hysteria and hysterical practices are normalized even mythologized as the very psychology upon which the patriarchal order stands. Masculinity constructs and maintains multiple scales moving from bodies and subjectivities, to interpersonal relations and families, and even larger scales of capitalism and international politics. These patriarchal scales work to construct gender boundaries and "produce differentiation and domination within them" (Connell 1983: 41). The scales of patriarchy reify a closed logical system that prescribes masculine domination and the subordination of the other. Scale, male hysteria and the maternal are related to desire and the anxiety men experience over *the* mother. For effeminate male hysteria, these issues coalesce in a bodily resistance to the gender roles placed upon them by patriarchy. This form of male hysteria is a confusion over the scalar relationship between the interior and exterior world and, as such, it exposes the unstable production of gendered identities, bodies, and patriarchal scale. For institutionalized male hysteria, the desire and anxiety for the mother works to construct a gendered hierarchy which reaffirms the masculine phallus and suppresses the effeminate other. Institutionalized hysteria and patriarchy enable space and scale to appear natural and allow the constitution of the gaze to remain masculine. We would like to now consider the way that one aspect of institutional male hysteria – the gaze – aids in the construction of patriarchal scale. A focus on the masculine gaze positions male hysteria within patriarchal culture and its inscription in popular motion pictures.

Van Sant, the murderous gaze and the thrill of the *mise-en-abyme*

Van Sant's passing reference to Hitchcock in *To Die For* is critical for two reasons. First, it helps situate his work squarely in the context of Laura Mulvey's (1975) well-worn arguments for the existence of a voyeuristic "gaze" in Hollywood cinema wherein the camera looks, the male subject looks, and the female is looked at. Second, while Hitchcock's work may be thrilling because of numerous disingenuous narrative plot twists, there are other forms of thrill that emanate from the seemingly scaleless patterns of fractal images.

362

According to Mulvey, Hitchcock's heroes are the bearers of the gaze in a narrative where women are the object of its sadistic impulses. By implication, Hitchcock offers women an exclusively masochistic relationship to his films. His articulation of "patriarchal consciousness" and institutionalized male hysteria constructs the female subject as simultaneously victim, predator, and monstrous other, who devours the male subject (Denzin 1995: 117). In *To Die For*, Suzanne is caricatured in precisely this way, to the extent that Van Sant is satirically re-compiling the layers of Hitchcock's voyeuristic desire. We interpret Suzanne's character as an ironic production of the masculine gaze: she wants to be looked at, but she also needs to be an integral part of an institutionalized news media that depends upon a male image of impartiality and objectivity (cf. Croteau and Hoynes 1992).

By constructing a sexualized gender at the scale of the body, Suzanne is willing to capitulate to large scale patriarchal constructions by conforming to the excesses of the traditional male gaze. Kathleen Kirby (1996: 126) suggests that this kind of gaze " . . . outlines a structure of immobility in which . . . women can attain no agency and are destined to perpetual violation." The power that Suzanne aspires to as a television journalist appropriates a different kind of gaze, one that Kirby (p. 126) describes as a plastic medium wherein "the power and the quality of the *gaze* determines the gender positionality of the participating *subjects*." In this case, Suzanne wields a gaze that is reversed and, by so doing, she is given power over a hapless husband and the three teenagers who become involved in the making of her documentary. Although there is apparent control offered by this formulation of the gaze, Kirby argues that it does little more than reverse gender polarities while maintaining patriarchal binary relations. With the maintenance of a male/female dichotomy, the gaze that Van Sant is suggesting here is distanced from the sex of the bodies that he represents. We argue that this formulation of the gaze is ultimately disempowering to women and men because it hides within the scales through which patriarchies are perpetuated. Suzanne's cold and calculated attempts to use the male gaze, and her self-aggrandizement as a sexual object, ultimately fail because the strictures of patriarchy refuse women the willful construction of their own images. Suzanne, nonetheless, is convinced that she is in control of her image:

> There are some people who never know what or who they want to be until it is too late and that is a real tragedy in my book because I always knew what I wanted to be, always. Okay, who wants to be on TV?

After the murder of her husband, Suzanne contrives a story that incarcerates the two male teenagers while she goes into hiding to avoid "bad press." Suzanne's ultimate demise is directly related to another gambit to further her career in television and her dismissal of responsibilities around husband and family. She is lured from hiding by a Mafia killer (contracted over the phone by her father-in-law with a knowing nod from her mother-in-law) posing as a television producer who is interested in making a documentary out of her story. It may be argued, then, that Suzanne is murdered by patriarchy through the scale of the body, the family, and the corporate news media.

It seems to us that the success of Van Sant's filmmaking derives in part from his cognitive mapping of the patriarchal/political consciousness (see Jameson 1984, 1992). At its best, we read Van Sant's work as an interpretation of the ways fractal geographies highlight the seemingly infinite and scaleless hysteria within patriarchy that is forced through the body and beyond in a seemingly unassailable *mise-en-abyme*. In contemporary cultural criticism, Andrew Benjamin (1991: 15–16) uses the *mise-en-abyme* as a metaphor which "serves to structure the possibility of interpretation in advance of the act of interpretation itself." This is similar to Diane Elam's (1994) use of the term as a literary device, but she expands it to suggest the issues of scale that we elaborate upon here. Rather than fractals, Elam highlights the *mise-en-abyme* as a metaphor for understanding the infinite possibilities for women. Elam's use of the *mise-en-abyme* as a literary device parallels our usage of fractals and male hysteria. She suggests that the *mise-en-abyme* inverts the relationship of part and whole, making the object untenable to the subject, which produces a parallel hysterical regression in the subject. Not only can the relationship and definition of the part to the whole not be grasped, but the

> *mise-en-abyme* . . . opens a spiral of infinite regression in representation . . . The subject and object infinitely change places within the *mise-en-abyme*; there is no set sender or receiver of the representation. The infinitely receding object in the *mise-en-abyme* closes down the possibility of a stable subject/object relation. On the one hand, the object cannot be grasped by the subject; it slips away into infinity.
>
> (Elam 1994, 27–8)

As we move up and down, or zoom in and out of, a socially constructed series of scales, patriarchy represents itself again and again but we learn nothing new of its construction. However, the *mise-en-abyme* of male hysteria begins at the scale of the body and moves outward to reify this psychology at larger scales.

In addition to its relation to fractals and male hysteria, we use the *mise-en-abyme* as a counter to the more familiar film concept, *mise-en-scène*. Whereas the latter is used to describe a stage setting, or the surroundings of an event, we use the *mise-en-abyme*, literally, as the context of the abyss. The notion of patriarchal constructions of scale, male identity and the hysteria of looking into the abyss (or how patriarchy hides the abyss through the contrivance of scale relations) come together in the *mise-en-abyme*. In patriarchal societies the masculine gaze is naturalized through a *mise-en-abyme* that subsumes the *mise-en-scène*, making the infinite regression in representation inevitable. Thus, patriarchy maintains the gaze as masculine as a means through which it may enforce and police spatial hierarchies. When male hysteria moves beyond these normalized bounds, transgressive spaces/subjects become "visible" signifiers of the other. For example, women's roles in popular film mirror the separation anxiety inherent in male hysteria: women are often either objects to be gazed upon and *desired*, or objects of *anxiety* and aggression. A *mise-en-abyme* is graphically presented in *To Die For* when Suzanne as a young child talks of her enduring need to control her own image while at the same time being video-taped watching herself on the television screen: what we see is an infinite regression of Suzanne's image. In addition to its use of the gaze, then, we contend that *To Die For* is "thrilling," but not in the Hitchcock sense of being shocked by dramatic and unknowable plot twists. The thrill we write about here is thoroughly known and knowable: images repeat themselves in a fractal geography whose patterns we are drawn to, and excited by, because they mirror deep seated, and perhaps repressed, desires and anxieties.

Beginning with Jean Baudrillard's (1988) unreflexive camera eye, some cultural critics and film theorists suggest that representations and images neither mimic nor mirror "reality" but, rather, they engage, resist, and re-vision "reality" to the extent that it is impossible to pry apart the image and its source. We believe that there are nonetheless vestiges of the mirror in contemporary film when a repetitive and obsessive fractal geography is used to thrill.[5] The thrill is encountered through a narcissistic gaze that does not reveal anything more about the seemingly scaleless representations encountered. Like Suzanne watching multiple images of herself on the television, the experience is like looking at repeated images in a hall of mirrors: intrigue stems in part from a regression that instills the sense of an infinity that is independent of scale, and in our wonder we fail to notice that each recurring image reveals nothing new about the image itself. Our obsession is derived from a continual mirroring that engenders a sense of liberating scale and transcending the finite.

We argue that despite the thrill of a *mise-en-abyme* that contrives multiple mirror images *ad infinitum*, our selves/bodies are grounded by scale relations within families, communities, cities and citizenships that are inscribed and controlled by patriarchy. We think that Van Sant's skill derives in part from his ability to turn the thrill of encountering the *mise-en-abyme* into a male hysteria through which patriarchy is unbound and laid open. In what follows, we use *Drugstore Cowboy* and *My Own Private Idaho* to illustrate some of the ways male hysteria is constructed and then hidden within the fractal geographies of the *mise-en-abyme*.[6]

Transgression through intimacy and the scale of the body

If the gaze and institutionalized male hysteria works to construct patriarchal scale, effeminate hysteria seemingly transgresses these boundaries. This is not an active, willful political protest, but rather an unconscious, bodily refusal to accept patriarchal strictures. We use transgression the way Foucault did, in the sense that it both "affirms the limits of being" and "affirms this limitlessness into which it leaps," thus allowing our "*sole* manner of discovering the sacred in its unmediated content" (Miller 1993: 88). Transgressive spaces then are the product of effeminate male hysteria, where liberatory masculinities may take shape and arise simultaneously from within and beyond the confines of patriarchal scale and the oedipal complex. They are safe spaces where identities are played with and difference is recognized as empowering rather than threatening.[7] We argue that Van Sant uses movies to create transgressive spaces that suggest an unmediated "reality" before the thrill of fractal geographies. From these sites, viewers and characters can interpret, and perhaps contest, dominant narratives and fractal geographies as they are woven into obsessively repetitive social patterns and naturalized interpersonal relations.

In *Drugstore Cowboy* and *My Own Private Idaho* transgressive spaces are fabricated through drug use, the open road, narcolepsy and intimacy. Set in Portland, Oregon in 1971, *Drugstore Cowboy* focuses on Bob, a drug addict who heads a "quartet of low-grade desperadoes who rob drugstores and filch pills and Dilausdid from hospitals" (Vineburg 1990: 27). The rewards of their heists enable the gang to spend the bulk of their days in a drug-induced transgressive space. Van Sant is not necessarily suggesting that drug-induced states of mind are liberatory but, nonetheless, he uses them to create a fluid space for the gaze. Following Anne Friedberg's (1993) and Giuliana Bruno's (1993) arguments, we contend that some forms of the gaze can accommodate difference and liberate sexuality through the practice of transgressive spaces. These spaces offer indifferent boundaries where genders come together and change one another which, in turn, allows the gaze to be reconstituted as a flexible medium. Thus, in *Drugstore Cowboy* and *My Own Private Idaho*, Van Sant uses mind-altered states to suggest the possibility of liberatory transgressive spaces that may be attained in many different ways: slumming, cross-dressing, journeys, intimacy. In Aitken and Lukinbeal (1997), for example, what we call a flexible mobile gaze in the road sequences of *My Own Private Idaho* symbolizes a movement through the layers of repressed sexual memories to a transgressive space.

Perhaps the most important aspect of Van Sant's transgressive spaces is that they contrive a medium for the gaze that is so fluid that we, as viewers, simultaneously feel like participants in, and voyeurs of, the intimacy of his characters. As a consequence, *Drugstore Cowboy* is not about celebrating the space of mind-altering drugs but, rather, it is about the nooks and crannies of our lives that we prefer to deny and hide. To this extent, Van Sant is expert in using intimacy to communicate the dangers of creating a space within which we artificially escape or deny feelings. Drug-induced states of mind are not necessarily liberating in themselves but they provide a metaphor and a medium for a flexible gaze, and it is out of this gaze that the contradictions in male hysteria find form.

Bob's gang is a makeshift and dysfunctional family comprising his wife Dianne and two younger followers, Rick and Nadine. Van Sant creates a bizarre nuclear family fantasy nested within a suburban landscape but there is clearly no idyll here to which anybody would want to aspire. The normality of the landscape is dissected by constant references to teenage drug addiction. While driving home on a suburban street, for example, cows, cars, and hats fly across Bob's field of vision as he indulges in some of the drugs from the gang's first heist. As soon as they arrive home, a teenage "TV baby" from across the street rushes over to buy drugs. In a particularly tense scene, Bob's phallic power is disrupted by Nadine and, after shaming her, Bob presages a central theme of the movie when he turns to Dianne with the comment, "Look at me babe, I'm hysterical."

It is important to note that a liberatory masculinity in Van Sant's work is not based upon mind-altered states *per se* but focuses, rather, upon fluid sexualities and how the gentleness of interpersonal space can usurp hard-edge hegemonic spaces. Gendered, interpersonal space is the medium which connects the individual to intimacy and the social. Van Sant's penchant for representing liberatory sexual identities is through a subtle engineering of a fluid gaze and intimate social spaces highlighted against an immutable political culture. When focusing upon intimate relationships, Van Sant best exposes transgressive spaces. Hysteria is part of this, but Van Sant's transgressive spaces are incomplete without intimacy and the scale of bodies.

The narrative of *My Own Private Idaho* revolves around the development of intimacy between Mike and Scott, two male prostitutes who journey on the roads of Idaho and live on the streets of Portland and Seattle. Throughout the movie, Mike's narcolepsy, as a transgressive space,

> enables an eerie, yet exhilarating sense of dislocation and narrative slipperiness. It's often unclear whether we're inside or outside Mike's skull, in real time or dream space. Whenever Mike falls out, Van Sant's camera winks out of existence, too. Events unfold in jump cut. We recurrently awaken with Mike as if the world were newly invented, under clouded circumstances in obscure locales.
> (Greensberg 1992: 24)

Through this disruption of narrative convention, Van Sant is able to open the possibility that sexualities are based upon different formulations of internal and exterior spatial performances. Sexual and political identities are formed in part through the negotiation and performance of interior, or psychic space (including the body) and exterior, or physical space (including the body). By re-configuring or deforming the relation between internal and external space, we may also change external reality. As one of us argues elsewhere, subjects (and subjectivities) do not necessarily lie deep within the interior spaces of the unconscious, but rather they are continuously formed and reformed on the surface, at a fluid boundary between interior and exterior spaces (see Aitken and Herman 1997). This boundary is porous and mutable, extending outward into external space, as space is produced, and retreating back to the body in times of attack.[8] It seems reasonable to argue, then, that the space of the subject is confounded by the enforcement of scale differences beyond the body. As the embodiment of scale differences, patriarchal structures may therefore bind and limit our bodies.

Van Sant's filmic genius continuously distracts us from Mike's mind-altered state. He uses Mike's narcolepsy to obsessively repeat images of mother love, home, family, the American "heartland," and so forth in a *mise-en-abyme* that diverts us from Mike's neurosis and leaves us concerned about his character and his relationships. Film critic David Denby (1991: 79) suggests that Mike is an "aimless young dish, passive, not too bright, he doesn't have a mean bone in his pretty body. Like Myshkin's epilepsy in Dostoevski's *The Idiot*, his narcolepsy is meant to be a sign of his essential innocence, more a poetic than a medical reality."

Whereas Mike's daily life is that of a male prostitute's and his fantasies mingle hysteria with mother love, Scott's day-to-day existence is not penetrated by soft-edged fantasies. Scott prostitutes his body as an affront to the upper-class spaces occupied by his father, the mayor of Portland. The space of his body interacts with the spaces within which he performs. His sexuality and social class affect not only his performances, but also the types of spaces that he routinely visits and occupies. The large inheritance he will receive on his twenty-first birthday, colors his street life. Nonetheless, because bodies are signifiers of cultural production and social practices, Scott is able to outrage and threaten his father by how he uses his body. Giuliana Bruno (1991: 36–7) argues that the relation between social practice and language, mediated by the body, is the scale at which a definition of ideology is produced. The scale of the body marks the site at which social signification makes its mark, it is a "reserve, an archive that informs the decoding of images . . . a social process where signs and object-signs are shaped by, and shape, social geography."

The first meeting of Mike and Scott signals their location within a social geography: an affluent woman picks out a few bodies from a row of rent boys lining a sleazy downtown street and then takes her purchases to a suburban mansion in an opulent Seattle neighborhood. The importance of this scene relates not to any kind of sexual bartering and promiscuity, but to the beginnings of an intimate relationship between Scott and Mike in a space that is marked by patriarchy. One of the most touching scenes in the movie is when Mike tells Scott of his love at a roadside campfire in Idaho. There is no sexual *mise-en-abyme* here, but rather we see Scott's concern mix with his dilemma over his own masculinity: Mike falls asleep comforted in Scott's arms.

In the end, Scott and Mike play out two forms of male hysteria caricatured respectively by obsessive father hate and duplicity, and mother love and lost innocence. The story is carried, however, by Mike and the liberatory potential of the transgressive space that Van Sant creates around him. Mike's fantasies and openness to intimacy contest the hard-edge of patriarchal capitalism, heterosexual norms and naturalized spaces to which Scott ulti-mately capitulates. It seems reasonable to suggest that Mike's innocence and openness to intimacy liberate him bodily from the impressions made by the patriarchal scale relations to which Scott succumbs. It is ironic, however, that power is accessible only to those who, like Scott, can move between the scales that are inscribed by traditional patriarchal structures. We now consider the *mise-en-abyme* of institutionalized male hysteria and the concept of jumping scale.

Scaling patriarchy: the *mise-en-abyme* and jumping scale

Like space, scale is materially established through social action and, as such, there are politics in its constitution that contrive a crucible for reproducing interpersonal and patriarchal relations. When scale is viewed as a "natural" ordering of the world the subjugation of social relations at each level closes avenues for political action, thus reifying scale as sacrosanct (Herod 1997). Dominant organizations and social structures control subordinate groups by confining their activities to manageable scales. It is this ability to harness power at other scales, or to *jump scale*, that enables individuals and groups to affect change:

> Scale now is causal in the realist sense that the constituent properties of social objects change, or are activated, as these objects stretch and contract across space. Neil Smith calls this stretching process the action of "jumping" between scales, and he rightly emphasizes that it is a process driven by class, ethnic, gender, and cultural struggles.
>
> (Jonas 1994: 258)

Within patriarchy's naturalized structure, jumping scale implies a bodily movement up or down a geographic spatial hierarchy. In *My Own Private Idaho*, Scott's embodiment is contrived in an oscillation between the scale of the public street, Pigeon's squatter community, and the patriarchal power base of his father's mayoral office. If Mike's battle is to regain lost mother-love, Scott's fight is over his resistance to the phallic power of his homophobic father. Scott's resistance is played out not only at the site of his body, which he sells, but also at the site of his chosen family in Pigeon's community. By positioning himself as "threat" to the phallic power of his birth father, Scott solidifies his alter street persona by accepting Pigeon as both lover and street-father. We focus here on how Scott is able to manipulate the social geography inscribed at the site of his body, which allows him to seemingly jump patriarchal scales.

Pigeon plays a Falstaff character to Scott's Prince Hal. Pigeon is Scott's teacher of resistance whom he will later denounce in a parody of Prince Hal's public castigation of Falstaff in Shakespeare's *Henry IV, Part One*. Van Sant uses Shakespeare to suggest an intriguing oedipal intertextuality. Both Scott and Prince Hal slum with street people to insure that their return to their father's court will be more momentous and spectacular. Prince Hal's marriage to a foreign woman and return precipitates his ascension to the throne as Henry V and, similarly, Scott's marriage to an Italian woman expedites his return to the upper-class life of his father. Van Sant's parody of Prince Hal's journey suggests an interesting commentary "on two very different cultures facing similar challenges relating to the shifting social ideologies of gender and sexuality" (Román 1994: 313). For example, throughout the film, Scott clearly is not interested in sex with males other than as a means of upsetting his father and staging his climactic redemption through hetero-sexual marriage. In contrast, Mike's love for Scott is confused with brotherly love and homosexual desire. This confusion between family, love and sexual preference establishes a complex space that is resolved for Scott with his choice of heterosexual privilege and patriarchal structure over the intimacy and indeterminacy offered by a relationship with Mike.

Once distanced from Pigeon's court, Scott confesses that he has always been planning to return to the world of his father's hegemonic norm. This conceit is spatially reinforced when Scott returns from Italy with his new fiancée. While driving through the streets of Portland, Scott's limousine passes by Mike's huddled body on the sidewalk. With this simple scene, Van Sant contrives the limousine as "a bubble of panoptic and classifying power, a module of imprisonment that makes possible the production of an order, a close and autonomous insularity" (de Certeau 1994: 111). Scott has now jumped scales; he has removed his body from the street and imposed a physical boundary between himself and Mike. For the rest of the movie, he inhabits sites of compliance rather than resistance: the limousine, a trendy private bar, the traditional, pompous site of the mayor's funeral.

The concept of "jumping down" economic and social scale is an old myth with a strong lineage. For example, Shakespeare's story of Prince Hal came during the age of discovery, when imperial expansion and colonization introduced England to a number of new cultures and social practices. Scott's jumping down suggests a control of scale that enables patriarchal culture an "opportunity to appropriate its marginal social elements so that like Hal, the different, strange, and gross could be extracted, consummately rehearsed and thus consummately foreclosed" (Román 1994: 325).[9] It seems reasonable to argue that this jumping down through scale is a variation of the hero's saga through which males find socially acceptable means to discover their manhood. Within a patriarchal society, the hero's tale enforces dominant behavioral roles that seek to naturalize culturally inscribed prototypes of masculinity. Taking this one step further, we argue that with the naturalization of mythic struggles, prescribed masculine modes of behavior are recycled in heroes' tales. Consequently, as a masculine *mise-en-abyme*, the relationship between the part (the specific actions that make a hero) and the whole (what constitutes a hero) become inverted. Mythic heroes, then, are institutionalized forms of hysteria – they provide established means through which men can attain, or regain, the phallus.

Cultural critic and gay activist, Michelangelo Signorile, criticizes Van Sant for his representation of heroic heterosexuality in *My Own Private Idaho*. Signorile claims that Van Sant embraces a logic wherein Scott, the hero, renounces gay, subordinate, lower-class life and celebrates the heterosexual, dominate, upper-class life (cited in Román 1994). On the surface, this may appear to be a reasonable criticism, but it misses the subtlety with which Van Sant portrays the sexual politics that surround Mike and Scott.[10] For example, Scott's capitulation to his father's patriarchy is not necessarily immutable. In his final castigation of Pigeon, Scott declares that he

> was [always] planning a change. There was a time when I had the need to learn from you my former and psychedelic teacher and although I love you more dearly than my dead father, I have to turn away. Now that I have, and until I change back, don't come near me.

Moreover, in the gaze that Scott and Mike exchange at the funerals of Pigeon and the mayor, there is a suggestion that Scott may at any moment choose to re-invent himself as a rent boy. More importantly, we argue that it is the contradictions between Scott and Mike that ultimately transcend the myth of hegemonic masculinity. Whereas Mike is an effeminate hysteric whose motherlove/motherland dissolves the "natural" scales imposed by patriarchy, Scott's institutionalized hysteria enables him to mimic the lack of the phallus. In the end, however, we agree with Signorile that Scott's hysteria is a performance of power in a *mise-en-abyme* that connects him to a grand lineage of male heroes. Scott's gender and social class enable him to mime homosexuality and lower social-economic class to dispose of desires and anxieties around economic and gender instabilities. Through a contrived hysteria, he empowers himself by weaving associations to a tradition that relates to the son's challenge of the father's power: a laudably heroic and institutionalized hysteria.

If we describe Mike's hysteria as effeminate, then it contrasts with Scott's institutionalized male hysteria because it does not fit well within the dominant order of masculinity. As we have described it in Mike's case, effeminate hysteria expresses itself more forcefully at the scale of the body while Scott's institutionalized hysteria extends outward from the body to the social and geopolitical realm. What we find particularly intriguing about Van Sant's work is a sensitivity to the liberatory potential of male hysteria, in combination with an uncanny sense of how to destabilize the space within which patriarchy works. Similar to the *mise-en-abyme* surrounding Scott's shifting masculine construction and scale jumping, territorial battles show the contention between various masculinities. Territorial battles are prime examples of where institutionalized male hysteria seeks to construct scale and subordinate the other through signifying them as effeminate.

Institutional hysteria: territorial battles and social control

In *Drugstore Cowboy*, the territorial battles between Gentry, the police detective/father figure, and Bob highlights the inherent power struggle between dominate and subordinate factions of male hysteria. At the time the movie was released in 1989, Van Sant's portrayal of drug-use was in stark contrast to the institutionalized hysteria that characterized the "war on drugs" and the insensitive "just say no" campaigns of the Reagan/Bush administrations. Mark Driscoll (1994: 113) puts it this way: "the hysterical war on drugs demands a cleansed social body practicing a clearing up of the nasal/market passage, a Dristan McCarthyism." Institutionalized hysteria around drug-use is about an assumed moral supremacy that highlights Foucault's (1965) point that society always needs a group to gaze upon and, thereby, to subjugate.[11] In *Drugstore Cowboy*, the philosophy and sentiment of using drugs as a foil against the rapier of society's hysteria is expressed by novelist William S. Burroughs who plays a cameo as an elderly addict priest:

> Narcotics have been systematically scapegoated and demonized. The idea that anyone can use drugs and escape a horrible fate is an enigma to these idiots. I predict in the near future right-wingers will use drug hysteria to set up an international police apparatus.

In *Drugstore Cowboy*, we witness the ways that institutionalized drug hysteria enforces control over the streets and poor suburbs, enabling police to disrupt and regulate private lives. On the front lines of this street battle is the panoptic gaze of the police who are ironically portrayed as inept hysterics. They treat Bob and his gang with contempt, beating them up and destroying their property.

In their first territorial battle, Gentry leads the police raid that destroys Bob's house. Bob asks Gentry, "Do you have a warrant?" To which Gentry asserts his phallic dominance by stating, "Yeah, I got a warrant, its pasted on the end of one of these slugs." During the raid, Bob's private space is destroyed and he is made a public, effeminized spectacle in front of his "family." In an exchange over Bob's golf clubs, Gentry asks where Bob plays. Bob says he's been playing at Mayfield, a public golf course. Gentry proclaims that "Mayfield is for pussies" and, by so doing, again asserts his phallic authority over an effeminized Bob. The police have intruded, marked, and destroyed Bob's private space and asserted their hegemonic power over Bob and his gang. The morning after the police raid Bob's "family" lies naked in the rubble of their living room as if their bodies were violated by the police.

When Bob and his gang find a new residence, the panoptic gaze of the police (and Gentry) emanates from a camper parked in a nearby street. Bob manipulates this gaze through a mischievous plot that makes him appear to be hiding his drugs in the house next door. In the police's zealous attempt to locate the drugs, an undercover agent gets shot and wounded by the owner of the house. The next day, Gentry advises Bob to "take a nice long drive somewhere far away from here," to which Bob retorts,

> Wait a minute Gentry, *I won the Goddamn war* not you. Who are you to dictate the terms? Why don't you go find some small town police department where you can step in as sheriff where all you have to worry about is the usual Saturday night drunks and kids fucking around.

In a sarcastic reversal of power positioning, Bob not only challenges Gentry, but crowns himself winner of the territorial battle. Two cops then grab Bob while Gentry punches him in the stomach and yells in his face: "So you think you won the war, huh? You are just a junkie that got one of my officers shot. And, soon as he gets a hold of you everybody is going to forget that you ever existed." Shortly thereafter, Bob packs up his crew and leaves on a road trip, a traditional cure to male hysteria.

The open road is the place where Bob, Dianne, Rick and Nadine can stay one step ahead of subjugation. The sedentarism in cities is inscribed by the police, parents, and the methadone program. Tragedy nonetheless strikes when Nadine overdoses on narcotics from a particularly successful heist and dies in the same motel that a group of sheriffs is using for a convention. After narrowly escaping from the sheriffs, Bob decides to leave the gang and to return to Portland to join a methadone program. Later, in a gentle scene in his Portland apartment, Bob says good-bye to Dianne. There are aspects of Bob and Dianne's love story that are touching. Although their relationship is largely unsentimental and asexual, it is, nonetheless, not without intimacy.

With the scene in the apartment, we are drawn to Bob's affection for Dianne when he tries to persuade her to join him in rehabilitation. Importantly, physical sex takes second place to Van Sant's portrayal of feelings and passion. In a wonderfully ironic play on his disregard for sex, as part of the rehabilitation program in Portland, Bob is given a tedious job drilling holes in machine parts: Van Sant's dwells at length on the phallic characteristics of this work (another interpretation of this scene is that Bob may be getting screwed by rehabilitation). As with *My Own Private Idaho*, clearly the liberatory masculinity in Van Sant's work is not based upon sexual identity *per se* but focuses rather upon how the gentleness of interpersonal space can usurp the hard edge of hegemonic space.

One evening, Bob returns home to find Gentry waiting in his apartment. Gentry again asserts his control over space and territory by first being able to locate Bob, and second, penetrating Bob's private space. Nonetheless, Gentry shows genuine concern: "I hope you make out on that job of yours, and I sincerely hope that you straighten up." Here, the father figure sympathizes with the son only after Bob seemingly renounces his transgressive lifestyle. The intimacy is genuine but the space is contrived: if Bob accepts Gentry's world it positions him in a structured patriarchal space.

The endings of *My Own Private Idaho* and *Drugstore Cowboy* are similar in the sense that a transgressive space remains open for the central protagonists. The credits begin to roll in *Drugstore Cowboy* after Bob is badly beaten up and shot by an old drug acquaintance, the teenage "TV baby." During the beating, Bob reflects in a voice-over that "there's nothing more life affirming than getting the shit kicked out of you. I know it by heart . . . I relaxed and gave into the notion that for the very first time in my life I knew exactly what was going to happen next." Bob knew that he was going back to the old territorial game and the beating was propelling him there; he was going to the one place that had the best drugs of all, the hospital.

Conclusion

Male hysteria's central and perhaps most promising feature is "its capacity for a deep relativizing of gender norms" (Micale 1995: 251). We argue in this chapter that the relativizing of gender norms commingles male hysteria and a fluid gaze in transgressive spaces that unhinge patriarchally constructed scales of being. These scales contrive a hegemonic masculinity that relies on performances of institutionalized male hysteria to provide continuity and social order. As James McBride (1995: 146–7) notes,

> like the endless wheel of life and death, discontinuity and continuity destine men to a life of repetition–compulsion in that they can do no other than to desire what is forbidden and to lose what has

been illicitly won. In this play between taboo and transgression, discontinuity and continuity, the masculinist imaginary conceives male activity as a game.

Reifying the obsessive repetition of the *mise-en-abyme* enables continuity to be maintained and allows sanctioned forms of institutionalized male hysteria to be performed and imprinted upon our bodies. At another level, however, male hysteria transgresses patriarchally based masculinities by uncoding men as men.

We use Van Sant's representations of intimacy and power to suggest that although scale may contrive subjectivities, transgressive spaces destabilize the continuities that delimit and spatially fix our bodies. For example, we are not drawn to Mike in *My Own Private Idaho* out of pity, but rather out of intrigue towards the illusive freedom of an unencoded body. Similarly, we are drawn to the intimacy between Mike and Scott as a site for transgressing patriarchal logic. Likewise, Bob and Dianne's intimacy in *Drugstore Cowboy* is not defined by sex and love but, rather, by a bodily struggle against the strictures of established patriarchal norms. In the end, Mike and Scott's transgressive space is disrupted by Scott's capitulation to his father's institutionalized hysteria. This hysteria is a revisioning of the heroic narrative as an acceptable means through which men from the upper class can mimic hysteria, jump scales, and return to the safe confine of the fatherland. In contrast, Suzanne's attempt to mime the hero's tale in *To Die For* results in her continual repression and eventual death.

We are drawn to these complex narratives because they highlight patriarchally-based fractal geographies that mirror thoroughly known social and spatial orders. Like fractals within fractals, the texts and subtexts in Van Sant's movies engage a patriarchal logic that is knowable but is mirrored continuously so that – like watching images in a hall of mirrors – we learn nothing new of its constituency. The patriarchal logic emanates, for example, from Mike's obsessive dependency on the lost mother, Scott's dependency on the homophobic father and his obsession with gaining power through jumping scale, Bob's obsession with narcotics and heists, and Suzanne's obsession with power, image, and jumping scale. The unsettling paradoxes that Van Sant exposes as self evident, in actuality enframe an institutionalized male hysteria that we would rather ignore or allow to persist as "natural" scales that constrain our lives. Van Sant creates narratives and a *mise-en-scène* that seemingly freeze these fractal geographies, and holds them in place, allowing us to examine the *mise-en-abymes* that surround our social space. It is as if we may find our own *private* Idaho, a landscape of uncontrived sexual and political identities, a space where the gaze is fluid and our bodies unconstrained.

The *mise-en-abyme* that constrains identities, bodies and the gaze is graphically depicted in the penultimate scene of *To Die For*. The female high-school student who was part of "Teens Speak Out" describes her new life of fame as a consequence of the affair with Suzanne:

> Suzanne used to say, "If people were watching it makes you a better person." So if everyone was on TV all the time, everybody would be better people. But if everybody was on TV all the time, there wouldn't be anybody left to watch. That's where I get confused.

At the word "confused," the teenager's image divides in two and then continues splitting as she talks about the invitations she has received to appear on various talk-shows. The cloned images of the girl are frozen in multiple takes, each a mirror image in time and space as she continues speaking: "But it really is something to think that I'm the one who is going to be famous. Suzanne would die if she knew." The multiple framing of the girl's image fades to Suzanne's sister-in-law skating on a frozen lake as the closing credits begin to roll and the soundtrack plays "Season of the Witch." Beneath the frozen lake lies the body of Suzanne. Another story ends with patriarchy firmly in control by first creating then destroying another witch/hysteric.

Acknowledgments

The authors would like to thank Christina Kennedy, Doreen Mattingly, Deborah Dixon, Larry Ford, and Sirena McCart for comments on earlier drafts and for their encouragement of our ideas.

Notes

1 Heidi Nast's (forthcoming) "Unsexy Geographies" similarly explores the fractal nature of the modern oedipal in heterosexism and racism at various national and international scales. Her work differs from ours in that we argue that male hysteria is an intimate accomplice in the construction of patriarchal fractals and their associated scales.

2 By masculine hegemonic structures, we mean the embodiment of ideal patriarchal orders that maintain a status quo by reinforcing the myth of the ideal man at all scales (from muscular bodies to political muscle). Hegemonic masculinities are continually reaffirmed in relation to what they are not, and so their relations to both subordinate masculinities and to women are important.

3 Jane Gallop (1988, in Weber 1994: 174) points out that the phallus "is neither real nor a fantasized organ but an attribute: a power to generate meaning. . . . To have a phallus would mean to be at the center of discourse, to generate meaning, to have mastery of language, to control rather than to conform to that which comes from outside, from the Other."

4 Propriety here refers to its two meanings of both proper social behavior for a given situation and its older seventeenth century meaning as both "property" and "knowing one's place." Effeminate hysteria upsets the notion of property as

the "physical nature upon which men play out their appointed roles" (Olwig 1996: 640).

5 In Aitken and Lukinbeal (1998), we discuss the repetitive violence, destruction and conflagration that characterizes many mainstream Hollywood "blockbusters" as precisely the kind of thrill that mirrors patriarchal desires and anxieties.

6 We discuss *My Own Private Idaho* at length in an earlier work which ties the notion of a continuously mobile male to disembodiment and male hysteria (Aitken and Lukinbeal 1997). In the current work, we elaborate on those arguments with a fuller appreciation of how scale is implicated in the continual transformation of sexual identity.

7 Giuliana Bruno (1993: 52) suggests that transgressive spaces enable "individuals from different backgrounds, classes, venues, and genders" to be "put in close touch, share intimacy in fleeting encounters." She argues that some public forums such as cinemas were transgressive sites in the first half of this century because they opened *flânerie* and voyeurism to women who would have been labeled streetwalkers if they had strolled alone on city sidewalks.

8 Internal spaces are penetrated, and sometimes attacked violently, on a daily basis (Kirby 1996). For prostitutes like Scott and Mike, the violation of the internal is a commodified penetration. With this in mind, Mike's narcolepsy is important because it highlights separation anxiety and desire for the mother rather than castration fear of the father.

9 Steven Mullaney (1998: 84) notes that the *mise-en-scène* of the theater (like today's popular movies) serves as an arena for the "rehearsal of cultures."

10 This criticism also contradicts Van Sant's personal sexual orientation and public activism:

> in the fall of 1992 [Van Sant emerged] as the leading force in Portland's gay and lesbian community to counter the perilous activities of the Oregon Citizens Alliance, an antigay group that put forth an initiative – which was defeated in the November election – that would have amended the state constitution to define homosexuality as "abnormal behavior" and would have forbidden state and local governments from including gays and lesbians in antidiscrimination laws.
>
> (Román 1994: 327)

11 Although the mad and hysterical roamed the streets free and were viewed with respect during the middle ages, Foucault argues that this changed after leprosy was all but eradicated and European society needed another group to subjugate. The gaze of society became encoded by social biases making madness "visible" in external space: hysteria, drug culture and prostitution came under sharp focus with this new gaze.

References

Aitken, Stuart and Herman, Tom. 1997. Gender, Power and Crib Geography: Transitional Spaces and Potential Places. *Gender, Place and Culture* 4, 63–88.

Aitken, Stuart and Lukinbeal, Christopher. 1997. Disassociated Masculinities and Geographies of the Road. In *The Road Movie Book*, edited by Ina Rae Hark and Steve Cohan. London: Routledge, 349–70.

Aitken, Stuart and Lukinbeal, Christopher. 1998. Of Heroes, Fools and Fisher Kings: Cinematic Representations of Street Myths and Hysterical Males in the Films of Terry Gilliam. In *Images of the Streets*, edited by Nicholas Fyfe. London: Routledge (forthcoming).

Baudrillard, J. 1988. *America*. London: Verso.

Benjamin, Andrew. 1991. *Art, Mimesis and the Avant-Garde: Aspects of a Philosophy of Difference*. New York: Routledge.

Bernheimer, Charles and Kahane, Claire, eds. 1985. *In Dora's Case: Freud–Hysteria–Feminism*. New York: Colombia University Press.

Bjørnerud, Andreas. 1996. Beckett's Model of Masculinity: Male Hysteria. In *Not I*, edited by Lois Oppenheim and Marius Buning, pp. 27–35. Madison: Fairleigh Dickinson University Press.

Blum, Virginia and Nast, Heidi. 1996. Where's the Difference? The Heterosexualization of Alterity in Henri Lefebvre and Jacques Lacan. *Environment and Planning D: Society and Space* 14, 559–80.

Bruno, Giuliana. 1991. Heresies: The Body of Pasolini's Semiotics. *Cinema Journal* 30, 29–43.

Bruno, Giuliana. 1993. *Street Walking on a Ruined Map: Cultural Theory and the City Films of Elvira Notari*. Princeton: Princeton University Press.

Caput, Mary. 1994. *Voluptuous Yearnings: A Feminist Theory of the Obscene*. Lanham, Maryland: Rowan and Littlefield.

Connell, Robert. 1983. *Which Way is Up? Essays on Sex, Class and Culture*. London: George Allen and Unwin.

Croteau, David and Hoynes, William. 1992. Men and the News Media: The Male Presence and Its Effect. In *Men, Masculinities and the Media*, edited by Steve Craig. Newbury Park: Sage.

de Certeau, Michel. 1994. *The Practice of Everyday Life*, translated by Steven F. Rendall. Berkeley: University of California Press.

Denby, David. 1991. *My Own Private Idaho*, a movie review. *New York Times*, 79.

Denzin, Norman. 1995. *The Cinematic Society: The Voyeur's Gaze*. London: Sage Publications.

Driscoll, Mark. 1994. James Bond and Immanuel Kant's War on Drugs: A Nosography and Nosegrammatics of Male Hysteria. In *Body Politics: Disease, Desire, and the Family*, edited by Michael Ryan and Avery Gordon, pp. 104–14. Boulder, Colorado: Westview Press.

Elam, Diane. 1994. *Feminism and Deconstructionism: Ms. en Abyme*. New York: Routledge.

Fischer-Homberger, Esther. 1969. Hysterie und Misogynie – ein Aspekt derr Hysteriegeschichte. *Gesnerus* 26 (1/2): 117–27.

Foucault, Michel. 1965. *Madness and Civilization: A History of Insanity in the Age of Reason*. New York: Random House.

Friedberg, Anne. 1993. *Window Shopping: Cinema and the Postmodern*. Berkeley: University of California Press.

Fuch, Cynthia. 1993. "Death is Irrelevant": Cyborgs, Reproduction, and the Future of Male Hysteria. *Genders* 18, 113–33.

Gallop, Jane. 1988. *Thinking Through The Body*. New York: Columbia University Press.

Greensberg, Harvey. 1992. Review of *My Own Private Idaho*. *Film Quarterly*, Fall, pp. 23–4.

Herod, Andy. 1997. Labor's Spatial Praxis and the Geography of Contract Bargaining in the US East Coast Longshore Industry 1953–1989. *Political Geography* 16(2), 145–69.

Irigaray, Luce. 1985a. *Speculum of the Other Women*, translated by Gillian C. Gill. Ithaca: Cornell University Press.

Irigaray, Luce. 1985b. *This Sex Which is Not One*, translated by Catherine Porter. New York: Cornell University Press.

Jameson, Fredric. 1984. Postmodernism, or the Cultural Logic of Late Capitalism. *New Left Review* 146, 53–92.

Jameson, Fredric. 1992. *The Geopolitical Aesthetic: Cinema and Space in the World System*. Bloomington: Indiana University Press.

Jonas, Andrew. 1994. Editorial: The Scale Politics of Spatiality. *Environment and Planning D: Society and Space* 12, 257–64.

Kahane, Claire. 1995. *Passions of the Voice: Hysteria, Narrative, and the Figure of the Speaking Woman 1850–1915*. Baltimore: John Hopkins University Press.

Kirby, Kathleen. 1996. *Indifferent Boundaries: Spatial Concepts of Human Subjectivity*. New York: The Guilford Press.

Kirby, Lynne. 1988. Male Hysteria and Early Cinema. *Camera Obscura* 17: 112–31.

McBride, James. 1995. *War, Battering, and Other Sports: The Gulf Between American Men and Woman*. New Jersey: Humanities Press.

Micale, Michale. 1995. *Approach Hysteria: Disease and Its Interpretations*. Princeton: Princeton University Press.

Miller, James. 1993. *The Passion of Michel Foucault*. New York: Anchor Books.

Mullaney, Steven. 1988. *The Place of the Stage: License, Play, and Power in Renaissance England*. Chicago: University of Chicago Press.

Mulvey, Laura. 1975. Visual Pleasure and Narrative Cinema. *Screen* 16(3), 6–18.

Nast, Heidi. Forthcoming. Unsexy Geographies. *Gender, Place and Culture*.

Olwig, Kenneth. 1996. Recovering the Substantive Nature of Landscape. *Annals of the Association of American Geographers* 86, 630–53.

Román, David. 1994. Shakespeare Out in Portland: Gus Van Sant's *My Own Private Idaho*, Homoneurotics, and Boy Actors. *Genders* 10, 311–33.

Saint-Aubin, Arthur. 1994. Testeria: The Dis-ease of Black Men in White Supremacist, Patriarchal Culture. *Callaloo* 17, 1054–73.

Showalter, Elaine. 1993. Hysteria, Feminism, and Gender. *In Hysteria Beyond Freud*, edited by Sander Gilman, Helen King, Roy Porter, G.S. Rousseau, and Elaine Showalter, pp. 286–344. Berkeley: University of California Press.

Showalter, Elaine. 1997. *Hystories: Hysterical Epidemics and Modern Media*. New York: Colombia University Press.

Smith-Rosenburg, Carroll. 1972. The Hysterical Women: Sex Roles and Role Conflict in 19th Century America. *Social Research* 39, 652–78.

Vineburg, Steve. 1990. *Drugstore Cowboy*, a movie review. *Film Quarterly*, Spring, p. 27.

Weber, Cynthia. 1994. Something's Missing: Male Hysteria and the US Invasion of Panama. *Genders* 19, 171–97.

Wood, Ann Douglas. 1973. "The Fashionable Diseases": Women's Complaints and Their Treatment in Nineteenth-Century America. *Journal of Interdisciplinary History* 4, 25–52.

Film References

Drugstore Cowboy. 1989. Director, Gus Van Sant. Avenue Pictures.
My Own Private Idaho. 1991. Director, Gus Van Sant. New Line Cinema.
To Die For. 1995. Director, Gus Van Sant. Colombia Pictures.

20

WRITTEN ON THE BODY

Eroticism, death, and hagiography

Giuliana Bruno

From the anatomy lesson to madness, from city views to the erotic–religious, the recognition, in the work of Elvira Notari, of an intertextuality with painterly, photographic and architectural discourses, as well as medical and religious representation, results from focusing on film stills, freezing filmic movement or analyzing extant frame enlargements and publicity stills.

I view the isolation of a still image as an act of participation in the intertextual process an interpretive intervention on the palimpsest – the network of quotations, hypotexts, paratexts, and peritexts that constitute Notari's lacunar textuality. Working in this way, a critic unavoidably returns to Roland Barthes's treatment of Eisenstein's stills.[1] Even if weary of a "third meaning," I subscribe to a Barthesean fascination with, and attention to, cinematic stills as a site of the articulation of the filmic. All the more so for Notari, whose filmic textual state is often only composed of surviving photographs. Given their status within such a field of filmic rarity, stills are seductive for their paucity and the lack they expose.

I do not claim that stills are symbols of the (lost) films, or specimens of their substance, nor that they exemplify a film's ultimate meaning. The pictures for which I offer a reading are not samples or extracts, "an idea that supposes a sort of homogeneous, statistical nature of the film element,"[2] but rather "citations," fragments deployed within Notari's fragmentary macrotext. Their existence does not exceed the fragment, nor supersede the film, with which they enter into a palimpsestual relation.

In the process of isolating a still from a motion picture, meaning is both crystallized and disseminated. In Barthes's words, "The still [institutes] a reading that is at once instantaneous and vertical."[3]

. . . Still an erotic body

I die because I cannot die. . . . It is not bodily pain, but spiritual, though the body has a share in it – indeed a great share. So sweet are the colloquies of love which pass between the soul and God.

Saint Theresa

The still image, object of our erotic detour, is quoted from *Fantasia 'e surdato* (Soldier's Fantasy, 1927) and deployed in a palimpsest of other citations. The frame enlargement in question is isolated from a sequence at the very beginning of the film, when by chance the protagonist, Giggi, meets a young florist, Ninetta, who sells on the sidewalk "the most beautiful flowers of her garden and her passion."[4] He wanted to buy some flowers for his mother, but she offered him a rose for himself. "And so they loved each other passionately," the intertitles read, as the camera showed them embracing and kissing; "but, in the dark night to come, a predator passed by and stole his dream of love." Cut to a shot of Ninetta, who lies swooning in the arms of her new lover. The shot functions as a flash within the sequence, offering a glimpse into an erotic moment that is singular in the visual texture of the film. The image is a discrete and isolated occurrence, almost a fugitive insert.

Let us look at the composition. The background is completely obscure. In the foreground, centrally placed in the frame, almost sculpted out of the darkness, are two figures, a woman and a man. Her lover's body is very muscular and masculine, a prototype of a fashionable filmic male beauty, an erotic hero à la Maciste. He appears to be naked, and his face is half-lit. While the man holds her, the woman lies horizontally, completely abandoned in his arms, seemingly dead. The composition and interaction of the bodies creates a sort of reversal of Michelangelo's *Pietà* (1499). However, the body is not dead, or rather this is a particular kind of death. The woman's state is "the little death," a moment of loss experienced in the climax of love. Of this moment, Georges Bataille writes:

> The orgasm is popularly termed "the little death." . . . The expen-
> diture of energy necessary for the sexual act is everywhere enormous.
> . . . Whether obscurely or clearly this little death is what is feared.
> On the other hand it is also desired (within human limits at least).
> No one could deny that one essential element of excitement is the
> feeling of being swept off one's feet, or falling headlong. If love exists
> at all it is like death, a swift movement of loss within us, quickly
> slipping into tragedy and stopping only with death. For the truth is
> that between death and the reeling, heady motion of the little death
> the distance is hardly noticeable.[5]

Elvira Notari visualizes Ninetta in a pose of sexual abandonment, in the erotic explosion of the flesh. In her representation of female desire, the woman's erotic involvement adjoins a death, which is not expressed as the wish to relinquish life but rather to live "the little death." Underlying the imaging is a plethoric state: at the moment of keeling over, eroticism, an excessive assenting to life to the point of death, borders on the desire to die. This female desire is a desire to live to the limits of the possible and to cross those limits with ever-increasing intensity.

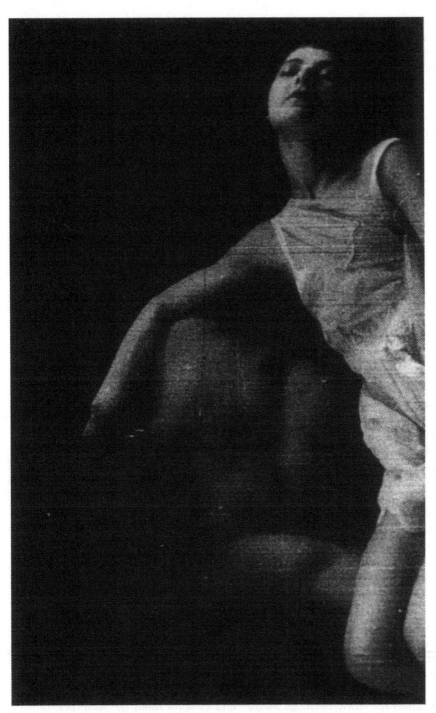

Figure 20.1 The dream of love, in *Fantasia 'e surdato* (Soldier's Fantasy, 1927) by Elvira Notari.

In her rendition of erotic relinquishing, Notari connects sensuality and death in a way that recalls the images of female saints, where that link is vividly exposed. The erotic death is conceptualized in the saints' own writing and visualized in the iconography that depicts them. Most appropriate to our image is Saint Theresa. Pushing herself to the limit, she spoke of a desire to live while ceasing to live, or to die without ceasing to live. Notari's image suggests this very desire of an extreme state, such that, according to Bataille,

> Saint Theresa has perhaps been the only one to depict strongly enough in words. "I die because I cannot die." But the death of not dying is precisely not death; it is the ultimate stage of life; if I die because I cannot die it is on condition that I live on; because of the death I feel though still alive and still live on. Saint Theresa's being reeled, but did not actually die of her desire actually to experience that sensation. She lost her footing but all she did was to live more violently, so violently that she could say she was on the threshold of dying, but such a death as tried her to the utmost though it did not make her cease to live.[6]

If Saint Theresa's words "I die because I cannot die" are appropriate to Notari's erotic imagery, it is because of the *fisicità*, the corporeal weight of the saint's vision. The body had a great share in Theresa's own imaging, and she, in turn, became an object of bodily representations, dense in "physicality," depicted as she was in the state of "little death."

One thinks in particular of Gianlorenzo Bernini's *Saint Theresa in Ecstasy* (1645–52), a sculpture of the saint and an angel who points the arrow of divine love at her. The angel's pose, his expression of satisfaction, and the arrow hint at a sexual vision. As in Bernini's *The Blessed Ludovica Albertoni* (1674), the spectacle of an erotic ecstasy is taking place. The sexual component of Theresa's fainting spell is evident: her abandoned languid pose and facial expression clearly indicate orgasm. As Pierre Klossowski puts it, "Turned completely about, her interior rendered exterior, the banners of the soul unfurled in the frozen swirls of the marble, the ineffable struggle mirrored in her rolling eyes, rising to the surface of her tautened lips, she sighs the bliss of her surrender."[7]

Looking more closely at the composition of Notari's shot, we may recognize an intertextual trace of Bernini's sculpture. Ninetta's swoon in the arms of her lover speaks of the iconography of Saint Theresa's ecstasy. Like the saint, Ninetta's body softly and lasciviously lies still in a languid pose. As in Bernini's statue, the male figure is erect, while she reclines, keeling over, with her head back, her eyes closed, her lips parted. Her dress falls in folds around her body, suggesting Bernini's play of veils. The texture of the clothing enhances the erotic twist. As Mario Perniola notes about the sculpture:

Figure 20.2 Michelangelo, *Pietà*, 1499

Figure 20.3 Gianlorenzo Bernini, *Saint Theresa in Ecstasy*, 1645–52

Figure 20.4 Gianlorenzo Bernini, *The Blessed Ludovica Albertoni*, 1674

The extraordinary erotic magic . . . does not depend simply upon . . .
the saint's lovely face, clearly indicating that she is about to faint, but
mostly on the fact that Saint Teresa's body disappears in the drapery
of her tunic. . . . While it still projects all the impetuous and vibrant
shuddering of a body in ecstasy . . . it accentuates the essential, the
transit between body and clothes, the displacement of what lies
beneath the drapery.[8]

In Notari's shot, as for Theresa's sculpture, the woman's dress covers and yet
uncovers, increasing the eroticism of the image. Unveiling is further rendered
as a contrast between light and shadow, and, in the play of (in)visibility,
the bodies of Ninetta and Theresa are likened. This erotic swoon, suspended
between life and (little) death, is a saint's "temptation": a "desire to fall, to
fail, and to squander all one's reserve until there is no firm ground beneath
one's feet."[9]

Hagiographical tales: seduction and the *corps morcélé*

Eroticism and religion are connected in a physical way. . . . The
dream of erotic love takes us to such heights only so that we can make
that jump, from the sky down to earth, down below.

Lou Andreas Salomé

Besides this instance of erotic imaging, many of Notari's stories of women's
vicissitudes recall the popular tales of the martyrdom of female saints.
Along with the operatic and the popular, Notari participated in a filmic
remaking of the hagiographic. Hagiography, the body of literature inspired
by and destined to promote the worship of the saints, has produced a network
of diverse forms of expression and support. As hagiography circulates in
Notari's texts, they participate in its process of social dissemination and
proliferation.

The intertextuality of hagiography and Notari's fictional world extends all
the way from thematics to the very form of narrative articulation. Similarities
are found in the generative process of narrativization and in the form of
characterization. In hagiography, as in Notari's cinema, individuality counts
much less than typology. Both use narrative strategies in which similar
features, events, and experiences are passed along to different characters and
given different proper names. And it is the combination, the bricolage
of elements that gives form to a character. As in the case of Gennariello, the
character ends up being more a model than a singular individual with one
proper name. In this respect, we might say that saints are constructed in a
fictional process paralleling that of the Gramscian "heroes of popular
literature."[10]

Another site of this intertextuality concerns the relation between the private and public spheres. Hagiography, as Michel de Certeau shows, constructs narrative as a passage between these two realms, as the drama of the saint lies in the very transformation of private into public, and vice versa.[11] The life of a saint is a composition of places, marked by a predominance of space over time. The history of the saint is popularly rendered as a geography, one rich in changes of scene. Such a passage and friction is the pivot of Notari's narrative form. Her public melodrama, constructed on the borders of private space and urban site, with its *mise-en-scène* on the threshold of "inside and outside the home," is grounded in such a popular experience of spatial location.

A blatant case of hagiographical interaction is *Trionfo cristiano* (Christian Triumph, 1930), the last fiction film for which a censorship record exists.[12] This was part of Notari's immigrant cinema, commissioned by a group of Italian Americans who asked Dora Film to make a film about Saint Pellegrino, the patron saint of their place of origin, Altavilla, a small village in the Neapolitan region. An early Christian, Pellegrino was condemned by a pagan emperor to be sexually tempted by women and eventually died of this torture. Although the film is lost, a fictionalization, written in Notari's style, survives:

> Because of Emperor Aurelianus's despotism, and his vile and evil passions, Rome was feeding more and more on the blood of her sons, who were falling under the murderous weapon of pagan superstition. Pellegrino reeked of Christian virtues, and he lived by spreading the Truth of Religion. Courageous in peril, he would run around to support the believers who were languishing, chained in prison, and those who were chained to city walls, and those who were sent to forced labor, destined to perish. . . . Because he was opposing the arrogance of the emperor, Pellegrino, together with his very dear friends Vincenzo, Eusebio, and Ponziano, . . . was captured and severely tortured. . . . After the experiment of torture, another way was tried: the seduction of love. Insinuating temptresses were chosen. They were entrusted with the difficult task of conquering the chastity of the saints and instructed to use all possible means of seduction. The response of the saints was that their real love was Jesus Christ on the cross. The woman who was trying to corrupt Pellegrino's feelings and heart was in turn taken by the gentle figure of the saint . . . and converted. Pellegrino and the others were condemned to death and died by torture. But only their bodies were conquered.[13]

A number of transpositions circulate in this tale, a story of sexuality, power, and death. As Pellegrino fights against the arrogance and despotism of the emperor, the saint becomes a sort of "popular hero" who opposes the dominant power. Described with sexual metaphors, "reeking" of virtues and "spreading" the Truth, he thrives in the company of tormented bodies, chained to city walls. The story exposes a Sadean fascination with torture, and seduction is used as a way to torture the saint. Edoardo Notari, who played Pellegrino, recalled that the film was shot with many nonprofessional actors, and that a great many naked women performed as temptresses.[14] A play of sexuality and sexual difference takes place in the seduction–torture scene. The ultimate torment is to be seduced by a woman, and a homoerotic subtext emerges, as the four men resist the women's sexual advances, persisting in their love of another man, Jesus Christ tortured on the Cross. Meanwhile, the temptress is herself seduced by the "gentle figure" of the saint. Ultimately this temptation reads as a battle between the sexes and sexual identities. The religious subject becomes a safe way to exhibit sexuality and its dynamics, one that, in fact, was able to circumvent censorial mechanisms. The film was shown during holy weeks, both in Italy and the United States.

The tale of Saint Pellegrino is a local version of the story of Saint Anthony, the saint who triumphed over his temptation by the devil. It is a microstory when compared to Anthony's legend, which has been interpreted in literature and art since the beginning of Christianity. A revival took place around the turn of the century, when Gustave Flaubert and Paul Cézanne, among others, offered versions of the temptation of the saint.[15] In Naples Domenico Morelli painted *The Temptation of Saint Anthony*, a picture of intense erotic charge that acquired great local fame.[16] Like Notari's Pellegrino, Anthony's iconography similarly speaks of seduction, exposing in some of its versions the temptations of sexuality and the war between the sexes.

Trionfo cristiano is an example of hypertextuality, in which a hagiographical tale is claimed as the hypotext of the film. However, hagiography also enters Notari's texts at a more subtle intertextual level without necessarily being acknowledged. The extant film *'A Santanotte* (The Holy Night, 1922), for example, can be read as a case of the superimposition of hagiography on popular cinema. As in *E' piccerella*, the credits describe *'A Santanotte* as a "popular passional drama by Elvira Notari" and credit the author with the direction and *mise-en-scène* as well. Characters are introduced in close-up: Nanninella, the main female character; her father, Giuseppone; Tore Spina, the man Nanninella loves; Maria, his mother; Carluccio, played by a "student of Scuola di Arte Dora Film," a man who wants to marry Nanninella; and Gennariello, the "good-hearted boy."[17] The *mise-en-scène* of the interiors, like those in *E' piccerella*, is crudely naturalistic, and the exteriors were shot on location in Naples.

The opening sequence is a striking commentary on battered women in the poor strata of society. Nanninella works hard in a café and, at home, attends to her drunken father. The omniscient narrator, sympathetic to the woman's hard life, describes Giuseppone as "a drunk who torments his daughter, who is a poor servant in a café."[18] As usual, he asks for money to buy liquor, but

Nanninella does not have enough, so he hits her and throws her down the stairs, hurting her. Tore and Carluccio watch the scene, and Tore helps her, begging her not to return to such a home. But Nanninella, "the Madonna of the Chains," behaves like a true female martyr and returns to her father to help him get into bed. The two young men are now both pursuing her. And Nanninella, "as beautiful as she is virtuous," goes on with her life in the café, where she is seen defending herself against the sexual harassment of its customers. To make the situation even worse, Carluccio approaches Giuseppone and promises to pay for his liquor if he lets him have his daughter. The father betrays her: "Nanninella shall be your wife. It is nice to have a son-in-law who ensures the liquor for my throat." Notari comments on the woman as a mere object of exchange: marriage seals the pact of her passing from her father's possession into her husband's. Nanninella attempts to protest, but her father's answer is more physical violence. At her plea – "Father, why would you kill my heart?" – he batters her again. Tore, whom she loves, confronts her father, but during their verbal fight, Giuseppone accidentally falls down a cliff and dies. The innocent Tore, accused by Carluccio of having murdered him, goes to jail.

Nanninella's Madonna-like character reinscribes in social terms a common story of violence against women – a story that the hagiographical tradition renders as violence against a virtuous virgin martyr. The comparison with the life and death of the saints increases as the story progresses. Nanninella is tormented by the doubt that Tore may actually have killed her father. But Gennariello, the angel, "brings her the light." He reveals to her that Carluccio sent the innocent Tore to jail in order to have her. The last part of the film enacts Nanninella's voluntary sacrifice. Pretending to be in love with him, she simulates to perfection the part of the enamored fiancée, hoping to get at his secret. Carluccio believes her and proposes marriage, while Nanninella continues the pretense.

It is now the night before the wedding. Tore's mother knows the truth and asks Gennariello, "Are you going to permit that saint to sacrifice herself to such a degree?" Cut to Nanninella's squalid home. She is alone, staring at her wedding dress. In a subjective shot, the empty white wedding veil is moved by the wind in the half-light. The dress is the last costume of her act. The woman looks at herself in a mirror and cries out, "Here comes my death sentence." The wedding dress, her prop, will kill her. Marriage is a tomb.

The wedding sequence is a strong moment in the drama. Throughout the last part of the film, Nanninella constantly stares into the void, as if living in a detached, contemplative state. During the wedding she is catatonic. Her performance – the theater of hysteria – almost turns into catalepsy. Carluccio practically carries her down the aisle of the church. The camera follows the wedding march as if it were a funeral march. The couple reaches the priest. Distanced from her body, Nanninella marries. During the entire wedding party, which follows, Nanninella sits motionless on a chair while everybody is singing and dancing. She stares out, abandoned, her wedding dress disheveled.

The wedding sequence is shot in long takes to render the woman's catatonia, but is intercut with an action sequence –Tore attempting to escape from jail to save Nanninella from marrying. He succeeds in fleeing from prison and goes first to his mother, who assures him that Nanninella is "a saint." But Tore does not get to perform the predictable last-minute rescue. On the edge of the paternal house, Nanninella, like Margaretella, will die, knifed by a man. For the story to exist in melodramatic form in the narrative of excess, the woman must be killed, or self-destruct. Like the saint in the hagiography and the woman in opera, both sources of melodrama, she must die. As French feminist theorist Catherine Clément comments, writing on operatic narratives:

> Dead women, dead so often. There are those who die disemboweled . . . there are those who die for having embodied too well the false identity of a marionette-woman or for having affirmed that they are not there where the men are looking for them. . . .
>
> In other times, in serious structuralist conclaves – in days now known as bygone – you would have been entitled to a clean blackboard, just to classify them all, these dead women, according to the instrument of her death or the guilty party. . . . The frightened, pathetic exercise of taxonomic intelligence, reassuring itself by filling its sensible little categories. But no matter how hard I laugh, there is always this constant: death by a man . . . the infinitively repetitive spectacle of a woman who dies, murdered.[19]

According to this well-rehearsed operatic plot, death for the woman is predetermined. Tore gets to the wedding, the scene of the crime, too late.

It is the wedding night. This, we must assume, is "the holy night" evoked in the title. Like those saints from the Catholic hagiography who figuratively become the bride of Christ, Nanninella, too, must give herself to her master. And she must give up her virginity. But she pursues her own plan, and seeking to know what the man is hiding, she teases Carluccio to tell her, as a test of love, the truth about her father's death. Caught in his pride, sure of his power, Carluccio confesses that Tore was innocent. Nanninella attempts to escape with her secret, but he stops her, and they fight. Carluccio takes a knife and stabs her. On the wedding night the woman's body is violated, as the man's knife penetrates her white virginal attire.

At this point, Notari's camera, which had used a long shot for the fight, cuts to a medium close-up. Approaching the spectator, Nanninella's hand tears at the wedding dress, showing us her wound. By way of displacement, we are reminded of the exposure of blood-stained wedding sheets. With blood on her white wedding dress, her head down, she descends the stairs, continuing to expose her wounded breast. The wound is deep, and it looks as if the breast has been severed. Then Tore arrives, a last embrace, and Nanninella dies – a virgin and a martyr.

Nanninella's wound and her death are topoi circulated in Italian hagiography. Both the narrative development and the visual rendition of Nanninella's life and death recall the story of the torture inflicted upon Saint Agatha, who is believed to be a woman who lived in Catania, Sicily, in the third century. She was pursued by Quintinianus, the prefect of Sicily, who had fallen in love with her and wanted to marry her. She refused because she was a Christian and he was not. Quintinianus tried to have her against her wishes, using his power to compel her to marry him. However, since she continued to spurn him, he inflicted a final torture upon her: he had Agatha's breasts severed.

Saint Agatha's story is depicted in Italian art,[20] and the iconography generally portrays her holding different objects such as a lamp or a book, seen as symbols of martyrdom and virginity. In many cases Agatha is represented with the instruments of torture or pointing to her severed breast, or holding her shorn breasts in her hand. Sometimes the actual excision of the breast is not hidden, and the scene of torture is visualized in progress. Agatha is then shown as her breasts are being cut off, while a core of observers watches the spectacle of her mutilated body.[21]

Portrayals of Saint Agatha are found in Neapolitan painting. The closest to Notari's representation is a painting by Francesco Guarino (1611–54), dated 1637. This is among the best-known images of the Neapolitan Baroque, wrongly attributed for a long time to Massimo Stanzione who also painted the saint.[22] Like the image constructed in Notari's medium close-up – that cut made to show the wound's cut – the Sant' Agata of Guarino is caught in a sensually dramatic pose, on the threshold between eroticism and death. Guarino placed the figure at the center of the composition, her head slightly turned, with her dress open showing some flesh. As in Notari's shot, Agatha's right arm and hand hold a piece of fabric, the piece that touches the wound. Here, too, the woman is alone, against an empty background. The mutilation takes place elsewhere, off-screen, as it were. In the absence of torturers, props of torture, or symbolic objects, we perceive only the (in)visible effect of the violent act. Although Guarino does not show the slash itself, his metonymical visual signs are all the more powerful: blood seeps out of the woman's body, and bright red drops seep through the cloth, spotting her white dress throughout. The chiaroscuro of the painting enhances the uncannily erotic portrayal.

In both Notari's and Guarino's representations, the cut clearly marks an erotic zone – the woman's breast. The wound and the suffering have sexual overtones. This is often the case in painterly iconography of religious subjects, as the martyrdom of both female and male saints is grafted onto the flesh, and, likewise, the agony of Christ dying on the cross is eroticized.[23] In the crossing

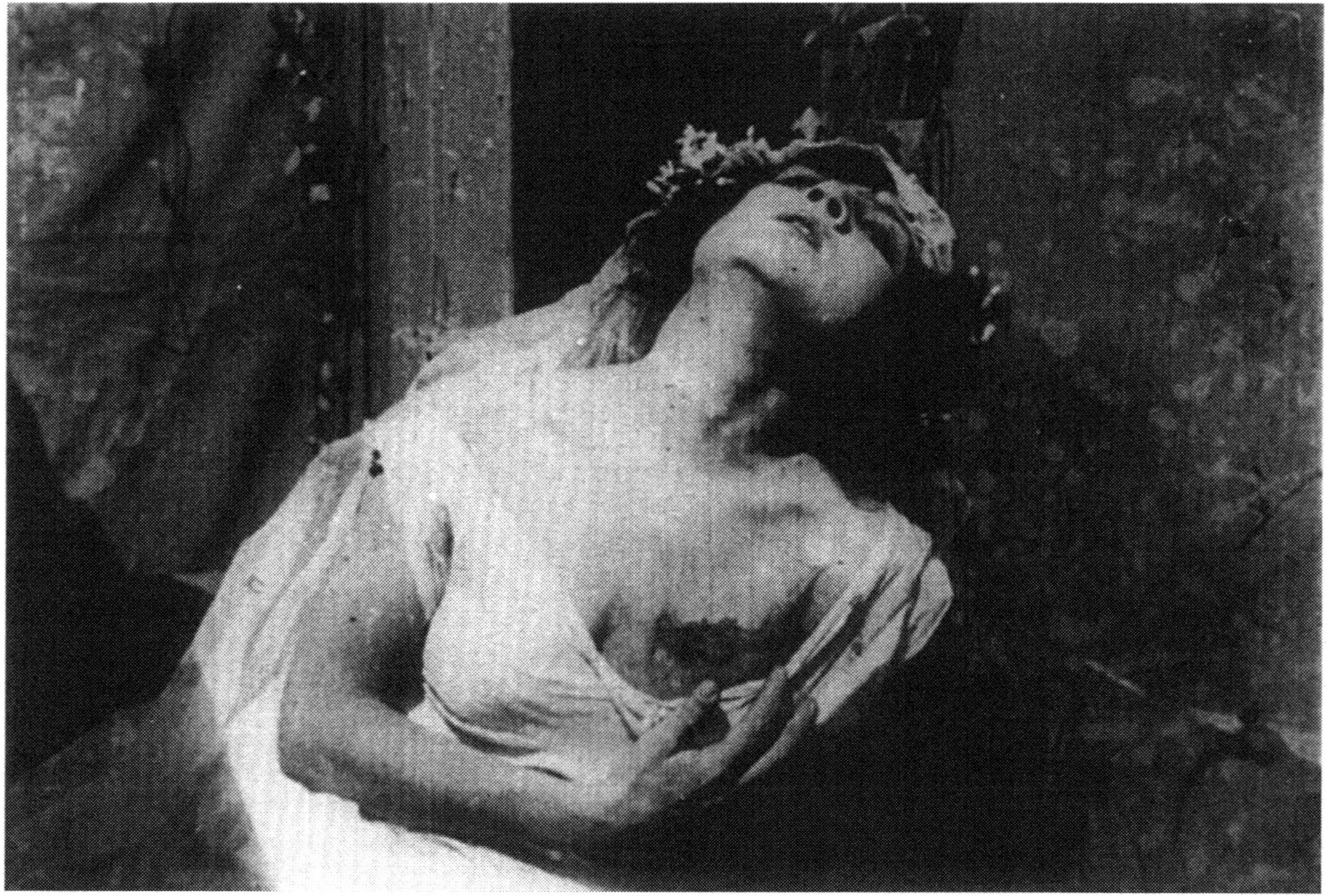

Figure 20.5 '*A Santanotte* (The Holy Night, 1922) by Elvira Notari

Figure 20.6 Francesco Guarino, *Saint Agatha*, 1637.

of death and sexuality, as art historian Leo Steinberg claims, the Image and the Word converge, and the word "passion" seems most appropriate when applied to Christ's sufferings. Jesus' wound was in the side, but the painted blood often drips down into the groin. Determined by a force other than gravitational, the "blood hyphen" is created. "Now last and first wound are connected, as though the graph of Christ's lifelong passion were traced on the chart of his body."[24] This chart is not always genitally defined, for Christ's body, often the site of a more diffused eroticism, is at times treated as female.[25]

Notari's cut on the woman's breast maps a space of female sexuality, as does Guarino's extraordinary representation. Several clusters of fantasy, anxiety, and fear are delineated by this wound. And we can still feel remnants of the horror that drives the traditional iconography of Saint Agatha. The visual construction of that scene of mutilation is sustained to a great extent, though not exclusively, by a threat of castration. There the "blood hyphen" connects two wounds, the woman's primary and her last, now adjoined in death. The cut evokes that primary lack marking a woman's body – a body seen as castrated, wounded. And the mutilated piece is, in a way, fetishized, displaced into a chain of diverse object signifiers (the breast itself, a book, a lamp, etc.) standing for what is now a missing part.

However, if some of the horror of the Guarino–Notari image evokes the figure of castration, the scene here is different from Agatha's usual depiction. In a peculiar visualization of the mutilation, and in the absence of displayed fetish objects, we have a rarefied yet denser representation of female sexuality. Metonymically the wound draws a complex sexual topography. A clear cut, it speaks of the dark areas of sexual difference. A nuanced fantasmatic scenario is visualized, where sexuality and violence, eroticism and death, sadism and masochism, mutilation and self-mutilation intersect.

If one looks at the two images as complementary, as a double-sided representation, it becomes apparent that the "blood hyphen" is not genital. This particular "hyphen" is primarily oral, as it connects, on the topography of the breast, milk and blood. The white (of the women's dresses) and the red (of the blood) are the colors of the female fluids. Color conveys a relation of signifiers. In Notari's shot, red blood oozes out of the woman's breast in place of white milk. Guarino's image is all the more explicit and suggestive. Constructed as it is on the semiotics of "seeping," it creates a link between blood and milk – two fluids that seep. Blood seeps through the cloth, just as if it were milk seeping from the breast in nursing a baby. In this way Guarino literalizes in the image the oral hyphen, as he shows, charted on the breast, both the act and the effect of seeping blood-milk.

The wound on the breast, whether suggested by Guarino or openly exhibited by Notari, is being touched by the woman's hand. Holding a piece of cloth, the hand points to the breast. This, a cathectic object, the site of a prime oral desire, has been attacked and injured. We are reminded of that "infamous" scenario delineated by Melanie Klein.[26] In the view of the female psychoanalyst, the child's relation to the mother's breast departs from an idyllic scene of pleasure: the desire for the breast is coupled with sadistic fantasies. As it fails to give, and to be good, the breast becomes the internalized prototype of all bad objects. In a complex dynamics of projection and introjection, it is also perceived as dangerous, and the child projects its own aggression against it, often biting it. Destructive impulses are developed.

In fantasies of an oral-sadistic nature, the breast is attacked, devoured, cut off, or cut to pieces. The oral desire manifests itself as a cannibalistic impulse directed against the mother's bosom and, consequently, "also against the inside of her body: scooping it out, devouring its contents, destroying it by every means which sadism can suggest."[27] This fantasy of the mother's breast as a primary site of pleasure and aggression defines it and then her whole body as a *corps morcélé*. Is "woman" part "Agatha" – an erotic representation, with breast severed, bitten off, body mutilated, cannibalized? The Notari–Guarino image does at least suggest this when read together with Melanie Klein. The fantasy described by Klein takes a material form in the twin representation, literalized as it is in a female corporeality. As Guarino shows us the blood dripping from the injured mother's breast, Notari exhibits the wound provoked by the oral-sadistic attack.

The oral-sadistic desire for the breast is an important feature of psychological development, for it stimulates, as Annette Michelson puts it, the formation "of a horror feature, the longest-running one known to us."[28] The cannibalistic desire does not terminate, but rather undergoes transformations. During the second oral stage this impulse is also directed onto the penis, the other object of oral desire. In Klein's view, a hyphen breast–penis is established on the basis of orality and sadism. This implies a reading of castration that is particularly enlightening for Agatha's iconography. In this view, the breast is found at the root of the castration complex, and a wish to bite off the penis(–breast) is given form. This hyphen is evoked in the traditional iconography of Saint Agatha, for the horrific threat of castration is indeed grafted onto the breast. As a development of this imaging, the Guarino–Notari imago bears some traces of the threat, but the scene here is more distinctly oral. It is the oral sadism of the early phase, directed to the breast, that reemerges here and, fully embodied, becomes the protagonist of the horror feature.

An infantile scenario is reenacted by these pictures, or, we may say, it is the return to the infantile scene of part object that is "taking place." In terms of an infantile fantasy, the *corps morcélé* is not an enclosed temporality or an exclusive geography. Such fantasy "travels" and, in turn, is re-traversed. It transfers to the hysterical scene, for the hysterical body may also take the form of a *corps morcélé*, in the fantasmatic regression to the infantile scenario of mangled body parts. Fragmentation and mutilation inhabit this anatomical construction – a borderland topography. In this way, Agatha enters a well-traveled bodily archive, the anatomic–analytic, where she adjoins the mangled female corpses of the anatomy lesson.

Thus the story of the saint whose breasts were shorn continues to circulate in fantasmatic as well as literal forms. As a sexual topography, Saint Agatha's body quite literally inhabits the streets of southern Italian towns. In Naples, *vicoli* are filled with images of saints, Agatha included. Little altars are set up at street corners: statuettes, paintings, and prints depict a saint's life and death in a manner that denotes a pagan subtext. The altars are often quite elaborate, including all kinds of fetish objects. Ex-voto are exposed: images of body parts (feet, hands, hearts, breasts, etc.), in silver or cast metal, testify to the saint's specific area of corporeal corcern, which includes healing power over that body part. Saints are often portrayed in a way that speaks quite clearly of the link between sexual and religious imagery, as seen in high art. In its low-culture version, this imaging is sometimes obscene and even blasphemous. The sexually defined language of the representation is recurrently gender specific.

Saint Agatha's gender fable returns in various disguises at various street corners. A telling gender tale is revealed to the curious *flâneuse* who "street-walks" in Catania in search of food or some other kind of pleasure. Window-shopping in the saint's hometown, where she is patron saint, one discovers that the city's excellent *pâtisserie* features pastries in the form of Agatha's severed breasts. The theory of part object inscribed in Agatha's story has been transformed into a consumable commodity. These ricotta cheese cream puffs are not just elusively reminiscent of the female form. The definite shape, the consistency, and the white icing with the red cherry nipple leave no possible doubts. These are unmistakably breasts – Saint Agatha's breasts. So shamelessly breasts are exposed in the pastry shops. What kind of erotic gratification do they offer? Is this only a fetishistic ritual? These object parts are not there simply to be looked at or touched. These breasts were made to be eaten. Hence, one (literally) gets a taste for a "scene" of this nature: the cannibalistic impulse directed to the breast is reincarnated via pastry, and it returns, by way of food, to the mouth – back to the oral. At any time one may go and eat the mother's breast, whenever one feels that oral urge. And customarily on Sunday the Catanese family eats the motherly breast of the saint. They perform their oral-sadistic ritual, perhaps buying this convenient commodity on the way back from church, where, during the Mass, every good Catholic devours Christ's body in the form of the host. Or they may eat the pastry-breast during the usual stroll along the *corso* (avenue), on the way to a movie, where one, indeed, devours images, absorbing them while being absorbed by them, eating up fantasmatic imaging.

In this way, Agatha's story keeps on being reembodied, surfacing once again in the domain of microhistory. Agatha is not a saint who has made History, but one who has made stories. She is not a "king" of saints. Unlisted in the twelve volumes of the *Encyclopedia of Catholic Saints*,[29] she is marginalized in *The Penguin Dictionary of Saints*, where her story is ruled out as "a worthless legend":

> There was certainly' a virgin martyr named Agatha at Catania in Sicily. . . . Her worthless legend, of which there are many versions, tells us that . . . Agatha was tortured in various ways, and . . . eventually died of her sufferings. . . . St. Agatha was said to have been subjected . . . [to] the cutting off of her breasts.[30]

These kinds of "worthless legends," embedded in the popular, happen to tell women's stories. Clara Erskine Clement, the author of *Saints in Art*, written in 1899, acknowledged a display of female physicality and its vicissitudes in her sympathetic account of Agatha's iconography:

> The story of *St. Agatha* is so painful, and her martyrdom included such horrors, that it is not necessary to recount them. The picture by Sebastian del Piombo in the Pitti, and others which represent the tearing of the breasts, are too revolting for description; in truth, I could never so study them as to be able to write of them.[31]

As Clement's reaction exemplifies, the representation of Saint Agatha and Notari's version are precisely the kind of women's stories that generate forms of resistance. A mixture of fascination and horror is provoked by such excessive and ephemeral tales, as these figurations of female sexuality and gender difference expose a quotidian and familiar image in uncanny forms. The breast is wounded. The female body turns into a ruined map. Laceration; the male, a child, the woman herself may cause injury. The act of severing leaves a gap, a gaping wound. Woman's sex becomes lacking. Her lack, a lacuna – the horror vacui. Powers of horror. Should we be horrified? Moved by a "passion" for this lacuna, fond of this *vacuum*, I insist on the scene of the analytic void.

Let me offer one last example: a citation drawn from *E' piccerella*. In this fragment, as in *'A Santanotte*, a woman is being knifed. This scene is represented in two out of three of Notari's extant films; in both cases, the cut is aimed at her breast, and the act has the same narrative motivation: a woman is punished by the hand of a man. In the case of *E' piccerella*, a female sexual topography is mapped by the syntagmatics of editing and its system of suture. It is now the final sequence. Tore, the man who has knifed the unruly Margaretella, is in jail. In a subjective shot from his point of view, two medium close-ups of the woman are shown. The first image materializes in front of Tore's eyes, while, after a few seconds, a second image begins to superimpose itself. Together with Tore, we first see Margaretella in the form of a beautiful seductress. Then we see, again as the man's hallucinatory fantasy, Margaretella, whose dress is now torn, and who exhibits the open wound on her breast. And it is while the man, moved by desire, leans toward the beautiful woman that she turns into the wounded body. Tore is repelled. The dynamics of editing reproduce a psychoanalytic dynamic: the female image is double-sided. The woman's body is both desirable and repulsive. Man is attracted by her beauty and repelled by what she represents. As her desirable body bears the mark of a lack, read as the missing phallus, woman becomes a signifier for the threat of castration; the gaping wound, her vagina, provokes fear and anxiety. As a mother, woman also embodies a split. Her breast is perceived by the child as a double signifier: it is good and giving, but also bad in its rejecting aspects. As these splittings take filmic form, projections of male fantasy in female imaging are revealed: Margaretella's beautiful body is unveiled, and her anatomy revealed – the body marked by a wound.

The psychoanalytic underpinnings of the cut are grounded, in Notari's case, in the socio-cultural terrain of "low" forms of discourse. Knifing belongs to the realm of the "material bodily lower stratum" and speaks for the strong material presence of the body in popular culture. As a popular form of spectacle, fights with knives can also be interpreted as a reenactment of delivering a child,[32] and, in such a reading, the hyphen blood-milk is again rewritten on the female body. Knives in general, and fights involving knifing, are elements widely circulated in Neapolitan popular culture. The act of knifing, often directed toward a woman, was so popular that it has acquired its own name in the Neapolitan milieu; *sfregio* has produced an extensive "slasher" fiction, ranging from actual street-life cases to popular songs, novels, and films.[33] It is such a local, everyday figuration of the body's cut that Notari transposes in her narratives when remaking Saint Agatha or exposing the sexuality of Margaretella in the form of *sfregio*.

A shared popular component, then, fosters the intertextuality of popular cinema and hagiography, a *transito* grounded on the socially repressed terrain of sexuality. Like popular cinema, hagiography is itself an expression of popular fiction. Grafted onto folklore, hagiography represents the outer edge of historiography. The hagiographer, as de Certeau remarks, is traditionally condemned, but while orthodoxy represses fiction, fiction reenters through the back door, penetrating the official culture through effraction: this heretical literature, intended for the people, is nonetheless tolerated for its usefulness among the people. It is therefore not surprising to find that hagiography continues to circulate in the popular realm.[34] The southern Italian story of Saint Agatha – an expression of a local popular culture – migrates in another popular form of expression: cinema. Agatha's dark, excessive woman's melodrama continues, via street life and pastry, its hagiographical journey of dissemination. And in its transit through popular culture, it becomes the hypotext of a popular cinema manufactured by a southern Italian woman.

Ultimately, we may conclude, the strongest point of contact between popular narrative and the hagiographical discourse is a common concern for a spatio-corporeal language. Hagiography, as the sexualized vision of saints reveals, insists on a bodily geography and offers an encyclopedia of the body, constructing its somatic topography. It highlights "a topography of holes and valleys: orifices (the mouth, the eye) and internal cavities (the belly, and ultimately the heart), one favored in turn over the other, and written into the dialectics of inside–outside or inglobing–englobed, in order to embody a rich spectacle of entries and exits."[35] As we have aimed to show, topography and/as the language of the body are important features of Notari's popular fiction, both rooted in *il ventre di Napoli*. As a production of Naples' urban belly, this work expresses the dialectics of entries and exits at the threshold of private/public. Traversing the "exteriors" of the city, "interiors" such as the orifice-mouth are drafted, as for the uncanny laughter at "the table of the poor." Such concern for the interior landscape, expressed as an anatomy of the (in)visible, extends to the female subject. Representing her anatomy, in the form of a dissection, or exposing the wounded body, a cut of *sfregio*, Elvira Notari speaks a language that borders on the territory of the popular – that of *corpo*reality.

Notes

1 Roland Barthes, "The Third Meaning: Research Notes on Some Eisenstein Stills," in *Image, Music, Text*, ed. Stephen Heath, New York: Hill and Wang, 1977.
2 Ibid., p. 67.
3 Ibid., p. 68.
4 This and the following two citations are taken from the film's intertitles.
5 Georges Bataille, *Erotism: Death and Sensuality*, San Francisco: City Lights Books, 1986, p. 239.
6 Ibid., p. 240.
7 Pierre Klossowski, *The Baphomet*, Hygiene: Eridanos Press, 1988, p. 78.
8 Mario Perniola, "Between Clothing and Nudity," in *Zone 4, Fragments for a History of the Human Body: Part II*, ed. Michel Feher, with Ramona Naddaff and Nadia Tazi, New York: Zone Books, 1989, pp. 255–6.
9 Bataille, *Erotism*, pp. 239–40.
The relation between erotism and mysticism had been emphasized earlier by Lou Andreas Salomé in her 1910 essay "Die Erotik." Salomé claimed that "erotism is directly linked to a religious expression, and vice versa. The relation is established on the basis of an empowerment of life where interior and exterior come to consciousness with stimulating effects. During such a process, this unifying force, this increased pleasure of living and willing, directs itself toward a voluptuousness, which is more strictly corporeal." (Lou Andreas Salomé, *L'erotismo: l'umano come donna*, Milan: La Tartaruga, 1985, p. 65 [Italian trans.]).
10 On the "heroes of popular literature," see Antonio Gramsci, *Selections from Cultural Writings*, eds and with an intro. by David Forgacs and Geoffrey Nowell-Smith, London: Lawrence and Wishart, 1985, p. 350.
11 See Michel de Certeau, *The Writing of History*, New York: Columbia University Press, 1988, esp. pp. 269–83.
12 Censorship visa no. 24810, obtained 30 April 1930.
13 From a publicity flier for the film.
14 According to Vittorio Martinelli, who reports the testimony of Edoardo Notari. Martinelli, *Il cinema muto italiano*, vol. 4, Rome: Bianco & Nero, 1981, p. 403.
15 On the temptation of Saint Anthony, see Bram Dijkstra, *Idols of Perversity: Fantasies of Feminine Evil in Fin-de-siècle Culture*, New York: Oxford University Press, 1986, pp. 253–7.

16 See Maria Antonella Fusco, "L'arte (1860–1970)," in *Storia delle città italiane: Napoli*, ed. Giuseppe Galasso, Bari: Laterza, 1987, pp. 451–2.

17 All citations in this paragraph are from the film's credits.

18 From the film's intertitles. All following citations in quotation marks, up to the next footnote, are also from the film's intertitles.

19 Catherine Clément, *Opera, or the Undoing of Women*, Minneapolis: University of Minnesota Press, 1988, p. 47.

20 For an account of the iconography of Saint Agatha, see George Kaftal, *Iconography of the Saints in the Central and South Italian School of Painting*, Florence: Sansoni, 1965; Kaftal, *Iconography of the Saints in the Painting of Northern Italy*, Florence: Sansoni, 1982; Kaftal, *Iconography of the Saints in Painting of North East Italy*, Florence: Sansoni, 1978; Kaftal, *Iconography of the Saints in the Painting of North West Italy*, Florence: Le Lettere, 1985; and Kaftal, *Iconography of the Saints in Tuscan Painting*, Florence: Sansoni, 1952.

21 For example, *Saint Agatha in Arfoli* (Valdarno), a fresco of the fourteenth-century Tuscan School, represents the life and death of the saint. Its sequential nature narrativizes the painting. In this rendition of Agatha's torture, the mutilation is clearly represented: we see a man in the act of performing the operation. As in Nanninella's case, Agatha's dress is torn, and the blood of the cut marks an erotic zone. A painting of the Swiss School, *The Martyrdom of Saint Agatha* (1473) also shows the breast being shorn by the hand of a male torturer. Agatha, here impassive to torture, is semi-naked. A male public watches the scene.

22 On Guarino's *Saint Agatha*, see Raffaello Causa, *Pittura napoletana dal XV al XIX secolo*, Bergamo: Istituto italiano d'arti grafiche, 1957, p. 37; Causa, "La pittura del Seicento a Napoli dal naturalismo al barocco," in *Storia di Napoli*, part 5, vol. 2, 1972, p. 931; and Riccardo Lattuada, "Problemi di filologia e di committenza nelle opere di Francesco Guarino alla Collegiata di S. Michele Arcangelo a Solofra, *Annali della facoltà di lettere e filosofia dell'università di Napoli*, vol. 13, n. 11, 1980–1, esp. pp. 116–18. Raffaello Causa attributed the painting to Guarino in 1957.

23 See Leo Steinberg, "The Sexuality of Christ in Renaissance Art and in Modern Oblivion," *October*, no. 25, Summer 1983, pp. 1–222.

    On the subject of the representation of the human body in religion, see also Caroline Walker Bynum, *Fragmentation and Redemption: Essays on Gender and the Human Body in Medieval Religion*, New York: Zone Books, 1991.

    For a discussion of healing, suffering, and religious representation, see Andrée Hayum, *The Isenheim Altarpiece: God, Medicine and the Painter's Vision*, Princeton: Princeton University Press, 1989.

24 Steinberg, "The Sexuality of Christ," p. 160.

25 See Bynum, *Fragmentation and Redemption*. Bynum offers a number of readings in the area of gender and religion, including a reply to Steinberg's emphasis on Christ's male genitality. Her treatment of the female body and religious practice shows that Christ's body was also treated as female.

26 See Melanie Klein, *Love, Guilt and Reparation & Other Works 1921–1945*, New York: Delta, 1975.

27 Ibid., p. 262.

28 Annette Michelson, "'Where Is Your Rupture: Mass Culture and the

Gesamtkunstwerk,'" October, no. 56, Spring 1991, p. 51, Michelson mobilizes Klein's scenario in her mapping of art object and part object.

29 *Encyclopedia of Catholic Saints*, 12 vols., New York: Chilton Books, 1966.

30 Donald Attwater, *The Penguin Dictionary of Saints*, New York: Penguin, 1965, p. 32.

31 Clara Erskine Clement, *Saints in Art*, Boston: Page and Co., 1899, p. 281.

32 See Mikhail Bakhtin, *Rabelais and His World*, Cambridge, Mass.: MIT Press, 1968.

33 The knifing of a woman who is unfaithful to her lover or husband recurs in the popular culture of Naples. In a novel adapted by Notari, Francesco Mastriani feels the need to explain the meaning of *sfregio* to "his kind female readers" who might be unfamiliar with Neapolitan popular customs. Francesco Mastriani, *Ciccio il pizzaioulo di Borgo Loreto*, Naples: Lucet, 1950, p. 20.

   A *sfregio*, a cut on the face aimed at disfiguring the woman, appears in Gustavo Serena's *Assunta Spina* (1915). In this film of the Neapolitan milieu, based on a play by the Neapolitan author Salvatore Di Giacomo, the main protagonist Assunta, played by Francesca Bertini, is knifed by her lover and disfigured by the cut. The difference between this cut on the face and Notari's *sfregio* is that her cut is aimed directly at an erotic zone – the breast – and thus has a clear bearing on female sexuality and sexual difference.

34 For an analysis of the relation between hagiography and popular culture, see also *Les saints et les stars: le texte hagiographique dans la culture populaire*, ed. Jean Claude Schmitt, Paris: Beauchesne, 1985.

35 De Certeau, *The Writing of History*, p. 279.

Conclusion

21

EVERYDAYPLACESBODIES

Heidi J. Nast and Steve Pile

Perhaps you will allow us to start this conclusion with a narrative about everyday events, unexceptional except in their specificities. After this story, we will draw out some ways in which places are experienced through the body, and how the body is experienced through places. But, more than this, we will develop our concluding remarks around the social production of places-bodies.[1] It is not our intention to "conclude" this collection by closing down the indeterminacies of the bodies-places – that is, by ending the discussion – but by opening out the possibilities for thinking places through the body. In the text, there are always the ghosts of other stories . . .

* * * * *

Heidi is waiting for, no . . . hanging on, a phone call. But, she cannot call the person who has not called her. Steve is supposed to be calling someone. But, he cannot bring himself to call the person who wants him to call. The telephone has become the focus of a vortex of different feelings for both of them: guilt, anger, desire, loss and, most importantly, anxiety. So, they decide to do the only thing they can in the face of such whirling emotions – get away from the telephone altogether. Somewhat despairingly, Steve kicks the answering machine into action and they leave the flat to get something to eat from a nearby restaurant. As they turn into the street, Heidi asks whether the answering machine is on . . . again.

The sky is blue, as it rarely is in London; London always feels like another country when the sun shines, but the sun is not shining. The light is just thinking about fading, but has not yet made up its mind. The evening vacillates with the light, but decides it is not yet its place to cool. Heidi and Steve walk down the street and try to figure out how to write the conclusion to the book they're editing. They are keen to tie the threads of the book together, yet also to talk about the warp and weft of places and bodies, and also to speculate about what is commonly referred to as the "politics of location." They are not sure how and have no clear ideas. At one point, in frustration, they decide not to write a conclusion at all – and then they laugh with the freedom that this "decision" brings.

After getting to the cross-roads, they decide to turn left and eventually they end up at the Poori Hut vegetarian restaurant. Steve is familiar with this place and knows the score, but Heidi does not recognize many items on the menu. So, the woman who takes their order also takes Heidi's questions about the food. When they've finished eating, Heidi and Steve are offered a bowl of mixed spices, which act as a breath freshener, so Steve immediately pinches some and chews it greedily. Heidi then mischievously informs Steve that a recent study had found traces of urine on the complementary mints given out in some restaurants. Heidi laughs, while Steve does his best to ignore the comment, saying that he had thought that he had heard Phil Crang say something like that at some conference or other. Now, they both began to wonder what exactly was in the breath freshener (not including urine!) – they start guessing, but they don't know for sure; so a woman at the next table, who was from Madras (in Southern India), kindly helps them out.

As they leave the restaurant, they remember how sad they felt – and Heidi looks at the sky and says that it is smoggy. Steve denies this, but not convincingly enough. As they start to walk back to the flat, neither feel they can face the "no message" machine, and they somewhat reluctantly decide to watch the Hollywood blockbuster film, *Twister*, in the hope/expectation that it would take their minds off things for a while. If films can allow you not to think, then this wasn't one of them. The all too ludicrous plot kept tripping up their intention to suspend their disbelief. They enjoyed the movie, they think, but they still cannot escape the damned vortices of their inner worlds – so they start laughing at the movie: they wonder whether the film's scriptwriter, Michael Crichton, could do for cultural geography what he had just done for climatology, and theme parks and palaeontology before that. Westernized World? Juxtaposition Park? Twisted? And then they chat about the things they might write in the conclusion . . .

* * * * *

Academic writing often deals in extremes – the *most* significant sites, the *most* exceptional bodies, the *most* important social relation of power – yet it may well be that significant and exceptional things also happen in (y)our own backyard. On the other hand, rendering the familiar unfamiliar, through writing the body down, quickly seems painfully anodyne or perversely analytical or pretentiously animated (as Pratt argues, Chapter 16). It is easy to see, it seems, when the Emperor has no clothes (as the children's story goes). His naked body is there for all to see and it can be recognized as such: naked, stripped of power, laughable. So, one of the key problems in writing about the intricate and delicate interlacing of bodies, spatialities and places is that we all have bodies . . . and that they are so familiar . . . and that we know what they do and don't do . . . and that we inhabit them with comfort and difficulty as we do, and as others let us . . . and it is easy (though appearances can be deceptive) to spot a fraud . . . and that the exceptional is exceptional.

And our story is certainly not exceptional – not in terms of our own lives, nor in terms of other people's. But, then again, it is exceptional; or, at least, specific. Let us tease out a few geographies of the body. Actually, we are intending to do something more focused than simply list geographies of the body – after all, the list could well be endless. Specifically, like the contributors to this book, we are interested in the ways in which bodies are traced through and trace out networks of power. Corporeal geographies lie at the heart of geographies of lived (deathly?) power arrangements.

Our story began with the telephone – let us start our discussion (t)here. The history of the telephone has been bound up with moral panics and with utopic visions (see Stein 1996). Yet, technology is slow to develop and slow to embed in people's everyday lives (as Thrift shows, 1996). The telephone is an object with many components – including hands and ears. Seemingly, the telephone annihilates space, by allowing (almost) instantaneous connections between people, however far apart they are – and indeed, it is this sense of connection that much telephone advertising imagery draws on.[2] In our case, the telephone has created closeness only to invert it into distance – a sign of distance away from someone, a distance that cannot be bridged by the telephone (why else would there be answering machines, but to shift between distances: in and out? Be in when you're out, be out when you're in . . .). In this case, the telephone does not simply terminate distance, it is a technology of what we will call "proxemics." We use the term proxemics in the context of the telephone, partly because they – like other technologies of space such as cyberspace and trains (see Chapters 4 and 15) – not only play on distance and closeness, but also lie at the intersection of the matter-as-event and knowledge-as-practice,[3] and also "interface" the machine and the body (hands, ears and, soon, eyes too).[4] The telephone, like other "connections" to the world (whether through hands or computers, see Chapter 4), subsist in networks of power: through production, exchange, meaning, use, and so on. These networks can be fixed and relatively stable; they can be fluid and chaotic; and, they can be both – as any user of telephone exchanges knows only too well.

Proxemics teases out the simultaneously fixed and fluid nature of spatial arrangements by articulating the sense that networks shift, alter and stabilize around effects of power, meaning, subjectivity and objectivity. These effects are both bodily and spatial, but are rarely explicit or open or conscious. Instead, they are felt; thought through the body rather than the mind. Thus, proxemics describes an "unconscious" relationship between the body and spatialities. In this sense, bodily feelings constitute the telephone as much as the telephone is the context within which those feelings become possible. Proxemics simultaneously articulates the inside out and the outside in and the desire for, and fear of, closeness/distance – where "inside" and "outside" the body, the place, become less significant than "the near," "closeness," "distance" and "the far away." Proxemics simultaneously speaks of the double/d articulations of near/far, inside/out, and of desire/fear, duty/ guilt, and – as we describe below – of other spatial relationships. Like Moebius's famous strip, proxemics moves through a space where location is indeterminate, relational and multi-dimensional (see Grosz 1994). So, it is through this understanding of the term "proxemics" that we will map out the mutual constitution of places-bodies.

At this juncture, we would like to move our discussion of proxemics into another place through the body: the house. This will add a sense of *scale* to our discussion of proxemics, but "scale" in this case is not to be understood in the sense of "nested hierarchies" which become larger and larger (which, it should be noted, is commonly linked to greater and greater power, in a too quick association between territory and power: for a critique, see Massey 1993). Instead, we understand scale to involve specific relationships between bodies, places and power – as Smith's analysis of homelessness demonstrates all too clearly (1993). Just as telephone connections are apparently global, but actually involve specific connections between particular places, so too the seemingly small spatialities of the house need to be understood through other proxemics of space: involving not only different scale relationships – such as the body and/or the city and/or the globe – but also the power relationships that constitute and make those scales. Just as telephones lines scratch the flows of information into the face of the earth, so too the (Western, at least) house cuts up space into parcels (of property), which are differentiated both inside and outside by walls, fences, and so on. And these seemingly innocent spatial arrangements – of walls, of fences, of telephone lines – betray the territorializations of power relations (of capital, ownership, sexuality, nation, race, ablebodiedness, and so on) – at the scales of the body, and/or the city, and/or the global (see Chapters 2, 3 and 8).

As Apter (Chapter 7), Bermann (Chapter 10) and Dorn (Chapter 11) show, the small architectures of space are written through the body – through the confinement of women's bodies and the excesses of those bodies – such that the house becomes a place which simultaneously expresses and disciplines the individual, locking them into spatial practices which define and reproduce the body (see Chapters 6 and 16). The house becomes a site of intense significance (see also Nast and Wilson 1994): capturing and releasing desires, fears and fantasms; defining and making a permeable and shifting inside– outside; linked and crossed by multiple, variant connections to the world; a place where bodies stay, move, reproduce, stop; where memories are housed (see Chapter 5). Bodies and places become real through their physique, through their emotional lives, through their actions, through their thoughts. But we should pause for breath, at this point. For the time being, we can note that proxemics describes how bodies and places are corpo*realized* through the embodied spatial practices encountered up to now: the telephone, the house.

So far, this discussion of proxemics has – implicitly – dealt with specific aspects of spatial relationships: first, the near/far connections and exchanges of the telephone; and, second, the boundedness and the differentiation of space/place as represented by the house. It is now possible to move on to another aspect of spatial arrangements of power: networks.[5] We believe that, from the perspective of proxemics, networks cannot be understood to be fixed, stable, flat and passive linkages between people, places, and power relations. Nor are networks free-floating, fluid, without form, and subordinate to just anybody's will and whims. Instead, networks are multiple, dynamic and productive of the perpetually changing continuities that make up the proxemics of everyday life. In a different register, networks are not simply lattices of power, nor merely fixed coordinates on a social map, but also about the shifting paths, movements and sites of interaction that chart out – and assemble – bodily geographies. Let us return to our opening story.

The house is situated near a cross-roads, from where would it take only fifteen minutes to walk into an exclusive white middle class area or, in another direction, into a Greek and Turkish Cypriot community living in privately owned terraced housing or into a run-down housing estate where the tenants are mainly black people, or into a predominantly white working class neighborhood. It should be clear by now that housing and residence in London is graded through thoroughly racialized and classed grids (see, for example, Smith 1989): these urban outcomes are produced by dense, dynamic and interwoven networks of social relations. Our point here is not just that these networks produce specific racial and class patterns that mark, or scar, urban space, but that they also frame the feelings, encounters and actions that comprise everyday life, in place, and through the body.

Entering a restaurant can initially be slightly disconcerting (for example, does the waiter show you to a place or do you choose your own?), but only slightly, since the ways in which you are meant to behave – learnt through the years – quickly become apparent (whether you've entered a MacDonalds or some posh up-town restaurant). It is relatively easy to find a place and to get bodily needs – thirst, hunger, etc. – attended to. In our everyday story, we were getting a kind of fast food vegetarian Indian meal. If our path to the restaurant was somewhat arbitrary and indeterminate, the route the restaurant took to us was less so. It will not be surprising if we argue that this situation was made possible by the networks of social relations which were set up through British Imperialism. The extension of the British Empire throughout the world during the nineteenth century created and changed lines of connection between places – including India and Britain. Through these altered networks, many people (and not just slaves and slave masters) migrated or were forced to move for one reason and another (see King 1995). And these networks persist and channel flows of capital, people, information and value(s), despite the Empire's demise in the latter half of the twentieth century. The story of Empire underlies the story of our meal, undeniably. But the story of Empire cannot simply be read as a one way street, which does not allow for encounters, exchanges and meetings of all kinds (see for example

Pratt 1990 and Low 1996). As Nast argues, we are always everywhere negotiating different worlds and worlds of difference (see 1994).

So, our meal is enabled by migrancy in a context of specific networks of power, but where that power does not necessarily play out as fixed dichotomy between the powerful and the powerless, nor as a fixed gradient differentiating between higher and lower, but as a set of spatial practices through which people "learn" and "take up" their place in the world. These places can be more and less circumscribed, and the boundaries marking off specific people and specific behaviors can be more and less permeable. So, in understanding the seating (of power) arrangements in a restaurant, it is necessary to think beyond a sense of social position in which people are situated within differential axes of powerfulness or powerlessness, in which there are clearly marked and uncrossable divisions between people, and where everything that is done is necessarily at someone else's expense – even while barbaric things can also happen (as Bermann poignantly shows, Chapter 10).

Through specific linkages, an Indian restaurant is set up in North London, and a North American woman and an English man eat there. One (India) or more places (including America, at least) are translated into another, England. This "translation" (or, in Martin and Kryst's phrase, "place contagion," see Chapter 12) is not always easy, not always safe and not always rewarding. But the idea of proxemics asks us to consider the other sides of these relationships: our meal was not simply born of domination, oppression, tyranny, exploitation, and so on. We enjoyed our food, and our interactions with each other and the others we met. We learnt some things (however true or false) as we talked to strangers with different experiences, knowledge. Such things may not be so easily confined to, reduced to, understood as nor explained by the latitudes and longitudes of an Imperial legacy of spatialized power relationships. There are always possibilities for enjoyment and pleasure that lie in other bodies, other places (as de Lauretis argues in another context, Chapter 13).

Movements through networks can involve pleasure, nevertheless they can also be terrifying. To take a journey from one place to another sometimes involves knowing where you are going, and what you will find, but sometimes networks can enforce paths which have no clear destination, which cannot be reversed and cannot be changed, and sometimes networks can jump you into unexpected situations, into unfamiliar lines of power and to places that have no map (as Blum evocatively suggests, Chapter 15). Crossing boundaries can be liberating and confining and both at the same time. Movement proxemics are as ambiguous and dynamic as any other spatiality.

While networks can provide people with a clear sense of where they are in the world, and where they are supposed to be, those locations are "hairy" – hairy in the sense that any location will network out to other locations in specific ways; hairy also because these become entangled with other networks. These entanglements show the individual that, while they may feel at the

410

center of their world, they are not the center, cause, and master of the world of social relations – such that perhaps they are not welcome in that place (see Chapter 9); that they may feel "in place" or "out of place" (see Cresswell 1996); but that might allow them to negotiate these relationships, even to create new places and bodies (see Chapter 14). This hardly exhausts an analysis of networks and movement, but we would nevertheless like to move our discussion of proxemics onwards – on to the next aspect: representation.

In our story (which you might have forgotten by now!), we ended up going to see a movie. The film depicts two competing groups of scientists – one group, amateur, honest and good; the other, corporate, dishonest and bad – attempting to release data-collecting balls into the middle of a tornado. It may be too much to argue that the release of these objects, which have tails, into the tunnel of tornadoes (which are called at one point "sisters"), enacts and allays a sexual anxiety on the part of the filmmakers. Nevertheless, sexual scripts do underlie and give meaning to many narratives (and these scripts are not innocent of power relations, as shown in different ways in Chapters 18 to 20). Unsurprisingly, this film is also a love story. However, it is not the plot that concerns us here: it is the constitution of proxemics through real fantasies and fantasized real. It is common to find in discussions of place a clear distinction between the real and the imagined, yet this book, this film and our story, demonstrate that what we *know* as "the real" and "the imagined" are mutually constitutive, where neither one nor the other are determinate in the last (or even the first) instance (see Part 4; and also Chapters 5 and 7). So that, for example, things that have gone can stay "present" in our minds and in our hearts, partly because we become thinking, feeling and acting individuals through our real-and-fantasized experiences of the things in our (social, personal) worlds (see also Chapters 5 and 15).

Further, discussions of "Orientalism" have shown how fantasies become real in the practices of Imperial power, both in terms of the constitution of places and of bodies (see Bhabha 1994 and Pile 1998; see also Chapter 17). Here, the film transports us into a different place, with entirely different emotional intensities – including responding to the exciting action sequences, but also laughing at the dialogue, and wondering if there wasn't a better way for the scientists to go about the modest task of discovery. We are not in America exactly, but then again we are not in North London either – but somewhere "between." In many ways, of course, all places are "between," and it is in the specific ways that they are "between" that makes those places unique. And much the same can be said of bodies. In the space of the film, places and bodies are projections (see also Chapters 18, 19 and 20). From this perspective, both bodies and places are "between" an extra-discursive reality and the ways that people understand that reality (see also Grosz 1994: xi). To add to this sense of betweenness, it can be suggested that bodies and places are also locations in intersecting webs of social relationships, where bodies-places are always between one position and another.

While bodies are "meaty" and places are "some where," they are not reducible to their physiques, but neither can the physicality be simply cauterized from the ways in which the body is understood, practiced and lived. The problem is how to think about this relationship *between* the meaty reality of bodies and places and their various kinds of fantasized representation. One way might be to think of this "betweenness" as both real and fantasized; and neither fully real nor fully fantasized; and really fantasized and fantastically real. Another formulation could be that the proxemics of everyday life involves a double imagination that is attendant to the changing relationships between a real that cannot be stripped of its cultural meanings and an imaginary that cannot be isolated from its material contagions (see, for example, Chapters 12 and 13).

The conclusion has almost turned full circle, back to places-bodies, via proxemics. Back to an element of the story we could easily forget – the sky, our emotions. As we walked through London streets that glowed darkly, so did our moods; the skies, too, saturated in the colors of our feelings. The atmospheres of climate and emotion seem to lie outside the field of social analysis: it would appear that the skies are the stuff of science and fiction, while feelings are personal and not (to be) shared. Yet, the purpose of this conclusion has been to show that everyday stories are constituted by the proxemics of places and bodies; that places and bodies are socially and spatially produced through proxemics; and, that proxemics are implicated in both the forms and relationships of power – and, to move the story on again, struggle and negotiation. So far, in this conclusion, we have evoked a never less than double imagination by deploying the idea of proxemics – its distances, juxtapositions, boundaries, networks and movements. It is this understanding of proxemics that we would like to offer to conceptualizations of the politics of location.

There has been widespread use of spatial metaphors in critical theory of late. We believe that "space" has been drawn on for two significant reasons: first, in order to talk about the material contexts of oppression and resistance; and, second, in order to establish the grounds on which communities of resistance are to be constituted.[6] These grounds have often been established through a coarse division between "us" and "them," where the basic question is whether someone is on the right side of the fence, whether this is constructed in terms of class, race, gender, sexuality or some other privileged form of oppression. Instead, there might be other ways of conceptualizing the spaces of power, which do not presume that power is only about domination by the few of the many. Instead, it might be possible to think of fine tuned latitudes and longitudes, where locations are ambiguous and indeterminate – where, indeed, the grounds in this cartography of power are variable and uneven, but are nevertheless firm enough to recognize not only oppression and marginalization but also the capacities and resources that located communities of struggle can draw on.

412

From this perspective, proxemics suggests that a politics of location would involve accounts of places-bodies in *changeable* relationships to others, where those relationships are still mapped through positions in two-dimensional X and Y grids of domination, oppression and marginalization, but also through the people's n-dimensional (A, B, C . . . ?) capacities to produce directions and orientations, to negotiate their connections and dislocations, to determine their (channeled) mobilities and (temporary) fixities. In this sense, what might be reactionary in politics will be the permanent, undisclosed and unreflective stabilization of politics around a fixed, static and undialectical sense not only of space, but also of places and of bodies. The "radical" move, then, is to think through the implications of spatializing the politics of location, to animate the surfaces of politics in ways which enable "space" to be seen as constitutive of political struggles, to see struggles as located within/against/for dense, shifting and intense relations of power: that is, to see communities of struggle formed through the proxemics in place, through the body.

In place, through the body . . .

You are walking down the street, the wind is blowing, freezing, you pull your overcoat around you and look down at the pavement, too cold to look up.

You walk down the street, the sun is warm, a light breeze fans your clothes, and you think you might see the sights today.

You are walking, then stop and look up, the skyscrapers crowd overhead and your head begins to whirl, you are not sure where you are, nor where the silent aeroplane is, going; you feel lost.

You walk down the road, the heat is intense, you are assailed by the noise of the traffic, almost unable to hear the hawkers' demands, smells occupy your mind, you feel hungry, but do not know where to go.

You go shopping, but cannot find anything that you like – actually, you decide that you have not got the money to buy the things you do like.

You push your way down the street, you are tired, but no-one notices, every inch of the pavement is a hazard, an obstacle course – you hurt.

You march down the street with many others, you wave your banner (which you secretly suspect won't actually change the world), but the people who watch you shout and jeer, you feel simultaneously amongst friends and enemies, simultaneously strong and frightened.

You stand in front of the law accused of a crime you did not commit, but it does not matter because you will be punished anyway.

You stand at the foot of the bed, and realise that you cannot bear to look at the dead body anymore, that there is always a last time.

You sit in a café, with a cold cup of coffee and a slice of cake that has seen

413

better days, and type words that appear on a computer screen thousands of miles away, and you are delighted to be communicating with friends.

You see another tragedy on TV and you don't care – nothing is shocking.

You look in the mirror and see yourself dying – you are shocked.

You cough, and wheeze, and try to move your head, but you do not have the energy to get up, to get out of the house, to go to work.

You have been locked out of your workplace in a labor dispute and you fear that you will never work again.

You are laughing and dancing and chatting and laughing – "a little bit" drunk – wondering how you will get home in this state.

You walk through the alley, the night is clear, the dawn seems to be breaking somewhere near, but not that near, and you feel afraid: are those footsteps behind you, should you turn round?

You hide, but think that you've been betrayed by friends, and you wonder whether the others have been found – hide and seek?

You turn your head as you are attracted by someone you desire, you wonder: is it OK to talk to them in this place?

You turn your head and feel yourself caught by the stare of another, you turn away, but still feel their eyes crawl over your body: do you run?

You take a train, but get on the wrong one, frightened and lost, you get off in a place which you do not recognize – and, shaking off your sense of dislocation, you decide to stay a while and enjoy the sights.

You stop to look at a picture, and someone joins you, you both look at the picture, but neither of you says a word, occupied in the silences and noise of thought, both together and miles apart.

You are doing some research, but cannot get access to the people you want to talk to . . .

You sleep and dream of other worlds, of places you could not imagine, of people you have never met, and wake and remember nothing: where have you been?

You are tired and lie down, you close your eyes, and you never move again because your body has frozen.

You are confined in a prison, you don't know how you got there, no-one seems to have locked the door, you want to get out, right now, but it is more than walls that pen you in, and you think about how others might be feeling, you think about the endless ways in which you can experience the body through places, and everyday you wonder about the endless possibilities for realizing/releasing places through the body.

Notes

1 It is important to note, here, that we are not thinking of social production only in terms of political economy at this point, so that social production includes exchanges, circulations, consumings, refusals, and reproducings of all kinds, and

that social production is constituted through specific relations of time and space (following Massey 1994). The term we have settled on to describe this relationship between the spatialized practices of the body and the bodily production of space and place is "proxemics" (which is discussed in the main text). It, implicitly, relates both to de Certeau's argument about embodied spatial practices (1984) and to Grosz's discussion of the Moebius strip-like spatialities of the practiced body (1994).

2 Indeed, Steve got a junk letter from British Telecommunications PLC that day which began simply, "Dear Dr S. Pile, Connections."

3 See Bruno Latour (1991), on this.

4 We wouldn't want either the word "machine" or "body" to be taken too narrowly. Machines for living might include for example houses and cities (see Nast and Wilson 1994), while bodies are made up of many kinds of fluids as well as the meaty bits (see Grosz 1994).

5 The significance of networks in structuring both human relations and the relationship between the human and the non-human world have been given a variety of treatments. Two distinctive approaches are exemplified by Latour's account of the ways in which knowledge and power are practiced through the production and maintenance of networks (see, especially, 1987) and by Castells' sense that the logic of networks determines how social relations are structured (see, especially, Castells 1996).

6 See, for example, Rich 1986, Mohanty 1991, and Frankenberg and Mani 1993; for a discussion, see Pile 1997.

References

Bhabha, H., 1994, *The Location of Culture* (London, Routledge).

Castells, M., 1996, *The Information Age: economy, society and culture, Volume 1. The Rise of the Network Society* (Oxford, Basil Blackwell).

Cresswell , T., 1996, *In Place/Out of Place: geography, ideology, and transgression* (Minneapolis, University of Minneapolis Press).

de Certeau, M., 1984, *The Practice of Everyday Life* (London, University of California Press).

Frankenberg, R. and Mani, L., 1993, "Crosscurrents, crosstalk: race, 'postcoloniality' and the politics of location" *Cultural Studies* 7(2) pp. 292–310.

Grosz, E., 1994, *Volatile Bodies: toward a corporeal feminism* (Bloomington, Indiana University Press).

King, R., 1995, "Migrations, globalization and place" in D. Massey and P. Jess, eds, *A Place in the World? Places, Cultures and Globalization* (Oxford, Oxford University Press in association with the Open University) pp. 5–44.

Latour, B., 1987, *Science in Action: how to follow scientists and engineers through society* (Cambridge, Mass., Harvard University Press).

Latour, B., 1991, *We Have Never Been Modern* (1993, London, Harvester Wheatsheaf).

Low, G. C.-L., 1996, *White Skins, Black Masks: representation and colonialism* (London, Routledge).

Massey, D., 1993, "Power-geometry and a progressive sense of place" in J. Bird, B. Curtis, T. Putnam, G. Robertson, and L. Tickner, eds, *Mapping the Futures: local cultures, global change* (London, Routledge) pp. 59–69.

Massey, D., 1994, *Space, Place and Gender* (Cambridge, Polity Press).

Mohanty, C. T., 1991, "Cartographies of struggle" in C. T. Mohanty, A. Russo and L. Torres, eds, *Third World Women and the Politics of Feminism* (Bloomington, Indiana University Press) pp. 1–47.

Nast, H. J. and Wilson, M. O., 1994, "Lawful transgressions: this is the house that Jackie built . . . " *Assemblage: a critical journal of architecture and design culture* 24 pp. 48–56.

Nast, H. J., 1994, "Opening remarks on 'women in the field' " *Professional Geographer* 46(1) pp. 54–66.

Pile, S., 1997, "Introduction: opposition, political identities and spaces of resistance" in S. Pile and M. Keith, eds, *Geographies of Resistance* (London, Routledge) pp. 1–32.

Pile, S., 1998, "Freud, dreams and imaginative geographies" in A. Elliott, *Freud 2000* (Oxford, Polity Press).

Pratt, M. L., 1990, *The Imperial Eye: travel writing and transculturation* (London, Routledge).

Rich, A., 1986, "Notes towards a politics of location" in A. Rich, *Blood, Bread and Poetry* (London, Virago) pp. 210–231.

Smith, N., 1993, ""Homeless/Global: scaling places" in J. Bird, B. Curtis, T. Putnam, G. Robertson, and L. Tickner, eds, *Mapping the Futures: local cultures, global change* (London, Routledge) pp. 87–119.

Smith, S., 1989, *The Politics of "Race" and Residence* (Cambridge: Polity Press).

Stein, J., 1996, "The ideological telephone: the politics of an urban technology in Victorian and Edwardian London." Paper presented at the Social and Cultural Geography session on "Virtual geographies," at the Royal Geographical Society (with the Institute of British Geographers) annual conference, Glasgow. Copies available from author.

Thrift, N., 1996, "New urban eras and old technological fears: reconfiguring the goodwill of electronic things" *Urban Studies* 33(8) pp. 1463–93.

INDEX

Page numbers in *italics* indicate
illustrations

abjection 7, 188, 254, 257–8
ableism 10–11, 183–8, 192–3, 195
Aboriginal spatial history 316, 322
absolute space(s): apparition sites as
224, 225; scientific conception 53
Aeschylus 141
Agatha, Saint, sexuality and death in
iconography of 391–8
agoraphobia 80–1, 83; Freud's Hans
263, 270–8
Aitken, S. 361, 366, 368
Algeria 8, 129, 131
Alice (*Alice in Wonderland*) xvii–xviii, 4
Allen, K.R. 98
Alloula, Malek 129
Alma-Tadema, Sir Lawrence 119
Altvater, E. 33
Anderson, Benedict 324
Anderson, W. 286, 296
anthropology: Beothuk culture in
Canada 311–23 *passim*; of Marian
apparition sites 11, 207–29;
reflection in 94, 96
anxiety about place 74, 80–1; *see also*
agoraphobia
apartheid 9, 153–64
*Apartment in the Harem of Sheik Sâdât in
Cairo* (Dillon) *130*
architecture: harem as architectural
fantasm 119–32; *see also* modernist
architecture
artificial intelligence 50, 201n10

arts, harem in the 8, 119–32
Assunta Spina (film) 401n33
ataxia, spatial dissidence of Patty Hayes
190–9
ataxic mapping 201n11
attachment theory 271
Awatere, D. 338

Bachelard, Gaston 72, 75
Bailey, S. 97
Bakhtin, Mikhail 55, 56, 57, 58, 59,
60, 187, 188
Balzac, Honoré de 121, 125
Banks, Joseph 308
Barker, Frances 57
Barrès, Maurice 126
Barthes, Roland 60, 381
Bartky, S. 287
Bataille, Georges 382, 384
Baudrillard, Jean 58, 286, 200n2, 365
Bausted, Suzanne 302n9
Beavon, K. 157
behavioral geography, and disablement
201n10
Bell, Catherine 208, 224, 225, 226n2
Bell, L. 346
Benjamin, A. 364
Benjamin, Walter 65, 178
Beothuk cartography 13–14, 305–36
Beradt, Charlotte 165
Bernini, Gianlorenzo 384, 385, 386
Bersani, Leo 237, 239
Berton, Pierre 311, 312, 313, 318
Bhabha, Homi 328, 331, 411
binary opposites, model of body/city 45

417

vision, and VEs and body *see* virtual
environments
visions, Marian apparition sites 11,
207–29
voyeurism, and harem 125, 126, 131

Waldinger, R. 34
Walker, C. 153
Walker, R. 338, 339
Wark, M. 29, 30, 33, 35
warrior body, Maori identity in *Once
Were Warriors* 337–55
Weber, C. 359, 361
Weeks, J. 162n2
Weiss, Edoardo 271
Well of Loneliness, The (Radclyffe Hall)
11, 232–42
When the Cathedrals Were White (Le
Corbusier) 135–51
White, A. 187
whiteness, Le Corbusier's Radiant City
134–6, 147–51

Whitford, M. 39n4
Wigley, Mark 147, 152n7
Williams, R. 60
Wilson, M. 199, 221, 408
Winichakul, Thongchai 324
Winter, Keith 311, 317, 318, 321–2
women: and difference in Nigeria
102–10; Le Corbusier and American
140–3; and location 2; reproductive
power 186; *see also* female . . .
Wood, A.D. 359
Woolf, Virginia 2
work *see* Filipina domestic workers
Worth, H. 256
Wright, E. 256

Yoon, H. 338
Young, Iris Marion 187, 188, 249, 257
Young, Robert 318

Zubaida, Sami 323
Zulak, G. 256